PLANT SCIENCE RESEARCH AND PRACTICES

QUERCUS

CLASSIFICATION, ECOLOGY AND USES

PLANT SCIENCE RESEARCH AND PRACTICES

Additional books and e-books in this series can be found on Nova's website under the Series tab.

PLANT SCIENCE RESEARCH AND PRACTICES

QUERCUS

CLASSIFICATION, ECOLOGY AND USES

BENT J. STEFFENSEN
EDITOR

nova science publishers
New York

Library of Congress Cataloging-in-Publication Data

Names: Steffensen, Bent J., editor.
Title: Quercus : classification, ecology and uses / [Bent J. Steffensen]
Other titles: Plant science research and practices.
Description: Hauppauge, New York : Nova Science Publishers, [2020] |
Series: Plant science research and practices | Includes bibliographical references and index. |
Identifiers: LCCN 2020021683 (print) | LCCN 2020021684 (ebook) | ISBN 9781536180268 (hardcover) | ISBN 9781536180374 (adobe pdf)
Subjects: LCSH: Oak. | Oak--Identification. | Oak--Ecology. | Oak--Utilization.
Classification: LCC SD397.O12 Q847 2020 (print) | LCC SD397.O12 (ebook) | DDC 634.9/721--dc23
LC record available at https://lccn.loc.gov/2020021683
LC ebook record available at https://lccn.loc.gov/2020021684

Published by Nova Science Publishers, Inc. † New York

CONTENTS

PREFACE

Quercus: Classification, Ecology and Uses introduces a literature review of research conducted on *Quercus* analysed by species, country and topic, and the main biological processes related to the oak's response to global change are reviewed.

An extensive review on the uses of acorn oils, flours, extracts, and beverages is presented. The nutritional composition and bioactive compounds naturally present in acorn are described, as well as the effects of their bioactivities on human health.

Some research on pollination characteristics and acorn production in determined species of this taxon is examined, highlighting aerobiological data as an important aspect to be considered in reproduction studies.

Additionally, deciduous species of the *Quercus* L. genus are studied during their introduction in the botanical garden of Southern Federal University in Russia. The ecological parameters of 17 species were evaluated, including winter hardiness, drought resistance, phenology, genetic data, and resistance to diseases and pests.

The authors aim to determine whether *Q. crassipes* and *Q. castanea* can be recognized as foundation species, denoting that both species are dominant oaks of Mexican temperate forests.

Next, this compilation discusses the most widely explored phytochemical group in oak leaves, polyphenols, of which condensed and hydrolysable tannins have been reported in high concentrations in different *Quercus* species.

The authors consider the habitat and species diversity of oak forests in Bulgaria. The flora of studied oak forests, their structure and functionality, production, chemistry, intensity of biological turnover, and caloric content of phytomass fractions are presented in detail.

An extensive study is conducted to access the population status of *Quercus leucotrichophora* seedlings in a part of Garhwal Himalaya Oak forest through the establishment of permanent plots in different altitudes.

Chapter 1 - Iberian oak forests and woodlands are vital providing a variety of ecosystem services. They provide raw materials like cork, wood and fuel, and fodder (acorns) for wild and domestic animals. Oak trees provide a climate regulating service by their capacity to sequester carbon and therefore to mitigate the effects of climatic change. There is an increasing demand for cultural services provided by oak woodlands, mainly for recreation and ecotourism. However, some global-change drivers are negatively affecting oak forests, therefore diminishing their ecosystem services and in consequence undermining human well-being. The main drivers affecting Iberian oaks are: land use changes, introduction of exotic pathogens, air and soil pollution deteriorating oak health, and climatic change, in particular the combined reduction of rainfall and the rise in temperature.

In this chapter the authors focus on the ecology of Iberian oaks and how they are affected by global change. Firstly, the authors introduce a literature overview of the research conducted on *Quercus*, analysed by species, countries and topics. Secondly, the authors review the main biological processes related with the oak responses to global change: i) genetic diversity, climatic refuges and adaptation to climatic change; ii) ecophysiological responses to stress and disturbance, functional diversity, growth patterns, and scaling up from leaf to ecosystem; iii) demographic responses, causes of regeneration failures, spatial patterns of recruitment, and drivers of decline and dieback. Thirdly, the authors review the management strategies in three case studies: the agro-silvo-pastoral systems (*dehesas* and *montados*), the cork oak woodlands, and the afforestation of a polluted-land for phytostabilization of contaminants.

Chapter 2 - Acorns from the *Quercus* spp. trees can be found in copious amounts in the tropical and temperate regions and have a significant role in gastronomy, folk medicine and animal feed in these regions. Erstwhile, these nuts were an excellent food source during scarcity times, but nowadays acorns are largely perceived as animal feed and are far from being as widely used as other common nuts. Despite their large availability, only reduced quantities of acorns are curretly being used for human nutrition, so a new management approaches and usage of the produced wordwilde acorns would foment and contribute positively to the planet's sustainability. These nuts have an excelent nutritional profile and are a good source of bioactive compounds with numerous health benefits and relevant biological activities, which can contribute to health maintenance and well-being. In other to take advantage of these benefits, various technologies have been used to extract several compounds from different acorn species and several acorn-based functional products are under development, among which some are already found in today's market. The incorporation of acorn in food matrices has shown to improve food properties and nutritional profiles without affecting negatively the sensory characteristics and consumers' perception of these new foods. A brief description of the acorn history, production, and sustainability was done in this chapter. The core of this chapter involves the acorn transformation and use for human consumption. Thus, an extensive revision about the study and the use of acorn oils, flours, extracts, and beverages was done. The nutritional composition and the bioactive compounds naturally present in acorn were described, as well as the effect of their bioactivities on human health. Some green processing and extracting technologies were also discussed. Additionally, a section about tannins and acorn toxicity was also presented and discussed.

Chapter 3 - *Quercus* genus includes around 600 species of trees and shrubs that can be found in a wide variety of habitats, including temperate deciduous forests, temperate and sub-tropical evergreen forests, sub-tropical and tropical savannah and subtropical woodland in North Hemisphere. *Quercus* comprises arboreal dominant species of many different ecosystems.

Their ecological but also economic importance is unquestionable, mainly regarding to timber industry in temperate forests, and livestock feed and cork management in Mediterranean areas. All these species are monoecious with unisexual wind pollinated flowers. Each spring male flowers placed in catkins, produce large amounts of pollen grains to ensure fertilization. This anemophilous characteristic allows to study airborne pollen as a bioindicator of the reproductive phenology of this taxon and as a valuable source of information for clinicians and population allergic to this pollen type. In this chapter the authors review main literature related to *Quercus* aerobiology and phenology, and their contribution to a deeper understanding of the ecological behavior and the complex reproductive performance of these species. In this sense, the authors analysis some researches relating pollination season characteristics and final acorn production in determined species of this taxon, also pointing out *Quercus* aerobiological data as an important aspect to be considered in reproduction and ecological oaks studies.

Chapter 4 - Deciduous species of *Quercus* L. genus were studied during their introduction in the Botanical garden of Southern Federal University, Russia. Ecological parameters of 17 species and varieties of *Quercus* L., including winter hardiness, drought resistance, phenology, genetic data, resistance to diseases and pests were evaluated in a complex research. In addition, the hyperspectral survey data were tested as a tool for remote identification of oak species. *Q. macrocarpa*, *Q. hartwissiana*, *Q. iberica*, *Q. robur var. tardiflora* Czern. were shown to be the most promising for the regional (North-Caucasus, Russia) cultivation. The results have suggested that hyperspectral vegetation indices REP_Li and CL may be successfully applied for the remote identification of *Q. robur* species using unmanned aerial vehicles. It has also been shown that *Q. robur var tardiflora* samples of different origin have a high degree of genetic similarity when studied using the ISSR method.

Chapter 5 - Theory ecology is applied to comprehend how genetic variation within tree dominant species may be an important driver of ecological processes. In this sense, studies performed suggest that the genetic diversity of the foundation species can have robust organizational influence at the population, community and ecosystem levels. Foundation species are a small subset of the total species in an ecosystem; it has been suggested that they should capture most of the variation of the community structure and ecosystem processes. The trees of forest ecosystems are excellent candidates to be considered as foundation species because their architectural, functional, and physiological characteristics define the structure of the forests and can influence the microclimate; their biomass contributes significantly to ecosystem processes. Species of the genus *Quercus* (Fagaceae) are one of the most important tree canopy groups, due to its diversity and dominance in temperate forests in particular. In ecological terms, oaks are important for their ecological functions, also their canopy function as habitats for a wide variety of biological groups, including epiphytic plants, mammals, birds, fungi and arthropods. In particular, the gall-forming wasp belonging to the Cynipidae family (Cynipini Tribe) associated with the genus *Quercus* are considered one of the most specialized groups of herbivorous insects. The galls produced by cynipids form a microcosm with great ecological activity, harboring a closely associated community of guest organisms (including non-gall-inducing cynipids, flies, moths and beetles) and parasitoids. The parasitoids associated with cynipids are specific to them. It has been reported that their communities may be affected by the genetic attributes of plants. Therefore, oaks and the insects associated with the galls oak canopy represent an excellent system to study the effect of genetic variation of foundation species on cynipid communities, their associated parasitoids, and secondary fauna. Mexico is considered one of the most important centers of diversification of the genus *Quercus*, with 161 species (30%). *Quercus crassipes* and *Q. castanea* are two species of red oak, which are a canopy dominant elements of different forest types that support a wide variety of associated species. Therefore, the aim of this chapter is to determine if *Q. crassipes* and *Q. castanea* can be recognized as foundation species. This review denotes that both species are dominant oaks of Mexican temperate forests. They have wide geographic distribution and both species act as important reservoirs of canopy arthropod fauna. Also, the authors have shown that genetic diversity in both oak species results in extended phenotypes that extends to the community and ecosystem levels, in particular, on the associated community structure of oak – gall – inducing wasp –parasitoids and secondary fauna associated with oak canopy.

Chapter 6 - In recent years, interest in the study of bioactive compounds extracted from natural sources has increased. Customarily, oak leaves (*Quercus* spp) have been used as food for livestock and as traditional medicine thanks to their nutritional and phytochemical composition, respectively. Different extracts of oak species have demonstrated antioxidant, anti-inflammatory, antimutagenic, anticancer, antitopoisomerase, gastroprotective, hepatoprotective and genotoxic potentials, in addition to promoting the enzymatic inhibition of cholinesterase and α-glucosidase. These biological effects have been directly related to the phytochemical richness of oak leaves, which can be affected by environmental factors and ethnobotanical origins. The encapsulation of oak infusions and its fermentation with the Kombucha fungus have proved to be an alternative for the improvement of organoleptic and functional properties of the drink. In addition, emerging technologies such as PEF (pulsed electric fields) have shown to enhance their phenolic composition and their biological activity. The most widely explored phytochemical group in oak leaves is polyphenols, of which condensed and hydrolysable tannins have been reported in high concentrations in different *Quercus* species. This chapter includes a summary of the findings reported over the last years in different oak species, including a description of their phytochemical composition and their biological effects.

Chapter 7 - The chapter deals with considering the habitat and species diversity of oak forests in Bulgaria. The flora of studied oak forests, their structure and functionality - production, chemistry, intensity of the biological turnover, caloric content of the phytomass fractions, etc. are presented in detail. A summary of ecosystem services provided by oak ecosystems has been analysed. The attention was paid to the state of oak forests in the country and the sustainability of provided services. The focus of the discussions is also on the protection of habitats and the opportunities for their sustainable use as a natural resource.

Chapter 8 - An extensive study was conducted to access the population status of *Quercus leucotrichophora* seedlings in a part of Garhwal Himalaya Oak forest. Permanent plots were established in different altitudes and aspects. Among all disturbance stands the seedling density of *Quercus leucotrichophora* was found higher in the north aspect at moderately disturbed stand followed by highly and undisturbed stands. The *Quercus leucotrichophora* seedlings were best represented in all root collar diameter classes at north aspect as compared to south aspect. In the south aspect under root collar diameter class of 7.5 – 10.0 mm in undisturbed stand, < 10.0 mm root collar diameter class in moderately disturbed stand and 0 - 2.5 mm root collar diameter class at highly disturbed stand there were no seedling of *Quercus leucotrichophora*. In the rest of seedling species, among all disturbed stands, the seedling density was found higher again at north aspect in most cases.

Chapter 9 - An observable global trend since the mid-20th century is that anthropogenic pressure and anthropogenic transformations in the biosphere and its transformation into the noosphere are tangential to its degradation. Technology development, including innovative technologies related to the consumption of resources cannot be stopped and limited, as it is embedded in the development of human knowledge. Human society's dependence on ecosystem services, including natural resources, is steadily increasing, while practices for managing and using these services and resources are leading to the environmental degradation, the ecosystems and the biodiversity loss.

In: *Quercus*: Classification, Ecology and Uses ISBN: 978-1-53618-026-8
Editor: Bent J. Steffensen

Chapter 1

IBERIAN OAKS COPING WITH GLOBAL CHANGE: ECOLOGICAL PROCESSES AND MANAGEMENT STRATEGIES

Teodoro Marañón[1,*], Ignacio M. Pérez-Ramos[1], Rafael Villar[2], Vanda Acácio[3], Ismael Aranda[4], J. Julio Camarero[5], Enrique G. de la Riva[6], María T. Domínguez[7], Pilar Fernández-Rebollo[2], José L. Quero[2], José A. Ramírez-Valiente[4], Jesús Rodríguez-Calcerrada[8] and María Valbuena-Carabaña[8]

[1]IRNAS, CSIC, Seville, Spain
[2]University of Cordoba, Cordoba, Spain
[3]CEABN-InBIO, ISA, University of Lisbon, Lisbon, Portugal
[4]CIFOR, INIA, Madrid, Spain
[5]IPE, CSIC, Zaragoza, Spain
[6]Brandenburg University of Technology, Cottbus-Senftenberg, Germany
[7]University of Seville, Seville, Spain
[8]Universidad Politécnica de Madrid, Madrid, Spain

[*] Corresponding Author's Email: teodoro@irnase.csic.es.

Abstract

Iberian oak forests and woodlands are vital providing a variety of ecosystem services. They provide raw materials like cork, wood and fuel, and fodder (acorns) for wild and domestic animals. Oak trees provide a climate regulating service by their capacity to sequester carbon and therefore to mitigate the effects of climatic change. There is an increasing demand for cultural services provided by oak woodlands, mainly for recreation and ecotourism. However, some global-change drivers are negatively affecting oak forests, therefore diminishing their ecosystem services and in consequence undermining human well-being. The main drivers affecting Iberian oaks are: land use changes, introduction of exotic pathogens, air and soil pollution deteriorating oak health, and climatic change, in particular the combined reduction of rainfall and the rise in temperature.

In this chapter we focus on the ecology of Iberian oaks and how they are affected by global change. Firstly, we introduce a literature overview of the research conducted on *Quercus*, analysed by species, countries and topics. Secondly, we review the main biological processes related with the oak responses to global change: i) genetic diversity, climatic refuges and adaptation to climatic change; ii) ecophysiological responses to stress and disturbance, functional diversity, growth patterns, and scaling up from leaf to ecosystem; iii) demographic responses, causes of regeneration failures, spatial patterns of recruitment, and drivers of decline and dieback. Thirdly, we review the management strategies in three case studies: the agro-silvo-pastoral systems (*dehesas* and *montados*), the cork oak woodlands, and the afforestation of a polluted-land for phytostabilization of contaminants.

Keywords: climate change, dehesas, ecosystem services, functional diversity, genetic diversity, growth, hybridization, Mediterranean, *Quercus ilex*, *Quercus suber*, regeneration ecology, resilience

Introduction

The genus *Quercus* has about 450 species distributed throughout the Northern Hemisphere, from boreal regions in northern Europe to tropical forests in Southern America (Colombia) and Southeast Asia (Indonesia) (Manos and Stanford 2001; Hipp et al. 2019; Kremer and Hipp 2019). Oak trees occupy a wide range of habitats on a variety of soil types and often dominate the forest biomass (Cavender-Bares et al. 2016). Based on its high species diversity and abundance, and its wide geographic range the genus *Quercus* is considered an "evolutionary success" (Kremer and Hipp 2019).

The Iberian Peninsula, in SW Europe, harbours ten oak species with a variety of life forms: deciduous, marcescent and evergreen trees, an evergreen arborescent shrub and a rare prostrated shrub (Table 1). Iberian oak forests and woodlands are vital providing a variety of ecosystem services. They provide raw materials like cork, wood and fuel, and fodder (acorns) for wild and domestic animals (Bugalho et al. 2011). Oak trees provide a climate regulating service by their capacity to sequester carbon and therefore to mitigate the effects of climatic change; contribute to soil fertility and to the regulation of air, soil and water quality (Marañón et al. 2012). There is an increasing demand for cultural services provided by oak woodlands, mainly for recreation and ecotourism, as well as for their aesthetic and spiritual values (Leroy et al. 2019). However, global change is affecting negatively oak forests and therefore diminishing their ecosystem services and in consequence human well-being. The main drivers are: land use change, introduction of exotic pathogens, air and soil pollution deteriorating oak health, and climatic change, in particular the combined reduction of rainfall and the rise in temperature (Valladares 2004; Valladares et al. 2014).

Oak trees have experienced large climatic changes in the past. Part of their past and present success relies on the characteristics of their mating systems, that represent the natural mechanisms by which species maintain the genetic variability needed to adapt to environmental conditions along time and space (Boshier 2000). Extant oak species avoided extinction and colonized new areas through rapid migration and adaptation, extensive gene flow and hybridization (Kremer 2013). The genetic diversity, at species and population levels, is the base for a successful response of the genus *Quercus* to the global change. To cope with climate change threats, the Iberian oaks should combine a rapid response by phenotypic plasticity and the long-term local adaptation to new environmental conditions (Ramírez-Valiente et al. 2010).

Table 1. Iberian species of *Quercus* with their phylogenetic group (according to Denk and Grimm 2010), ecology and life form. For each oak species the research output (number and percentage of publications during 2000-2019 in the WOS) is indicated

Species	Phylogenetic Group	Ecology and Life form	Scientific publications	
			Number	(%)
Q. ilex L.	*Ilex*	Mediterranean, evergreen tree	1040	20
Q. coccifera L.	*Ilex*	Mediterranean, evergreen shrub	142	3
Q. suber L.	*Cerris*	Mediterranean, evergreen tree	699	14
Q. robur L.	*Quercus* (white oaks)	Temperate, deciduous tree	1576	31
Q. petraea (Matt.) Liebl.	*Quercus* (white oaks)	Temperate, deciduous tree	981	19
Q. pubescens Willd.	*Quercus* (white oaks)	Sub-Mediterranean, deciduous/marcescent tree	319	6
Q. pyrenaica Willd.	*Quercus* (white oaks)	Sub-Mediterranean, deciduous/marcescent tree	247	5
Q. faginea Lam.	*Quercus* (white oaks)	Sub-Mediterranean, marcescent tree	103	2
Q. canariensis Willd.	*Quercus* (white oaks)	Sub-Mediterranean, marcescent tree	46	1
Q. lusitanica Lam.	*Quercus* (white oaks)	Sub-Mediterranean, marcescent, prostrated shrub	2	0.04

The evolutionary success of Iberian oaks is partly due to their features shared with other trees: longevity, great size, high reproductive output and large seed mass, large populations and wind pollination (Petit and Hampe 2006; Savolainen et al. 2007). However, oaks have also singular traits that confer them higher fitness under Mediterranean conditions. Among them, their resistance to drought associated with water use efficiency, root biomass allocation and stomatal regulation, and their resilience to frequent disturbance related to sprouting ability and storage of nonstructural carbohydrate, and nutrients (Gil-Pelegrín et al. 2017; Rodríguez-Calcerrada et al. 2017).

The impact of global change may affect oak populations by increasing the mortality rate and reducing the recruitment rate. Early stages of the oak life-cycle are particularly vulnerable to global change drivers (Pulido and

Díaz 2005; Pérez-Ramos et al. 2012). Several factors contribute to the failure of oak regeneration: i) seed production can be scarce, with the exception of "masting years," ii) the large and nutritious seeds (acorns) are heavily consumed by many animals, and iii) the few emerged seedlings suffer high mortality by drought stress, herbivory and soil pathogens (Pérez-Ramos et al. 2012; Pérez-Ramos 2014). Climate change, especially the reduction of rainfall, will aggravate the pressure on these demographic bottlenecks. The suitability of micro-habitat for oak recruitment will depend on the net result of the multi-stage demographic outcome (Clark et al. 1999; Pérez-Ramos et al. 2012).

The aims of this chapter are to review key aspects related to the ecology and management of oak forests coping with global change. In particular, we review: i) genetic diversity, climatic refuges and adaptation to climatic change; ii) ecophysiological responses to stress and disturbance, functional diversity, growth patterns, and scaling up from leaf to ecosystem; iii) demographic responses, causes of regeneration failures, spatial patterns of recruitment, and drivers of decline and dieback.

Oak forests have a long history of human management shaping current landscapes. Three case studies of oak management are also examined in this chapter: the agro-silvo-pastoral systems (*dehesas* and *montados*) in West and South Spain, the cork oak woodlands of Portugal, and the afforestation of a polluted-land for phytostabilization of contaminants in the Guadiamar Green Corridor (SW Spain). We conclude with recommendations to assure the persistence of Iberian oaks and sustainable management of oak forests and woodlands.

SCIENTIFIC RESEARCH ON *QUERCUS*

The scientific publications on different topics related to the species of the *Quercus* genus reflect the interest of the society in oaks and the consequent investment in research (Marañón et al. 2014). The global databases and new informatic tools offer the opportunity to build a metaknowledge, identify and measure the complex processes of production and the use of knowledge (Evans and Foster 2011).

The search for the words "oak" or "*Quercus*" in the "Topic" field in the global database Web of Science (WOS) matched with 10,295 publications during the twenty-year period of 2000-2019. The publishing growth was exponential from the year 2000 (300 documents) to 2019 (776), with an annual rate of 3%, like other scientific disciplines (Larsen and von Ins 2010). The scientific research on Iberian oaks is outstanding, being Spain the second country (only after USA) and Portugal the seventh in the ranking of oak publications (Figure 1). To illustrate this exceptional research on *Quercus* by the Iberian countries their ranking position can be compared with the 10th place of Spain and 25th of Portugal in the ranking of general scientific publications (King 2004) and, if we assume a relationship with the country´s wealth, with the 13th of Spain and 49th of Portugal in the gross domestic product (GDP) world rank (www.statista.com).

The research effort is not evenly distributed among the ten *Quercus* species inhabiting the Iberian Peninsula (Table 1). The temperate oaks (*Q. robur* and *Q. petraea*) are the most studied and have received the highest number of publications (first and third in the rank, respectively), likely due to their wide European distribution. Holm oak (*Q. ilex*) is the second *Quercus* species receiving a higher research effort, in accordance with its dominance in evergreen forests and wood pastures in Mediterranean countries. The special research interest in cork oak (*Q. suber*, the 4th in the rank) might be explained by its economic relevance in the cork production (Aronson et al. 2009; see the case study section below), despite its relatively restricted distribution. At the bottom of the rank, the dwarf oak (*Q. lusitanica*), endemic of Iberia and North Africa, has received very little scientific attention.

Looking at the world top scientists who have published most about *Quercus*, it is illustrative to find a tree geneticist from France (Antoine Kremer), a global ecologist from Spain (Josep Peñuelas) and a forest-products scientist from Portugal (Helena Pereira). Thus, there is a variety of approaches and a geographical concentration in Southern Europe in the research excellence about *Quercus*. Another indicator of the attention devoted to the Iberian *Quercus* and the different topics investigated is the list of most cited publications in that 20-year period (refining the search by "Spain" or "Portugal"). Among the studies on *Quercus* with highest impact there are diverse approaches: genetic and phylogeography (Petit et al. 2002; Brewer 2002), sustainable management and restoration (Bugalho et al. 2011, Pausas et al. 2004), ecophysiology (Valladares et al. 2000), seed evolution (Gómez 2004) and soil biology (Sardans and Peñuelas 2005). In a previous survey of key words in publications about *Quercus* in Spain, the most frequent words were "climatic change," "regeneration," "drought," "disturbance" and "species richness," reflecting the dominant research topics (Marañón et al. 2014)

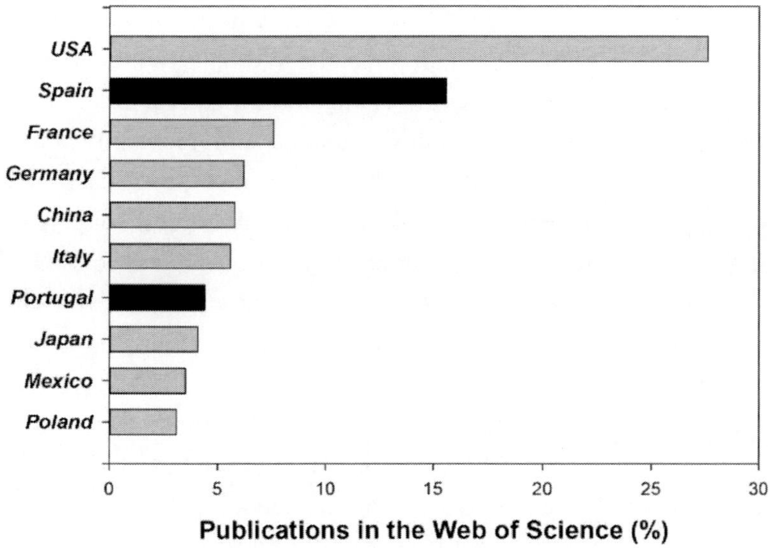

Figure 1. Rank of countries according to the percentage of publications about *Quercus*, for the period 2000-2019. The two Iberian countries (Spain and Portugal) are marked in black.

This bibliometric analysis is illustrative. We should have in mind that scientist´s mission is to advance in the knowledge (about oaks in this case) and not to publish the highest number of papers or get the maximum number of citations (Parker et al. 2013).

Genetic Responses to Global Change

Genetic Diversity and Climatic Refugia

In the Iberian Peninsula, a diversity of oaks including evergreen Mediterranean species and temperate deciduous oaks coexist in contrasting edaphoclimatic conditions (Table 1). Species of the genus *Quercus* are very abundant and widespread in Iberia, frequently forming mixed stands. Part of their past and present success relies on the characteristics of their mating systems, which represent the natural mechanisms by which species maintain the genetic variability needed to adapt to environmental conditions through time and space (Boshier, 2000). In *Quercus* species, hybridization (and subsequent processes of introgression), long-distance dispersal and asexual reproduction are the main traits that have allowed oaks to spread and adapt to environmental and anthropogenic changes during millennia. Given that population dynamics leave genetic traces among individuals, the analysis of neutral genetic variation is a useful tool to infer populations' past biological and demographic events regardless of the effects of natural selection. Also, according to the life-history traits of long-living social tree species (Petit and Hampe 2006), studies performed in oak species commonly show high levels of intrapopulation genetic diversity widely shared among populations at different spatial scales depending on species, population fluctuations and the type of molecular markers used. Most studies show that oaks are able to maintain high genetic diversity, highlighting the long-term stability of populations and their outstanding genetic resilience.

Contrasting differentiation between chloroplast (maternally inherited) and nuclear genes (biparentally inherited) is common in oak species, as a result of differences between seed and pollen dispersal (Petit et al. 2005; Petit and Excoffier 2009). All over the world, oaks present a strong geographical distribution of maternal lineages shaped by Quaternary colonization processes. Information from chloroplast DNA markers combined with palaeobotanical records have been used to locate glacial refugia and to trace postglacial colonization dynamics. In the Iberian Peninsula, Mediterranean evergreen species (*Q. ilex*, *Q. coccifera* and *Q.*

suber) show relatively limited expansion from multiple glacial refugia. *Q. suber* shows at least one glacial refugium in southwestern Iberia, *Q. ilex* and *Q. coccifera* multiple ones in central and northern Iberia, and all three oak species present numerous areas of secondary post-glaciation contact in the Peninsula. Moreover, hybridization explains the current sharing of organelle DNA among species. For instance, in calcareous eastern Iberia, extensive introgression of *Q. suber* with *Q. ilex* reveals potential refugia for the acidophilus species in scattered siliceous patches, where hybridization with *Q. ilex* allowed it to withstand arid glacial conditions (López de Heredia et al. 2007). Interspecific crosses with *Q. suber* as pollen donor produced a total replacement of *Q. suber* chlorotypes, and subsequent backcrosses with parental *Q. suber* diluted the nuclear contribution of *Q. ilex* by pollen swamping (López de Heredia et al. 2005).

The Iberian Peninsula also played a relevant role for temperate oak species as their European Southwestern most refugia during the last glacial period, from which postglacial colonization produced fast and large-scale expansions (Brewer et al. 2002; Petit et al. 2002). Frequent hybridization between white oak species (including sub-Mediterranean and temperate species, Table 1), facilitated in refugia by their small effective population sizes, explains the strong geographic structuring of chlorotypes, which are commonly shared among sympatric species. The Iberian Peninsula represents a primary glacial refugium and an important genetic reservoir for the clade (Figure 2), harbouring four of the six European maternal lineages– one (lineage B in yellow and brown colours) widely distributed in north-western Europe (from Great Britain to southern Scandinavia) and another one (lineage D in green colours) restricted to Mediterranean Spain, southwestern France and northern Africa (Petit et al. 2002; Olalde et al. 2002). This strong geographic structuring into well-defined biogeographic regions (with Atlantic and Mediterranean influences) probably was present before the last Ice Age, suggesting the existence of multiple primary glacial refugia in southern and also in northern Iberia, including north-eastern Spain (Olalde et al. 2002). In this area, where no clear pollen evidence prior to the Last Glacial Maximum has been found (Brewer et al. 2002), two other maternal lineages -lineage A (blue colour) predominant in the Balkans, and lineage C (red colour) located from southern Italy to Scandinavia- are present but controversy about their survival during the full glacial period exists (see Olalde et al. 2002).

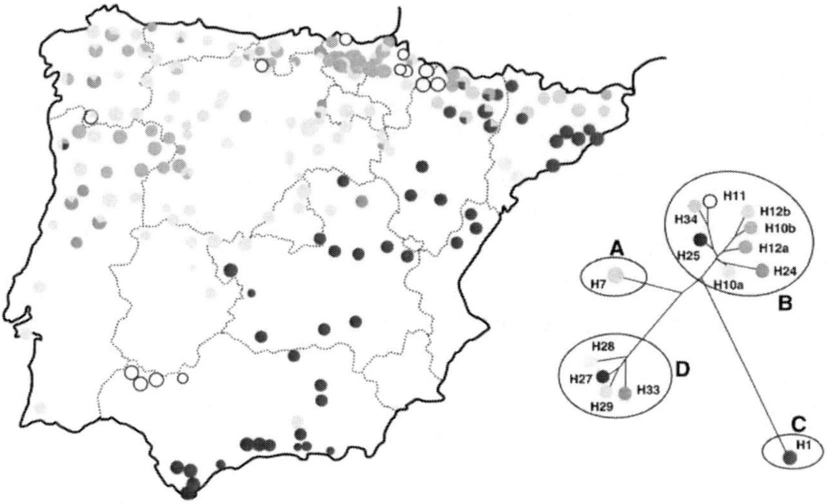

Figure 2. Chloroplast phylogeography of the white oaks in the Iberian Peninsula (from Olalde et al., 2002). Distribution of 14 cpDNA haplotypes grouped in four different maternal lineages (letters) as shown by the Neighbour-Joining tree. Reprinted with permission from Elsevier.

In this regard, an unpublished survey on the scarce white oak populations in the Balearic Islands located more than 200 km from the north-eastern coast of Spain, shows the same cpDNA haplotype of the maternal linage C, supporting the existence of primary refugia for this lineage in both regions (Valbuena-Carabaña, unpublished), and not just in southern Italy as suggested by Petit et al. (2002). The study of nuclear markers reveals high genetic differentiation among Balearics populations and the existence of bottleneck processes that suggest a former abundance and an ancient and natural origin of white oaks in the Balearics. Their long-term stability and marked genetic resilience provide oak species relevant conservation value, not only in continuous but also in marginal populations, especially in insular ecosystems where their scarcity often makes them be considered as introduced. The example of *Quercus suber* L. in the Balearic can also illustrate this point: just 67 survivors hold most of the genetic diversity found in the western Mediterranean distribution of the species, including Tyrrhenian and Iberian elements (Lorenzo et al. 2009; Jiménez et al. 2004; López de Heredia et al. 2005).

Anthropogenic clearing of forest structures seems to assure pollen dispersal and acorn production of different oak species traditionally managed as open woody pastures. These populations have maintained high genetic diversity and important levels of gen flow at intrapopulation scales leading to the formation of different kinship structures depending on the species and stand tree-density (Valbuena-Carabaña et al. 2005; Soto et al. 2007; Ortego et al. 2014). In addition, the widespread resprouting ability of oak species confers them important adaptations to natural disturbances and also resilience to the combination of human-induced fires, grazing and other traditional management practices like coppicing. This ecological trait is also one of the bases of the past ecological success of Mediterranean and sub-Mediterranean *Quercus* that explains forest dynamics along the Quaternary vegetation history of the Iberian Peninsula (Carrión et al. 2010). Despite the widespread belief that continuous asexual reproduction of coppice systems over centuries has caused losses of genetic variability, different surveys highlight the great genetic resilience of resprouting species facing anthropogenic disturbances (Valbuena-Carabaña and Gil 2017; Ortego et al. 2010). Although coppices and open woodlands harbour outstanding levels of genetic diversity, they often present important degradation states related to stand regeneration. To ensure the persistence and functionality of oak species in these managed woodlands, given the challenges posed by global change, it is necessary to adopt conservation and management measures to allow seedling establishment and sexual regeneration on these woodlands.

Overall, *Quercus* species present a great natural capacity to preserve high levels of genetic diversity under adverse conditions, which gives them the ability to cope with natural and human disturbances and their characteristic resilient nature. Intraspecific genetic diversity is of particular importance in processes determining the maintenance of fitness and the evolutionary potential of species in changing environments (Reed and Frankham 2003; Jump et al. 2009; Pautasso 2009). As temperatures increase and precipitation declines in the northern hemisphere due to climate change, conservation measures applied to southern boundary populations which are adapted to local climates are essential for the future preservation of species across their entire geographical ranges. In this sense, the preservation of intraspecific genetic diversity and the species' potential to interact should be crucial components of management and conservation programmes.

Adaptation of Oak Species to Climate Change

Adaptation to different environments is the main process driving the evolution of inter- and intraspecific functional diversity (Alberto et al. 2013; Cavender-Bares and Ramírez-Valiente 2017). Precipitation regimes and water availability, in particular, are probably the most important selective factors explaining the diversification of oak species but also the adaptive differentiation of populations within species. In fact, populations coming from areas with different precipitation have differences in traits such as leaf mass per area (LMA), nitrogen concentration and leaf thickness, and also in water use strategies (drought tolerance vs. drought avoidance) (Gratani et al. 2003; Ramírez-Valiente et al. 2010, 2014; Ramírez-Valiente and Cavender-Bares 2017). Other factors, such as low temperatures in winter, have also driven the adaptive differentiation among populations of some evergreen oaks (Aranda et al. 2005; Ramírez-Valiente et al. 2015). This adaptive diversity has been also documented at molecular level. For example, Sullivan et al. (2013) observed population differences in expressed sequence tags (EST)-SSR loci related to candidate genes of abiotic stress tolerance in red oaks. Gugger et al. (2017) reported differential transcriptome response in climatically-contrasting populations of *Q. lobata* when they were exposed to dry conditions. Besides natural selection, neutral evolutionary processes such as genetic drift can also originate population genetic divergence in functional traits. However, comparing quantitative adaptive variation and neutral molecular differentiation, Ramírez-Valiente et al. (2009, 2018) provided strong evidence that natural selection, rather than neutral processes, has driven the divergent evolution of traits such as LMA, leaf thickness and leaf size in *Q. suber* and *Q. oleoides*.

Oaks usually also exhibit high genetic variation in functional traits at intrapopulation level and consequently have a good evolutionary potential of traits under new selective pressures. Wind pollination, long life cycles and large population size are probably the factors that favour high levels of genetic variation within the populations in functional traits (Savolainen et al. 2007). Despite these general patterns, recent studies indicate that the potential to evolve under natural selection is trait- and population-dependent, and that particular populations can have low genetic variation in some traits that limit their potential to evolve in response to natural selection (Ramírez-Valiente et al. 2011, 2019). These studies suggest that intrapopulation genetic variation in traits should be considered as a key component of oaks for their potential responses in a climate change context.

The potential of oak tree populations to respond to new environmental conditions will depend not only on the genetic variation, but also on the phenotypic plasticity of traits important for fitness. Oaks usually exhibit high phenotypic plasticity to a variety of environmental factors such as light, water, nutrients and CO_2, which will guarantee a rapid response in the short term. In the long-term, some authors have suggested that plasticity might have a secondary role to face new selective pressures due to its limits and costs (Valladares et al. 2007). However, some studies challenge this idea and point out that plasticity of traits might have a role in facing climate change in the long term through its transgenerational component (Polle et al. 2001). In summary, high evolutionary capacity based on local adaptation and phenotypic plasticity of key functional traits could both guarantee oak species to cope with the challenges of the ongoing climate change.

Finally, it is worth mentioning that despite the high intra-specific variability, differences among *Quercus* species and co-ocurrence of multiple species inhabiting the same site is also common for *Quercus* species. The most striking case would be North and Central America where a high number of species co-exist in different sites, and numerous oak forests respond to the assemblage in a same community of evolutionary distant species and even from different phylogenetic origins (Cavender-Bares et al. 2004).

Ecophysiological Responses to Global Change

Functional Traits in *Quercus* Compared to Other Woody Species

Quercus species differ in several functional traits compared to other woody species (Escudero et al. 2017; Villar et al. 2017). In general, certain constancy of the morphological and phenological variables exists within the *Quercus* genus. For instance, oaks are small or medium-sized trees, monoecious with anemophilous pollination (Pulido 2002; Villar et al. 2017). In addition, *Quercus* species frequently show leaves with low specific leaf area (SLA, leaf area per leaf dry mass, the inverse of LMA) and high carbon construction cost compared to other woody Mediterranean species (de la Riva et al. 2016a; Escudero et al. 2017), which is related to long-lived tissues commonly associated with a more conservative use of resources (de la Riva et al. 2016b; Escudero et al. 2017). However, a certain degree of functional flexibility exists within the *Quercus* genus (Aguilar-Romero et al. 2017; Escudero et al. 2017; Villar et al. 2017). For instance, the evolution of different leaf lifespan habits within the *Quercus* genus has promoted diverse morphological adaptations associated with differential distribution across a wide range of habitats in Iberian Peninsula. In general, deciduous oaks tend to be more abundant in habitats with greater availability of water and nutrients, where the overstorey canopy is denser. By contrast, leaf traits of evergreen species allow tolerance to drought, but not to shade. So, they tend to dominate in dry and low fertile habitats, where the overstorey canopy is sparse. This spatial and geographical segregation suggests that both leaf habitats have different requirements (Escudero et al. 2017; Villar et al. 2017). In fact, deciduous oaks species differed in leaf structure and in physiological activities compared with evergreen oaks (Baldochi et al. 2010; Escudero et al. 2017; Villar et al. 2017). Thus, deciduous oaks from Mediterranean regions (such as *Q. canariensis*) have leaves with higher values of SLA and N concentration than evergreens (Figure 3) (Baldochi et al. 2010; de la Riva et al. 2014; Niinemets et al. 2004), favouring higher photosynthetic rates, stomatal conductance and relative growth rates at optimal environmental conditions. On the contrary, evergreen oaks (such as *Q. ilex*) have denser and thicker leaves with lower SLA, higher values of dry matter and the isotopic ratio ^{13}C to ^{12}C (Figure 3), lower mass-based photosynthetic rates and hydraulic conductance that promotes water saving and enhances tolerance during drought periods (Mediavilla et al. 2002; Niinemets et al. 2004). These results (Figure 3) show a high concordance with the axis of variation known as the "leaf economics spectrum" (Wright et. al. 2004; Niinemets 2015).

De la Riva et al. (2014) also found a negative relationship between SLA and wood density (Figure 3). This can be explained by the need for plants that have a greater SLA (with higher transpiration rates), to have stems capable of transport more water (that is, with greater hydraulic conductivity), which is related to a lower density of wood (Reich 2014). This coordination between leaf and stem traits was also found with roots traits, indicating that there is a common strategy in the acquisition of resources between the different organs (Figure 3; de la Riva et al. 2014). Similar results have been obtained in adverse conditions, such as sub-arctic (Freschet et al. 2010) or arid (Liu et al. 2010) environments, while contrasting with those found in tropical forests (Baraloto et al. 2010; Fortunel et al. 2012). This apparent contradiction suggests that the trait coordination is determined by the intrinsic limitations of each ecosystem (Kembel and Cahill 2011). In the specific case of Mediterranean ecosystems, the availability of water promotes the existence of a single axis of variation in which the different parts of the plant must be coordinated, due to the optimization that must exist between the investment to capture and that to conserve water for survival (de la Riva et al. 2016c). Therefore, the different *Quercus* species present a gradient in their functional strategies that are related to their distribution: from more conservative species in the use of resources, living in conditions of lower water availability, to the more acquisitive species in more humid areas.

Another clear difference between *Quercus* and non-*Quercus* species is that *Quercus* species have seeds between four to six times heavier than those of non-*Quercus* woody species with which they coexist (Villar et al. 2017); this large seed size has been suggested as an important contributor to the success of *Quercus* species (Larson-Johnson 2016). The size of *Quercus* seeds is positively related to a rich chemical composition and high carbohydrate and lipid contents (Fernández-Rebollo et al. 2017). These high seed reserves imply several advantages during the early stages of the seedling. For instance, seeds of larger mass originate seedlings of higher biomass (Quero et al. 2007; González-Rodríguez et al. 2010, 2012; Pérez-Ramos et al. 2010a), which may confer a resource uptake advantage because these larger seedlings have access to limiting resources (e.g., water, nutrients, light), promoting higher survival (Lloret et al. 1999). Also, a large seedling mass enhances seedling survival under deep shade situations, with a low carbon gain (Quero et al. 2007).

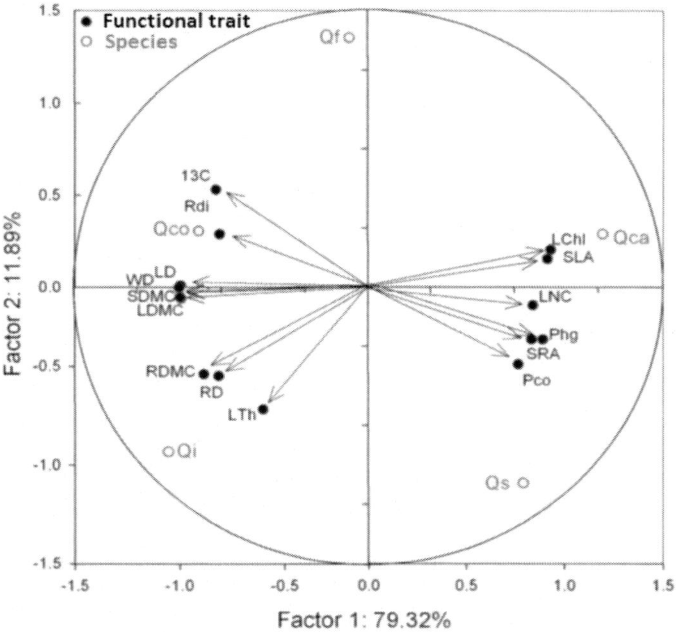

Figure 3. Principal Component Analysis (PCA) of 17 functional traits of leaf, stem and root of five *Quercus* species. Qi: *Q. ilex*, Qs: *Q. suber*, Qca: *Q. canariensis*, Qco: *Q. coccifera*, Qf: *Q. faginea*. Leaf thickness (LTh, mm), Leaf dry matter content (LDMC, mg g^{-1}), Specific leaf area (SLA, m^2 kg^{-1}), Leaf density (LD, g cm^{-3}), Stem dry matter content (SDMC, mg g^{-1}), Wood density (WD, g cm^{-3}), Root dry matter content (RDMC, mg g^{-1}), Specific root area (SRA, m^2 kg^{-1}), Root diameter (Rdi, mm), Root density (RD, g cm^{-3}), Height (Phg, m), Plant Area (Pco, m^2), Leaf Chloropyll (LChl, µg g^{-1}), Isotopic ratio of ^{13}C (13C, $\delta^{13}C$ ‰), Leaf Nitrogen Concentration (LNC, %).

On the other hand, these characteristics of *Quercus* seeds made them very palatable for predators (rodents, deers, pigs, birds, etc.) and compromise the regeneration of *Quercus* forests in many cases (e.g., González Rodriguez et al. 2012; Pérez-Ramos et al. 2012; see more details in the section on regeneration ecology below). Moreover, *Quercus* seeds are recalcitrant, so they only live for a few months (1-3 months), after which they lose the germination ability. In consequence, oak seeds cannot be stored and do not form seed bank, so the regeneration depends only on the current-year seed production, determinated by masting of species.

Quercus species also show a high biomass partitioning to roots compared with other woody species (Villar et al. 2014, 2017). The strong investment in roots makes the investment in leaves very low compared to that of non-*Quercus* species, explaining the low relative growth rate of *Quercus* (see growth section below). On the other hand, a greater proportion of roots may constitute an advantage to resist drought (Quero et al. 2006), allowing access to a greater amount of water in the soil, which can confer higher survival rates under adverse conditions (Lloret et al. 1999).

Ecophysiological Response to Stress and Disturbance

Functional and ecological variability across oak species is huge. As mentioned before, a diversity of leaf habits, growth forms and water- and resource-use strategies are found within the genus *Quercus* (Cavender-Bares 2019; Ramírez-Valiente et al. 2020). However, some ecological patterns emerge for evergreen and deciduous species in relation to their response to stress and perturbation (Figure 4). Oaks are relatively resistant to drought. A comparison of evergreen and deciduous oaks with other woody species revealed that oaks do not experience as low predawn and midday leaf water potentials during drought (Rodríguez-Calcerrada et al. 2017), probably due to their higher root development, water transport and water use efficiency. Moreover, stomatal conductance in oaks is reduced to half of its maximum value at approximately 1 MPa lower midday leaf water potential, which can be related to high osmotic adjustment in response to drought (Abrams 1990;

Rodríguez-Calcerrada et al. 2017; Forner et al. 2018). Altogether, these features reflect a remarkable capacity to tolerate low leaf water potentials and likely contribute to maintaining relatively high rates of net CO_2 uptake under drought conditions. Maintaining carbon gain under drought can be particularly important for oaks, which have comparatively high nonstructural carbohydrate (NSC) concentrations (Rodríguez-Calcerrada et al. 2017), and also probably high NSC pools due to high root development and root mass in proportion to plant mass (Villar et al. 2017). Storing NSC (and nutrients) requires perennial organs with living parenchyma cells. Indeed, some studies have found that 50% of sapwood is formed by parenchyma in some oak trees (*Q. pyrenaica*; Rodríguez-Calcerrada et al. 2015), and that median values of axial, radial and total parenchyma in secondary xylem are slightly higher in oaks than in other angiosperms (Robert et al. 2017). Another typical feature of oaks, the low growth rate, can also be related to their high carbon allocation to storage, given the trade-off between both sinks. According to Villar et al. (2017), high LMA and root mass fraction (RMF), and low leaf area ratio (LAR), would explain the comparatively low relative growth rate of oak seedlings (see more details in the section on growth, below). These features reflect a conservative use of resources and carbon which, in turn, constrain the competitive ability of oaks in undisturbed sites. In summary, high resilience to perturbations and drought (e.g., via resprouting), would be possible due to the relatively high leaf gas exchange rates at low plant water potentials and relatively low rates of growth which, together, would enhance carbon supply to reserve organs.

In the future, longer and more intense droughts are expected together with high flood risks due to torrential rainfall. Although there are oak species that inhabit bottomland sites prone to experience long floods (e.g., *Q. robur*), oaks are not particularly tolerant to waterlogging. The prediction of long-term responses to future climate projections, including longer droughts and floods, is uncertain partly due to few functional studies aiming to study the impacts of these perturbations in the long term.

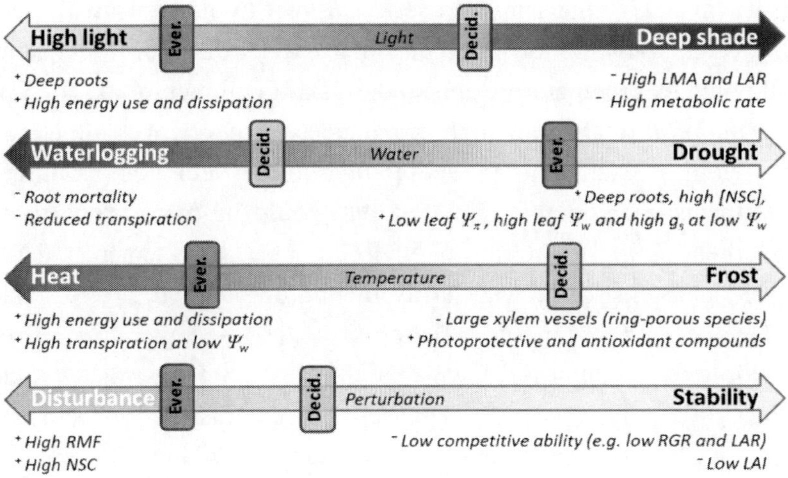

Figure 4. Functional bases of oak performance in relation to environmental factors. The vertical rectangles mark the position of evergreen (Ever.) and deciduous (Decid.) oaks in relation to each factor. Positive (+) and negative (-) symbols indicate positive and negative features in relation to stress and perturbation. We note that this is a simplified view of the vast functional and ecological variability of oak species. Abbreviations: leaf mass per area (LMA), leaf area ratio (LAR), nonstructural carbohydrates (NSC), water potential (Ψ_w), osmotic water potential (Ψ_π), stomatal conductance (g_s), root mass fraction (RMF), relative growth rate (RGR), leaf area index (LAI). Author: Jesús Rodríguez-Calcerrada.

Some of the above traits, such as the high LMA and RMF, and the low LAR contribute to explain the absence of oak saplings in heavily shaded understories. The abandonment of forest exploitation and the policy of fire suppression set up in some Mediterranean regions during the 20th century resulted in the replacement of some oaks by more shade-tolerant species. Future climatic changes will likely produce rapid and important changes in the understorey light regime that can affect oak recruitment dynamics. Phenotypic plasticity will be essential for plants to acclimate to new light regimes. Photosynthetic acclimation, involving increased nitrogen allocation to photoprotection and carbon fixation capacity, occurs within tree crowns and in shaded seedlings exposed to higher light intensities (Naidu and DeLucia 1998; Rodríguez-Calcerrada et al. 2008). In the longer term, light acclimation requires integrated changes in biomass partitioning and in the hydraulic system to meet new transpiration rates.

Heat waves are emerging as a serious threat to plant survival in recent decades. High transpiration rates of small leaves of some Mediterranean oaks alleviate leaf temperature and so the risk of cell membrane and protein damage under drought and high irradiance. Larger leaves of temperate species would reach higher, more favourable leaf temperatures for photosynthesis in temperate climates, but would be more prone to suffer from overheating in Mediterranean environments (Gil-Pelegrín et al. 2017). Moreover, oaks display a wide array of photoprotective, antioxidant and repair mechanisms to cope with both high and low temperature extremes. For example, the temporary down-regulation of photochemical efficiency after heat waves and frosts increases energy dissipation and avoids photodamage under concomitant high irradiance (García-Plazaola et al. 2017). The leaves of evergreen oaks are better prepared to frost than those of deciduous ones. In any case, early leaf-out, for example after a warm winter, increases the probability of leaf damage by a late frost. Carbon reserves are usually remobilized to form a new leaf cohort when the previous one has been scorched, but at the cost of carbon available for other metabolic processes, such as secondary growth. Live cells in stems and roots are protected by bark and soil, respectively, and so less vulnerable to extreme temperatures. Still, the embolization of the xylem can be large after a severe frost, or not so severe in the case of ring-porous oaks (Cavender-Bares 2005), and has delayed effects on tree metabolism and survival.

In a context of global change, oaks will modify their physiological response as a consequence of the progressive carbonization of the atmosphere (Norby 1996). A current hot topic is to know to what extent the fertilizing impact that emerges in most short-term experiments will be held in the longer term. The debate remains from the beginning of the studies analyzing the impact of atmosphere CO_2 enrichment on carbon uptake capacity and growth of oak species (Körner et al. 2005; Leuzinger and Körner 2007). In the short term, most studies show a general increase in growth and carbon-based processes such as leaf net photosynthesis for different oaks from North-America (Norby and O'Neill 1989), Europe (Scarascia-Mugnozza et al. 1996; Aranda et al. 2020) and East Asia (Chae et al. 2016). This trend is maintained in the long term although tuned by a down-regulation of the photosynthetic capacity, as observed for other forest tree species (Ainsworth and Rogers 2007); albeit some exceptions have been also described (see Killi et al. 2018 for contrasted response in *Q. ilex* and *Q. cerris*). Thus, a general CO_2 fertilization effect would be expected on growth and physiological performance of oaks under enriched atmospheres but with a moderate impact in the long term. This effect would result in an increase in biomass accumulation together with decreases in leaf stomatal conductance (reviewed in Frank et al. 2015), increases in water use efficiency (WUE; Norby and O'Neill 1989; Aranda et al. 2020), changes in light compensation point (Kitao et al. 2016), and modification in leaf composition (Tognetti et al. 1998; Shi et al. 2016). Nonetheless, all potential changes will be constrained by the responsiveness to other co-limiting factors such as light, water, nutrients or temperature (Lukac et al. 2010; Kitao et al. 2016). These changes can be modulated at different ontogenic states, and even be modified over the tree life time (Voelker et al. 2006). Along with alterations in the physiological performance regarding water and CO_2 economy, ecological interactions between oak species and other co-occurring species could be modified in the future. These ecological changes would range from variations in the inter-specific facilitation-competition relationships, to herbivory or pest response, or even interactions with the microbiome. Other unexpected potential changes boosted by growing under CO_2-enriched atmospheres could affect life-habit traits expression such as masting (after resource use modification), seed dispersal (from changes in quality of fruits) or recruitment success (Davis et al. 2007; Mohan et al. 2007). The impact of atmospheric CO_2 enrichment on these issues has barely been studied, but they need to be understood, considering the relevance of demographic and evolutionary processes in the future of oak woodlands under global change.

Growth Patterns of *Quercus*

Plant growth is a complex process that depends on intrinsic characteristics of the species and abiotic factors. Plant growth is defined as the increase in biomass over time and can be measured as the relative growth rate (RGR), which is the relative increase in biomass per unit of time (mg g^{-1} day^{-1}). RGR has traditionally been decomposed into two underlying components: the leaf area ratio (LAR, the ratio between the total leaf area and total plant dry mass) and the net assimilation rate (NAR, biomass gain per unit area and time; Evans 1972; Poorter et al. 2012). Thus, RGR = LAR × NAR. In turn, LAR can be decomposed into the specific leaf area (SLA, leaf area per leaf dry mass) and leaf mass fraction (LMF, leaf dry mass per total plant dry mass), being LAR = SLA × LMF.

As compared to other woody species, oaks are characterized by relatively low relative growth rates (Villar et al. 2017). The low values of RGR in *Quercus* species may be due to the low LAR of these species (Figure 5), as a result of both low leaf mass fraction (LMF) and low specific leaf area (SLA). SLA and LMF are important factors that contribute to the regulation of plant photosynthesis and growth (Poorter and Remkes 1990; Antúnez et al. 2001). Thus, lower values of these traits imply a low leaf area displayed per unit total plant mass invested and consequently a lower photosynthetic capacity per unit total plant mass implying a lower RGR (Reich et al. 1997; Ruíz-Robleto and Villar 2005). The low LAR of *Quercus* species can also be explained by the high proportion of biomass present in roots (around 40%) compared to other woody species (around 20%). The advantage of plants with an inherently low SLA is that it promotes long-lived foliage, which has longer time to pay-off the construction cost of leaves (Wright et al. 2004; Escudero et al. 2017). This strategy, which promotes the duration and nutrient use efficiency over potential growth rates, is energetically and competitively more efficient in stressed environments (Reich et al. 1997), such as most Mediterranean habitats.

Growth differences have been observed between different ontogeny stages (Villar et al. 2017). This may be due to several factors such as self-shading, tissue aging and increased allocation to structural components (Rees et al. 2010). For instance, oaks seedlings depend heavily on seed reserves (Quero et al. 2007; González-Rodriguez et al. 2010); this dependence is even more important for seedlings growing under low light conditions (Quero et al. 2007). However, the influence of seed mass becomes insignificant after the first year, increasing the dependency from the resource availability of the media after this initial stage (Pérez-Ramos et al. 2010a).

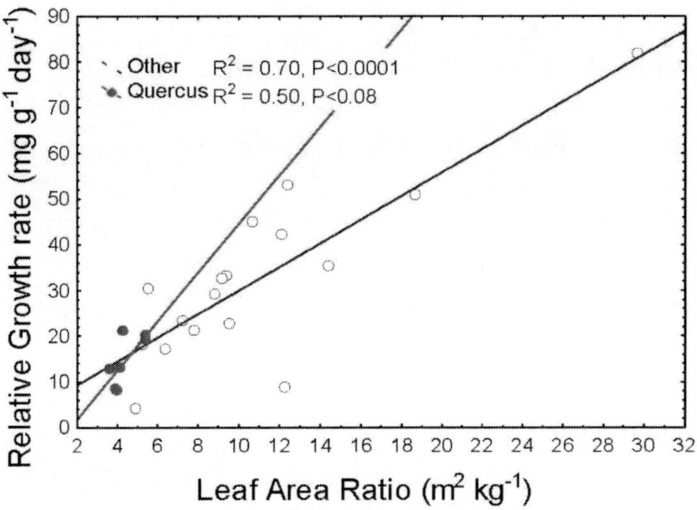

Figure 5. Relationships between RGR (Relative Growth Rate) and Leaf Area Ratio (LAR) for *Quercus* and non-*Quercus* woody species. The R^2 and the significance of the regression equation are shown.

The RGR decreases along the plant development, being negatively related with plant size (Evans 1972; Hunt 1982; Metcalf et al. 2003; Rees et al. 2010). Thus, climatic variables and resource availability are the most important factors for growth in Mediterranean oaks on the adult stage (Villar et al. 2017). However, the percentage of variation in growth explained by environmental variables (as temperature and precipitation) is lower than expected; for example, they only explain between 1 and 6% of the variation in RGR indicating that other factors, such as nutrient availability, competition and microhabitat differences, may explain the variation in growth under field conditions (Villar et al. 2017).

Field data of biomass allocation and RGR in Iberian oak trees are very scarce (but see Montero et al. 2005, Salomón et al. 2016 and Villar et al. 2017). However, they are needed to evaluate the potential for carbon sequestration of *Quercus* forests, as a measure to mitigate the increase of

CO_2 in the atmosphere. In addition, the knowledge of the factors affecting the growth of *Quercus* species may be useful to predict the variation in ecosystem services under different climate change scenarios.

Global Response - From Leaf to Ecosystem

Scaling-up exercises from organs, such as leaves, to higher organization levels in *Quercus* can be found in the scientific literature since the end of XXth century (Infante et al. 1997). These types of studies usually have been performed through either modelling (Zavala and Bravo de la Parra 2005) or experimental (Quero et al. 2008) approaches. Recently, leaf-to-ecosystem links have become more important in the context of forest decline, due to the utility of remote sensors for identifying pre-visual physiological-leaf symptoms in forest ecosystems. Calibrating remote hyperspectral images at ecosystem level with tree crown and leaf behaviour is a promising tool to detect early warning signals of forest decline.

Current trends of drought increase have led to diverse impacts in forest communities such as growth reduction, physiological changes, defoliation, plagues and diseases (Anderegg et al. 2015). Ultimately, water stress may increase embolism risk or reserves depletion, due to long-term stomatal closure (McDowell et al. 2008; Anderegg et al. 2015), conducting to oak tree mortality. This phenomenon has already been observed in *Quercus* forests at the Iberian Peninsula (Lloret et al. 2015), affecting gas exchanges between tree and atmosphere (Poyatos et al. 2013). These processes show a direct relationship between plant activity and aridity in *Quercus* forests (Peñuelas et al. 2018), mainly in their drier distribution limits.

The analysis of climate impacts on forest systems worldwide is conducted mostly by experimental plots, equipped with instruments to measure sap flow, soil moisture, leaf area, pigment content, water potential and photosynthetic efficiency. This type of infrastructure allows knowing in detail the processes of response to the climate at physiological level (Hernández-Clemente et al., 2017). Thanks to this kind of data, great progress has been made in the last decade developing quantitative models to estimate leaf variables at crown and stand levels, from hyperspectral and thermal imagery of high spectral and spatial resolution (Hernández-Clemente et al. 2011, 2012). On the other hand, early detection of changes in these variables may allow forest managers to implement more effective strategies in advance, when symptoms are pre-visual and still reversible.

Thus, using image analyses, the photosynthetic pigments such as chlorophyll (Ca+b) and carotenoids (Cx+c) have successfully been quantified from specifics absorption shown in the visible region (Hernández-Clemente et al. 2012) or using ratios between visible and near infrared (NIR; Main et al. 2011). Additionally, the xanthophyll cycle is related to the photochemical reflectance index (PRI), being employed in stress detection of leaf senescence and chlorosis processes (Hernández-Clemente et al. 2011). Regarding photosynthetic efficiency, the excess of photochemical energy to carry out photosynthesis is re-emitted as sun-induced chlorophyll fluorescence (SIF; Dobrowski et al. 2005). In summary, SIF is a weak emission produced by leaves and considered a powerful proxy of photosynthetic activity at stand level and therefore a non-invasive indicator to monitor vegetation status, resilience and recovery. On the other hand, water potential and stomatal conductance have been estimated from spectral information in thermal infrared (TIR; 8.0-14.0 μm). The differences between air and crown temperatures (Tc-Ta) are considered as an estimator of water pressure deficit (Jackson et al. 1981), suggesting that higher Tc-Ta values correlates to lower water potentials. Stomatal conductance is also used to relate stress level and canopy temperature (Moran et al. 1994). Stomatal closure modifies energy dissipation in the form of latent heat from regulating transpiration (Panda et al. 2014), being related to vegetation indices from hyperspectral data (Hernández-Clemente et al. 2011; Juntila et al. 2016).

Recent technological advances in remote sensing have developed hyperspectral sensors with higher spatial and spectral resolutions (Thenkabail and Lyon 2016). Several satellites are currently available (Landsat, Aster, Chris-Proba, Sentinel 2 A, etc.) that provide medium and low spatial resolution images and have been widely validated by the scientific community (Tang et al. 2016; Olsson et al. 2016; Sulla-Menashe et al. 2016). Thus, NASA is focused on the HyspIRI, ECOSTRESS and

THIRSTY (NASA-CNES) missions, the immediate future of hyperspectral remote sensing and high spatial resolution. It is worth highlighting the COPERNICUS Sentinel satellites from ESA, with the current Sentinel 2A sensor, which will provide images of the Earth with medium spatial resolution (10 m) through a wide range of sensors that cover the visible IR and thermal ranges. In the near future, the ESA program even includes a specific sensor for the estimation of photosynthetic efficiency, the Fluorescence Explorer (FLEX). These missions are designed, among other objectives, to improve the monitoring of terrestrial ecosystems, and will allow a better evaluation of the impact of climate change on them.

The use and application of these satellital sensors on *Quercus* forests for the diagnosis and evaluation of tree mortality processes will depend on the development of models that provide information about forest physiological status extracted from images of high spatial and spectral resolution (aereal sensor). On the other hand, these new technologies have initiated a change of scale in the generation of information on forest ecosystems, since they have opened the door to using high-resolution data on large areas (see Figure 4 in Hartmann et al. 2018). Several authors have demonstrated the utility of working with high spectral and spatial resolution data in the direct quantification of physiological variables, or the use of models that allow extrapolating data at larger scales (Zarco-Tejada et al. 2013). Such modeling becomes highly necessary for the study of *Quercus* forests that generally includes large areas of land. In this sense, the relationship between the real physiological status of the stand and the remote sensing estimates constitute the fundamental basis for developing predictive models that allow predicting the evolution of mortality processes over time. In fact, the link between remote sensing and physiology has recently been considered as one of the current priorities of global change research in Mediterranean terrestrial ecosystems (Hernández-Clemente et al. 2017). Their inherent heterogeneity and complexity require model validation from leaf-crown-stand to ecosystem level.

Demographic Responses to Global Change

Regeneration Failures under Different Climatic Scenarios

Quercus species are particularly vulnerable during their early life-history stages due to limitations imposed by several factors (Pérez-Ramos 2014). Since most of these life-history stages strongly depend on climate conditions (highlighting the influence of water availability), the rising aridity forecasted by climate change models for the Mediterranean area could potentially put in risk the persistence of Iberian oak populations and, consequently, trigger shifts in forest community dynamics and composition.

First, seed production is largely variable among years due to the broadly-known masting behaviour of oak species, and this temporal variability seems to be partially driven by precipitation patterns (e.g., Espelta et al. 2008; Pérez-Ramos et al. 2010b; Figure 6). A recent meta-analysis study of Mediterranean oaks (Pérez-Ramos et al. 2015) demonstrated that the influence of weather on mast-seeding is largely species-specific. Thus, the decline of seed production with drier conditions will be potentially greater in deciduous species, particularly those inhabiting wetter sites. In a wider global study (using time series for 219 plant species from all the continents) the masting phenomenon was associated predominantly to plant species with low foliar nutrient content, which are more frequent in resource-limited environments (Fernández-Martínez et al. 2019). These studies support that oak species inhabiting resource-poor sites (like most under Mediterranean conditions) would exhibit strong masting intensity, with relevant differences among species in their sensitivities to climatic variation. To what extent co-occurring species with contrasting functional attributes will respond to ongoing environmental changes in terms of seed output constitutes an exciting question to be further investigated not only in Mediterranean oaks but in other plant species world-wide.

Second, mature acorns are heavily consumed by different animals due to their large size and high nutritional content; therefore, seed predation is considered one of the most important bottlenecks for regeneration in oak species (Pulido and Díaz 2005; Pérez-Ramos et al. 2012). These regeneration failures over the seed-seedling transition are intensified in those ecosystems with super-abundance of wild and/or domestic herbivores, such as savanna-like ecosystems or *dehesas* (Pulido et al. 2010; Figure 6).

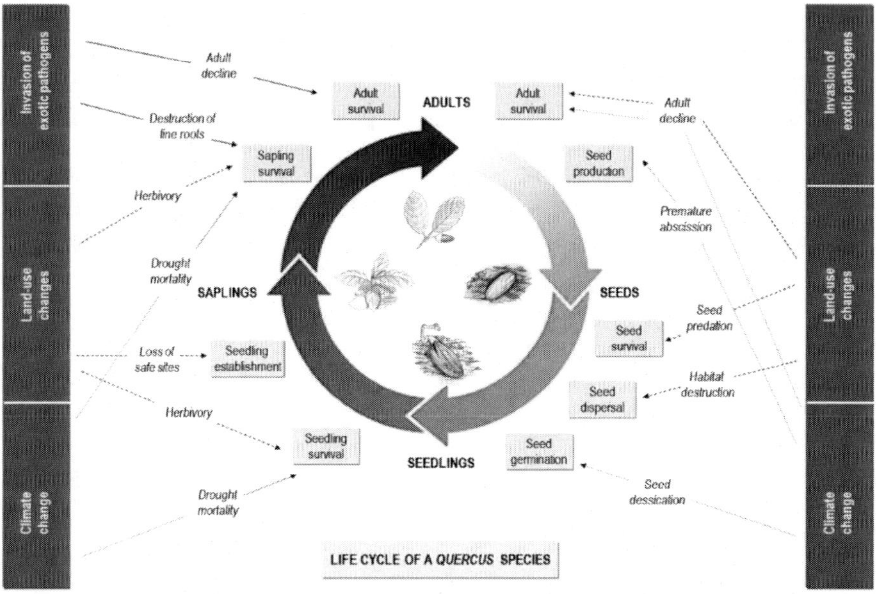

Figure 6. Summary diagram on the potential threats facing oaks over their life cycle due to three principal drivers of global change: climate change, land-use changes and invasion of exotic soil-borne pathogens. Author: Ignacio M. Pérez Ramos.

Third, oak seedling establishment is highly dependent on soil water content (Figure 6). On the one hand, acorns are strongly recalcitrant and require wet soil conditions for a successful germination (Finch-Savage 1992; Gómez-Aparicio et al. 2008; Urbieta et al. 2008). On the other hand, seedlings are particularly sensitive to water limitation and exhibit high mortality rates during their first years of life due to the summer water stress typical of Mediterranean climate (Pulido and Díaz 2005; Pérez-Ramos and Marañón 2012). Altogether, this implies that oak recruitment is commonly restricted to some sporadic years with the combination of large seed crops ("masting years"), where the large abundance of seeds may satiate the diverse assemblage of animals that feed on acorns (Kelly and Sork 2002), and high soil moisture ("wet years") when a higher proportion of the resulting first-year seedlings may survive the summer drought.

Finally, these regeneration problems in oaks could be aggravated due to the increasing attack of exotic soil-borne pathogens such as the oomycetes *Phytophthora* sp. and *Phythium* sp. (Gómez-Aparicio et al. 2012; Ibáñez et al. 2015; Figure 6). These soil pathogens seem to be involved in the extensive decline of some Mediterranean oak species (Pautasso et al. 2010; Garbelotto and Pautasso 2012). In particular, trees of *Q. ilex* and *Q. suber* species inhabiting forests and savanna-like ecosystems of the Mediterranean Basin are susceptible to this root disease (Sánchez et al. 2006; Sampaio et al. 2013). Although there is growing evidence that *P. cinnamomi* is potentially able to cause mortality by itself on healthy trees and seedlings, some studies have demonstrated that drought events might increase their susceptibility to suffer the disease (Roubtsova and Bostock 2009; Corcobado et al. 2014). However, from the pathogen´s perspective, results from a recent study under controlled conditions (Homet et al. 2019) suggest that an average drier climate might imply sub-optimal conditions for *P. cinnamomi* infections (and, therefore, a slower advance of the disease), due to the high water requirements of the oomycetes for the germination and dispersal of their zoospores (Desprez-Loustau et al. 2006). Therefore, understanding how the ongoing environmental conditions will influence this particular plant-pathogen interaction is not an easy task, since it will depend on complex responses of host resistance and pathogen growth to increasing levels of soil aridity.

Spatial Patterns of Oak Recruitment and Climate Change

It is broadly known that recruitment-driving processes in oak species are strongly context-dependent, appearing different types of plant–microorganism and plant–animal interactions as a function of the microsite where the seed is dispersed (Schupp and Fuentes 1995; Hulme 1997). On the one hand, seed rain varies spatially due to the non-random foraging patterns of seed predators and dispersers, which respond to preferences for certain microhabitats and rejection of others (Puerta-Piñero et al. 2010; Pérez-Ramos et al. 2013a). On the other hand, the spatial distribution of seed deposition across different microhabitats offers a wide mosaic of different micro-climatic conditions that will affect the probability of successful seedling establishment (Jordano and Herrera 1995; Rey and Alcántara 2000). Interestingly, suitability for recruitment of different microhabitats can also change over the life-cycle of the plant (concept of "ontogenetic conflict," *sensu* Schupp 1995) since the most favourable conditions for one stage may not be the most favourable for others (Pérez-Ramos et al. 2012). Due to the existence of spatially discordant patterns through plant ontogeny, a complete determination of microhabitat suitability for recruitment requires a multi-stage demographic approach where overall recruitment is quantified considering the largest possible number of stage-specific probabilities (Clark et al. 1999). Nevertheless, most studies aimed to identify safe sites for recruitment in oak species are only based on some early life-history stages (seed and seedling stages being often explored separately), potentially leading to incomplete or even misleading conclusions (Pérez-Ramos 2014). Information provided by these multi-stage demographic studies could be also used as a valuable tool for restoration and reforestation plans in degraded ecosystems dominated by oaks since they synthesize the most optimal conditions for natural recruitment of target species.

In a context of climate change, this complex network of interactions that governs both the magnitude and the spatial patterns of overall oak recruitment, could be seriously altered in the future by the expected decline in water availability. On the one hand, previous manipulative studies of rainfall exclusion have suggested that the ongoing increasing aridity might exacerbate recruitment limitation in *Quercus* species by aggravating the negative effects of water stress on different demographic processes (Lloret et al. 2004; Matías et al. 2011; Pérez-Ramos et al. 2013b). On the other hand, projected environmental changes might alter the spatial patterns of oak recruitment, as shown in a previous study where an experimental increase in drought displaced the peak of maximum overall recruitment towards microhabitats located under shrubs (Pérez-Ramos et al. 2013b). These authors suggested that the dependence on shrubs for an effective recruitment in some oak species could be intensified under future environmental scenarios of decreasing soil water. Understanding how plant communities will respond to projected changes of climatic conditions provides not only interesting basic knowledge for oak population dynamics, but also valuable applied information for ecosystem management since it will help us to implement new plans that contribute to forest resilience and foster plant adaptations to these ongoing climatic changes.

Responses in Terms of Growth, Vitality and Mortality

Recent changes in forest structure, growth and productivity, and in tree vigour (canopy dieback, defoliation) and mortality rates across global to continental scales have been attributed to rising temperatures and drought stress (Allen et al. 2010; Carnicer et al. 2011). Iberian oak forests are also vulnerable to these global-change drivers as several cases of drought-induced growth decline and dieback have demonstrated (e.g., Camarero et al. 2004, 2015, 2016; Gea-Izquierdo and Cañellas 2014). Usually, oak trees showing lower growth rates are more responsive to climatic constraints, and those producing vessels with smaller areas are more prone to drought-induced leaf shedding and dieback (Camarero et al. 2016). These are common responses to drought stress.

Radial growth of Mediterranean deciduous oak species is improved by wet-cool conditions in spring (Tessier et al. 1994), whereas growth of major evergreen species (as *Q. ilex*), which often start growing later as compared with coexisting deciduous species (e.g., *Q. faginea*), positively responds to high precipitation in spring but also in late summer and autumn (Gratani 1996; Montserrat-Martí et al. 2009). In *Q. ilex* trees, spring and late-summer to autumn rainfall enhance primary and secondary growth which explains its bimodal growth pattern, particularly in coastal sites under mild climate conditions (Campelo et al. 2009; Gutiérrez et al. 2011). In contrast, in continental inland sites, for instance in most of the Iberian plateau and mountains, winter coldness and summer drought constrain its growth and the corresponding growth pattern tends towards unimodality with a major peak in spring (Corcuera et al. 2004b). This plastic growth behaviour indicates that drought severity, duration and timing will differently affect *Q. ilex* stands depending on their location. First, the expected strongest impacts of winter-spring droughts on growth can occur on inland populations, whereas summer-autumn drought will impact more negatively coastal stands which usually receive precipitation due to convective Mediterranean storms during the late growing season. Second, drought timing is critical since it can determine the differential impact on primary (e.g., leaf production) and secondary (wood formation) growth (Misson et al. 2011; Camarero et al. 2015) so uncoupling tree phenological-constrained responses to water shortage and high temperatures. Therefore, growth duration is a major variable to better understand the long-term responses of Mediterranean oaks to drought (Lempereur et al. 2015), and to elucidate the potential roles, if any, played by acclimation (Barbeta et al. 2013).

Droughts are also associated to the formation of narrow xylem vessels in the evergreen *Q. ilex* which forms diffuse-porous or semi-ring porous wood (Villar-Salvador et al. 1997; Corcuera et al. 2004b), but also in deciduous, ring-porous species as *Q. faginea* and *Q. canariensis* (Corcuera et al. 2004a; Alla and Camarero 2012; Gea-Izquierdo et al. 2012). However, reductions in leaf area are more conspicuous and frequently observed in response to drought than changes in anatomical traits, at least in *Q. ilex* subjected to rainfall exclusion treatments (Limousin et al. 2010). Probably, it is more relevant to consider long-term changes in the ratio between transpiring and conducting areas or Huber values, i.e., the ratio sapwood area: leaf area. Anyway, drought and high temperatures in spring-summer (March, April, June) trigger a decrease in leaf production and transpiring area (Ogaya and Peñuelas 2006; Limousin et al. 2010, 2012), and an increase in leaf shedding, i.e., greater crown transparency and lower vigour associated to canopy dieback (Sánchez-Salguero et al. 2017). This explains that defoliation rose in Iberian oak species during the severe droughts (1994-1995, 2005 and 2012) which were accompanied by elevated spring temperatures (Lloret and Siscart 1995; Peñuelas et al. 2001; De la Cruz et al. 2014; Camarero et al. 2015, 2016). It is unclear how this loss in vigour associated with increased crown transparency or defoliation translates into increasing mortality rates in the long term. For instance, in a *Q. ilex* coppice forest monitored for the last 15 years, which was impacted by the 2005 and 2012 droughts, the mean mortality rate was only 0.2% yr^{-1} but rose to 1.3% yr^{-1} when considering only the dry years (Camarero et al. 2015). Forest inventory data will probably be unable to capture the localized negative impacts of drought on oak growth and mortality. Gathering this relevant information will require combining specific networks of field sites showing intense dieback (see for instance https://decaimientoencinar.wixsite.com/decaimientoencinar) with remote sensing information.

How global-change components such as climate warming and land-use changes interact in the long term is still understudied and will require additional and creative efforts. For instance, formerly coppiced *Q. pyrenaica* stands are nowadays overaged, and the competition between stools leads to reduced growth and decreased hydraulic conductivity making them prone to drought-induced dieback (Corcuera et al. 2006). Can thinning or restoring of coppicing systems in some Iberian oak forests provide them more resilience to drought? This question deserves to be answered by comparing thinning treatments in plots subjected to long-term monitoring (e.g., Rodríguez-Calcerrada et al. 2011). In addition, other resilience mechanisms related to shoot growth and canopy recovery are understudied. For example, buds in the evergreen *Q. ilex* may survive up to four years, while buds in the deciduous *Q. faginea* last for one year (Alla et al. 2013). Last but not least, forest densification in the Iberian Peninsula can lead to increased competition for water and wildfire risk exacerbating the negative impacts of drought on oak forests and decreasing their productivity making them more vulnerable to climate warming (Gómez-Aparicio et al. 2011).

Case Studies of Iberian Oak Systems under Global Change

Management of Iberian Agro-Silvo-Pastoral Systems (*Dehesas* and *Montados*)

Dehesas in Spain and *montados* in Portugal are the most spread agro-silvo-pastoral systems in the Iberian Peninsula (Fernández-Rebollo et al. 2014). These systems are the result of human intervention over natural oak woodlands, reducing trees and shrubs coverage to enhance pasture and oak productions and to enable occasional cereal and legume cropping. Hence, the vegetation in this ecosystem is characterised by a two-layered structure in the same unit of land: a tree layer in low density and an understorey of pasture or crop. The holm oak (*Q. ilex*) is the dominant tree species in the *dehesa*, while the *montado* is mainly occupied by cork oaks (*Q. suber*). Although both are multipurpose systems, livestock farming is the main source of income: flocks of sheep and herds of beef cattle, combined with Iberian pigs (in the case of *dehesa*), graze extensively, feeding themselves mainly on pasture, acorns and crops. *Dehesas* and *montados* constitute often small game reserves and, occasionally, are integrated into big game hunting grounds. Oaks also provide valuable forest products for the farm economy, such as firewood and cork.

Farming management of *dehesas* and *montados* encourages structural diversity at stand and landscape level, which combined with its wide distribution area and spatial continuity turn out in high levels of biodiversity. *Dehesas* and *montados* are listed under EU Habitats Directive (habitat 6310) and form part of the Natura 2000 network. Based on that, the conservation of biodiversity across large areas depends on the continuation of extensive farming management (Beaufoy and Cooper 2008), these systems have been classified as High Nature Value (HNV) by the European Commission. From a cultural and recreational point of view, *dehesas* and *montados* are valued for the aesthetics of their landscape and cultural heritage. The regulation services provided by these farming systems, such as soil and water conservation or carbon sequestration, are also noteworthy.

Oaks are a key element in the provision of ecosystem goods and services. However, their contribution to offer particular services is strongly affected by management practices and global change, in particular the expansion of pathogens and climate change (Figure 7). Hence, a major challenge is to encourage the most appropriate management practices to maintain oak trees and its functions while ensuring the profitability of the system. To achieve this goal, a first step is to understand better the trade-offs, synergies and feedbacks between productivity and the delivery of ecosystem services, in these changing circumstances.

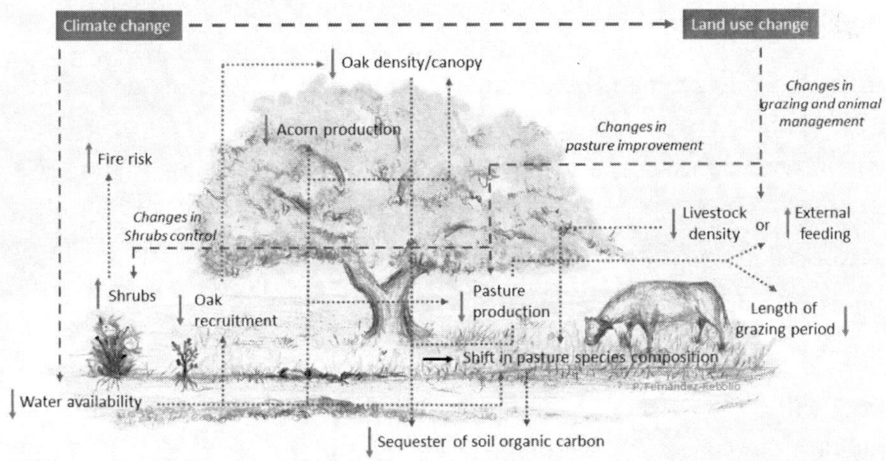

Figure 7. Interactions among the major components in the management of the agro-silvo-pastoral system *dehesa* and the effects of global change drivers (climate change and land use change). Author: Pilar Fernández-Rebollo.

The Mediterranean Basin is especially vulnerable to climate change, with pronounced increases in mean annual temperatures and reductions in overall rainfall, accompanied by changes in seasonal rainfall distribution and increased inter-annual variability (Bates et al. 2008). In these changing conditions, the agro-silvo-pastoral systems may not be sufficiently productive to support livestock and oak productions. Moreover, oak tree cover is likely to decrease as a result of climate change. Water and soil conservation within the farm boundaries will become key management points in the future. They will be achieved by reducing run-off and increasing soil water holding capacity (which in turn may imply at landscape level a decrease in water harvesting for other purposes, as irrigated agriculture or domestic water use), and by a proper allocation of water resource between pastures and oak trees.

In a predicted scenario of lower rainfall, farms should maintain shrubs control through a combination of agricultural practices as clearing, rotational crops, pasture sowing and grazing management to avoid the competition for water and nutrients between oaks and shrubs (Cubera and Moreno 2007; García-Moreno et al. 2014). Since there is general consensus that the nature and extent of interactions between shrubs and oaks are highly species-dependent, shrubs should not be eradicated from farms; thus, patches of this type of vegetation should be kept in some enclaves, especially highly diverse shrubs that enhance oak regeneration (Pulido et al. 2010; Rolo and Moreno 2011; García-Moreno et al. 2014). Moreover, the lack of appropriate control could promote shrub encroachment which, in turn, will increase the risk of fire, threatening the persistence of the oak. With regard to shrub control and the improvement of permanent pasture, there is increasing interest in the use of deep-rooting herbs, C4 grasses and legume mixes (Porqueddu et al. 2015). However, in spite of further experimentation it is required to provide information on the appropriate components of the mixture and its management; the effects of these grassland species on oaks (mature and seedling) should also be explored.

Livestock grazing regulates the competition for water between pasture and oaks. Pasture defoliation by grazers reduces leaf tissue and therefore transpiration, which increases, at any given time, the availability of water for the oak tree. Defoliation can also modulate the depth that herb roots can reach (Ingram et al. 2008), further separating the rooting-niche for the two functional groups, pasture and oaks (Moreno and Pulido 2009). In consequence, extensive grazing constitutes a valuable and suitable tool for maintaining healthy oak trees in those systems, improving the availability of water for trees, especially in late spring when water becomes scarce. Grazing also alters the nutrient cycle. In the Mediterranean permanent grassland, legumes increase with grazing; many of them have a small size and a prostrate growth habit; both traits promoted by the grazing pressure (Diaz et al. 2007; Tárrega et al. 2009). Their ability to fix atmospheric N increases soil fertility and improves the nutritional quality of pasture, reducing the need for feed concentrates and N-fertilization. In addition, grazing animals may accelerate nutrient recycling through the excreta, mainly by increasing the size of the inorganic N pool available in the soil (Augustine and Macnaughton 2006). The effects of grazing on soil nutrient availability have supported the hypothesis of the "grazing optimization theory," which suggests that extensive grazing may increase ecosystem production compared with non-grazing situation. Although this theory has been proved in different types of pastures, further research is needed to understand how grazing-mediated soil nutrient cycle affects the performance of oak trees and what consequences will have for them in a context of low water availability. Some studies reporting increased annual shoot growth, leaf N concentration and acorn production suggest that oak trees may also benefit from the effects of grazing on soil nutrient availability (Rolo et al. 2012; García-Moreno et al. 2014). The implementation of new grazing methods (as low or fast rotation methods) that encompass water conservation along with other management objectives, such as reserving pastures, allowing their seed bank to be recharged or maintaining them in good condition will result in a shortening of the period of water stress for oak tree.

Oak regeneration in *dehesas* and *montados* is negatively affected by livestock grazing (López-Sánchez et al. 2016). Livestock browse oak seedlings and consume acorns, thus interfering in the oak regeneration process. Other factors inhibiting oak regeneration are the high populations of rodents and soil compaction; moreover, with the predicted reductions in rainfall and the increase in inter-annual variability, the recurring droughts will arise as a main bottleneck for oak recruitment (see more details in the section on oak regeneration above). Regardless of the cause of the lack of regeneration, several management practices are applied to maintain the oak woodland; the more common is sowing acorns or planting seedlings at low density and protect them from the livestock. The use of healthy nursery stocks is vital to avoid the introduction of new tree pests or diseases into the farm, as it happened with the soilborne pathogen *Phytophthora cinnamomi*. Probably the initial source of infection by this soil pathogen was the large-scale afforestation program funded by the European Commission in 1992 (Fernández-Habas et al. 2019). In addition, a deep soil preparation of the planting site will be crucial in the future to improve initial seedling survival.

The future existence of the *dehesa* and *montado* and with them, the provision of ecosystem services, is linked to their ability to produce food and raw materials in a context of climate change. The redesign of these oak systems and the search for suitable management practices will be a major challenge for the Mediterranean-climate regions of the Iberian Peninsula.

Portuguese Cork Oak Land-Use Systems under Global Change

The cork oak (*Quercus suber* L.) is today restricted to the western part of the Mediterranean Basin, its largest extent being in Portugal (720,000 ha in 2015), about 30% of the species area worldwide (Pausas et al. 2009a; ICNF 2015). The cork oak woodland is a cultural landscape, created and maintained by human management where patches of changing tree density form a unique large-scale landscape mosaic (Pinto-Correia et al. 2011). This mosaic comprises, on one hand, cork oak agro-silvo-pastoral systems (*montados*) with an open tree layer, managed for cork production and in the understorey pastures used for grazing or crops, on the other hand, cork oak woodlands with higher tree density and solely managed for cork production (Figure 8). In most cases, and independently of the type of land-use system, cork is the most valuable product in the exploitation (Pinto-Correia and Mascarenhas 1999; Mendes and Graça 2009). Cork oak land use systems provide a large variety of ecosystem services (Bugalho et al. 2011) and are protected under the European Union Habitats Directive (92/43/CEE) and included in the Natura 2000 network (European Commission 2015). Portugal is the largest cork producer and exporter, with more than 50% of worldwide cork production. Cork exports represent approximately 2% of total Portuguese exports, with an increasing trend since 2009, having reached 986 Million euros in 2017 (APCOR 2019).

Figure 8. Cork oak woodland in Serra do Caldeirão, Algarve, Portugal. The reddish trunk indicates that the bark has been recently stripped off to obtain cork. They have the number nine to mark that they were harvested in 2019; the next harvest will be in 9-11 years. Author: Vanda Acácio.

According to national forest inventories, the cork oak area in Portugal remained relatively stable, with a minor decrease, from about 747,000 ha in 1995 to 720,000 ha in 2015 (ICNF 2015). However, cork oak agro-silvopastoral systems are threatened by lack of natural regeneration and insufficient tree recruitment for the renewal of stands (Pulido and Díaz 2002; Costa and Pereira 2007; Pausas et al. 2009b; Pinto-Correia et al. 2011). In 2015, about 50% of pure Portuguese cork oak stands presented sparse canopy cover (10-30%), 34% had more than 50 years old, and 33% presented low tree density (< 40 trees per hectare) (ICNF 2015). Furthermore, between 1995 and 2005, stands with less than 20% of crown cover increased in frequency (+9%) (Paulo et al. 2015) and cork oak defoliation rates registered a gradual increase (Sousa et al. 2007). In addition, 330,000 dead cork oaks were reported for an area of 712,000 ha in southern Portugal between 2004 and 2006 (Ribeiro and Surový 2008).

Cork oak decline in Portugal over the last decades has been associated with a complex interaction of human, climatic and biotic disturbances (Sousa et al. 2007). Mismanagement such as livestock overgrazing led to decreased cork oak natural regeneration, increased soil compaction, and oak root damage (Branco and Ramos 2009; Pinto-Correia and Godinho 2013; Godinho et al. 2016). Ploughing and shrub clearance with heavy machinery have also damaged natural regeneration and tree root systems, with further soil compaction and erosion, and intensive silvicultural practices such as pruning and cork extraction have weakened the trees (Cañellas and Montero 2002). Extreme droughts have also been associated with cork oak physiological stress that may have led to tree defoliation and mortality (Carnicer et al. 2011; Caldeira et al. 2015). Climatic disturbances may further enhance oak vulnerability to pests and diseases (Brasier 1996; Moreira and Martins 2005; Desprez-Loustau et al. 2006; Sousa et al. 2007; Branco et al. 2014; Tiberi et al. 2016). A major pathogen associated with increased cork oak mortality in Portugal in the last decades is the root rot oomycete *Phytophthora cinnamomi* (Brasier 1996; Bergot et al. 2004; Moreira and Martins 2005; Branco and Ramos 2009; Camilo-Alves et al. 2013).

Drastic changes in cork oak areas are also caused by land use changes. The expansion of fast-growing eucalypt and pine plantations supported by several afforestation policies have taken place in marginal lands with cork oak trees already under decay, in depleted soils or steep slopes with low cork production (Pinto-Correia and Godinho 2013; Costa et al. 2014; Acácio et al. 2017). On the other hand, land abandonment has led to the rapid expansion of native gum rockrose (*Cistus ladanifer)* shrubs in cork oak agro-silvo-pastoral systems (Acácio et al. 2009; Bugalho et al. 2011). These shrublands have formed dense and persistent canopies in southern Portugal, particularly in drier areas (Acácio et al. 2009; Costa et al. 2009), with invasive behaviour (Caldeira et al. 2015). The dominance of *Cistus* shrublands can be interpreted as an "arrested succession," hampering cork oak recruitment through different processes: they impede cork oak seedling establishment (Acácio et al. 2007), affect the physiological performance of adult cork oak trees (Caldeira et al. 2015), and act as reservoirs for oak diseases such as *P. cinnamomi* (Moreira and Martins 2005). In addition, competition between *Cistus* shrubs and cork oak seedlings tends to prevail over facilitation in harsh soil conditions of limited moisture and nutrients (Acácio et al. 2007; Caldeira et al. 2015).

Land abandonment with shrub encroachment, in combination with climate change, are associated with an increased burned area and wildfire severity over the last decades in Portugal (Marques et al. 2011; Moreira et al. 2011). In turn, recurrent wildfires may contribute to maintain and expand shrubland cover (Curt and Pausas 2010; Moreira et al. 2011; Acácio et al. 2009, 2017), since *Cistus* spp. are pyrophytic species that colonize extensive areas after fire (Corral et al. 1990). In spite of the relative resistance of *Q. suber* to fire, the increasing wildfire severity may increase oak tree mortality, negatively affect cork oak resprouting and seedling regeneration (Catry et al. 2012, 2013) and increase pest attacks on burned cork oak trees (Catry et al. 2017). Such vegetation shifts (from woodlands to shrublands) are also predicted by future climate scenarios that show a drastic reduction in cork oak area, especially in southern Iberian Peninsula, due to difficulties in cork oak regeneration and increasing mortality (Pereira et al. 2002; Correia et al. 2018), while shrublands are expected to expand in cork oak areas (Pereira et al. 2002; Ruiz-Labourdette et al. 2013; Caldeira et al. 2015; Baudena et al. 2020).

The long-term viability of the cork oak land-use systems in Portugal may be in jeopardy: concerted management, active restoration and sanitary actions are needed to release the arrested succession, protect natural regeneration and improve cork oak health, growth and recruitment towards forest recovery. Past programs of reforestation and afforestation have frequently resulted in high mortality of oak seedlings due to low genetic quality of reproductive material, inadequate planting techniques combined with summer drought (Almeida et al. 2009), and lack of proper management interventions and supervision during the first years. On the other hand, conservation and restoration of aging and degraded cork oak areas might be prioritized in terms of biodiversity preservation and carbon mitigation (Ovando et al. 2009).

New restoration techniques for degraded cork oak stands have been recently developed from several research projects, and include, among others, the use of good quality and mycorrhized seedlings, water harvesting with microcatchments, soil mulching and organic amendments, and the use of tree shelters to improve seedling survival and growth (Vallejo 2005; Cortina et al. 2009; Coello et al. 2015). More investment is also needed on resistant genotype selection (e.g., to higher temperatures and greater water stress; Rodriguez-Molina et al. 2002; Cheaib et al. 2012) and experimental treatments of infected oaks (Branco and Ramos 2009). At the landscape scale, restoration programs should include a combination of actions such as mixed plantations of oaks, small fruit trees and aromatic shrubs; increasing tree density and understory species composition in open woodlands; improving oak natural regeneration through rotating livestock systems; simulating natural fire breaks by creating a mosaic of forest gaps; controlling the dispersion of *Cistus* spp. through harvesting for *Cistus* distillation; and plantation of fruit and honey shrub species (Regato and Berrahmouni 2005). Novel management practices such as irrigation of cork oak stands are being tested by the cork industry in Portugal as a possibility to improve the productivity of stands (Correia et al. 2018). However, this option may be costly in regions of limited water availability under predicted climate change (Correia et al. 2018). Additionally, payment for ecosystem services, in collaboration with key-stakeholders at different levels of management, is a promising solution to produce novel economic incentives to promote the sustainable use, conservation and restoration of cork oak land-use systems (Bugalho et al. 2011; Mansourian and Vallauri 2012; Rosário et al. 2019).

The Guadiamar Green Corridor and the Phytoremediation of Contaminated Lands

Soil degradation is one of the main environmental problems in the Mediterranean countries (Jones et al. 2012). Intensification of agriculture, losses of organic matter, erosion and contamination are considered as the main stressors for European and Mediterranean soils, and it has been estimated that the impacts of these factors on soil biodiversity are potentially larger than the effects of climate change drivers on soil biota (Gardi et al. 2013). In the Iberian Peninsula soil contamination is an important issue in certain areas; mainly due to the geological context and the long history of mining activities. By the end of the 20th century, Spain was the European country with the largest numbers of abandoned mine sites (Martínez-Frías 1997). This situation has resulted in important cases of diffuse and point-source contamination by trace elements (TE) in soils from active and abandoned mining districts.

Trace element contamination may have severe consequences for soil and ecosystem functioning. In order to reduce the risks derived from this type of contamination and to restore soil functioning after pollution episodes several techniques have been developed over the last decades. Among them, phytoremediation is considered as one of the most feasible and cost-efficient alternatives (Garbisu et al. 2002). Phytoremediation includes the use of plants and their associated microorganisms to either remove TE from soil through the accumulation of these elements in harvestable plant biomass (the so-called phytoextraction technique) or to immobilize these elements in soil, by promoting a range of mechanisms that reduce the fraction of TE that is available to soil microorganisms and plants (the phytostabilization approach). The need for exploring native plant species with potential for the phytoremediation of contaminated Mediterranean areas has motivated a number of studies assessing the tolerance of different Mediterranean species to TE, the patterns of accumulation of these elements across species, and the major constrains for the establishment of Mediterranean plant species in contaminated sites. Here we resume the current knowledge about the use of Holm oak (*Quercus ilex* subsp. *ballota*) for the phytoremediation of TE-contaminated lands. Most of this information has been gathered during the monitoring of a large-scale phytoremediation program (the Guadiamar Green Corridor, in SW Spain) implemented in 1999 after one of the largest pollution episodes occurred in Europe: the Aznalcóllar mining accident (Madejón et al. 2018).

Holm oak is the dominant tree in many areas in the Mediterranean Basin. Its wide spatial distribution has promoted its use as biomonitor of air or soil pollution, and thus several studies have analyzed the spatial or temporal distribution of TE in Holm oak biomass. This tree has been used to infer the exposure to airbone contaminants in different regions in Italy (Maisto et al. 2004; Nicola et al. 2013). Bark collected from the annual segments of tree branches can be analyzed to evaluate the trends in air pollution going back 10 to 15 years (Drava et al. 2017). In the case of the Aznalcóllar accident, monitoring of *Q. ilex* leaves and fruits showed a reduction in the levels of As, Pb and Fe over the first three years after the mining accident, reflecting a decrease in the contamination of leaves by soil particles (Madejón et al. 2006).

If *Q. ilex* is going to be used in the phytoremediation of metal contaminated sites it is important to consider the patterns of uptake of the specific contaminants through roots and their translocation into the aboveground biomass. Holm oak accumulates Mn into leaves, a feature which is common to many species in the *Quercus* genus (Madejón et al. 2006). However, this accumulation is not high enough to justify its use for the phytoextraccion of Mn in soils contaminated by this metal. In addition, growth rates of this species in degraded areas are usually low, so that the removal of significant amounts of Mn or any other metal from soil could take, depending on the contamination level, a considerable amount of time. Therefore, this species is not suitable for the phytoextraction of contaminants from soil, as most of woody plants (Audet and Charest 2007).

In contrast, Holm oak can have a high potential for the phytostabilization of TE-contaminated soils. The main objective of a phytostabilization program is to maintain a low level of accumulation of TE in the aboveground biomass of the plants used, so that the risk of transfer of these contaminants to the food chain is minimized. Monitoring of TE accumulation in the leaves of afforested individuals in the Guadiamar Green Corridor showed that this species excludes TE from its aboveground biomass; TE concentrations in leaves were in all cases within the normal ranges, and did not respond to increases in the availability of TE in acidified soils (Domínguez et al. 2008). Under controlled greenhouse conditions the ability of this species to retain Cd in the fine roots and to exclude its accumulation into leaves was confirmed (Domínguez et al. 2009).

Tolerance of *Q. ilex* to high concentrations of Cd in the growing media was reasonably high, and although toxicity symptoms were developed at high doses (much higher than the level typically found in contaminated areas) it could tolerate soluble Cd concentrations of up to 20 ppm in the growing media. Cadmium is one of the most problematic pollutants, given its high toxicity to humans and its relatively high mobility in the soil-plant system, which favours its spread over the food chain in contaminated areas (Pérez de Mora et al. 2009).

Despite *Q. ilex* is a good candidate for the phytostabilization of TE contaminated soils in the Mediterranean area, the establishment of this species in such degraded sites is not an easy task. The negative effects that TE can have on soil biota and plant roots can reduce the success of *Q. ilex* plantations in contaminated areas.

High concentrations of TE in the soil might have negative impacts on several processes related to plant nutrient acquisition. Many works have shown that exposure to TE can alter the structure of the soil microbial communities, often leading to a decrease in microbial abundance and diversity (reviewed in Naidu et al. 2001), with consequences for key ecosystem processes like litter decomposition or the mineralization of organic matter (Helmisaari et al. 1995; Nieminen and Helmisaari 1996). In addition, exposure of roots to TE might inhibit the production or the elongation of fine roots, which limits the ability of plants to explore for water and nutrients. Also, some TEs compete with essential nutrients with a similar ionic ratio for exchange sites in the soil matrix and the rhizosphere. For all these reasons plants growing in contaminated soils often show some symptoms of nutritional deficiencies (Reimann et al. 2001; Mankovska et al. 2004; Jamnickà et al. 2007).

Impairment of nutrient acquisition could limit the establishment of woody plants after the restoration of contaminated sites in Mediterranean environments, which are often deficient in P and N (Sardans et al. 2004; Romanyà and Vallejo 2004). In a study conducted at the Guadiamar Green Corridor, the nutritional status of Holm oak saplings along a gradient of soil contamination was evaluated seven years after the plantation of this species in the area (Domínguez et al. 2010a). Results revealed some nutritional deficiencies, especially of P, in the young Holm oaks. These deficiencies were particularly strong in acidic soils; acidification is a common process in soils affected by mining activities, if pyrite is present in the ore (the oxidation of sulphides in the pyrite leads to acid drainage). Sixteen years after the plantation, the nutritional status of oak plants was again evaluated; this time the study also included the analysis of a wider range of functional traits at the root and leaf level, as well as the quantification of the taxonomic and functional diversity of the ectomycorrhizal fungi associated with oak roots (Figure 9; Gil-Martínez et al. 2018). Leaf phosphorus was related to ectomycorrhizal species composition; in particular higher leaf P was related to the colonization of roots by the fungi *Thelephora terrestris*. The transfer of TEs from soil to root was also explained substantially by the ectomycorrhizal composition, with the highest transfer of TE found in trees whose roots were colonized by *Hebeloma cavipes*. Therefore fungi-plant interactions play an important role in the ability of *Q. ilex* to acquire P and to immobilize TE at the rhizospheric level. The study also showed that contamination reduced the functional diversity of the ectomycorrhizal community (López-García et al. 2018), so it is likely that the nutritional deficiencies in Holm oak plants growing in contaminated soils are mediated by decreases in soil biodiversity in these soils.

The observed nutritional deficiencies in *Q. ilex* plantations could partly explain the low survival rates recorded for the Holm oak saplings over the first year after their plantation in the Guadiamar Green Corridor (Dominguez et al. 2010b). Monitoring of plant survival and growth over the 2003-2005 period (four to six years after plantation) showed that, on average, only 20% of *Q. ilex* saplings survived, and that there was a negative correlation between the survival rates at each monitored stand and the level of soil contamination of the stand. In the most contaminated and acidified plot, mortality rate was 97%. However, there were no significant differences in growth rates between the contaminated sites and a reference site, which was not contaminated by the pollution episode but was also afforested in the same period. These results suggested that once the plants were well established, the trace element content of the soil might have relatively lower importance for sapling growth than other environmental factors such as soil water availability, although P nutrition can be impaired by soil contamination as exposed above.

Figure 9. Holm oak saplings planted in trace-element contaminated soils of the Guadiamar Green Corridor (SW Spain). Sixteen years after the plantation (picture in the left) the transfer of contaminants from soil to Holm roots as well as P acquisition was explained substantially by the composition of the ectomycorrhizal community (López-García et al. 2018). Emergence and survival of seedlings in the most polluted soils was highly enhanced by the use of *Retama sphaerocarpa* shrubs as nurse plants (right) (Domínguez et al. 2015). Photos by Álvaro López-García (left) and María T. Domínguez (right).

The low survival rates in highly degraded soils could limit the potential of this species for the phytostabilization of contaminated areas. The success of *Q. ilex* plantations in these soils could be improved if they were designed considering the spatial heterogeneity of pre-existing vegetation, and taking advantage of the positive plant-plant interactions (facilitation effect) that can be particularly beneficial for late-successional species in highly disturbed environments. In another study conducted at the Guadiamar Green Corridor the usefulness of considering pre-existing shrubs as nurse plants for *Q. ilex* seedlings was evaluated (Figure 9; Domínguez et al. 2015). *Retama sphaerocarpa*, *Olea europaea* and *Phillyrea angustifolia* shrubs were used as planting microsites for acorns of *Q. ilex* along a gradient of soil pollution in the area. Seedling growth, survival, and chemical composition were monitored over three consecutive years, comparing the performance of the oak seedlings underneath the nurse shrubs and in open, unprotected microsites. Germination and seedling emergence was similar between both microsites, but seedling survival was greatly enhanced under shrubs; the shrub protection increased plant survival up to 85% (three years after planting), while in open microsites survival was lower than 30%. Shrubs attenuated light intensity at the ground level, reducing extreme fluctuations in temperature, and promoted higher soil water contents during the spring season, in comparison to open areas. Facilitation of seedling growth by shrubs increased along the gradient of soil pollution; therefore it is likely that the shading conditions provided by the shrub canopy allowed seedlings to cope with the toxicity provoked by the concurrence of low pH and high trace element concentrations in the most polluted sites (Domínguez et al. 2015).

In summary, Holm oak has a potential for the phytostabilization of Mediterranean contaminated sites, given its ability to maintain the concentrations of potentially toxic trace elements in its aboveground biomass at low levels. However, the feasibility of the plantations of this species for the phytoremediation on contaminated lands is much reduced by other physical and chemical constrains that often prevail in degraded soils, such as low water availability, acidification or nutrient limitations. The design of planting schemes in contaminated sites should consider alternatives to maximize the number of survivors during the first years after planting, when massive mortalities occur, for instance by using pre-existing vegetation as nurse plants.

Conclusion

The Iberian oaks belong to ten species with contrasting edaphoclimatic requirements having a widespread distribution and often dominating the landscape. Part of their past and present success relay on the characteristics of their mating systems which allow rapid migration and the maintenance of high genetic diversity levels promoting local adaptation and high rates of ecological divergence. They provide crucial ecosystem services but are threatened by several global-change drivers, including climate warming.

Quercus species differ in several functional traits compared to other woody species, as high seed mass, root mass fraction and leaf mass per area, and low relative growth rate. Deciduous oaks species differ in leaf structure and phenology, and in physiological activities compared with evergreen oaks being part of the causes of the different spatial distribution of both leaf habits. Oaks seedlings depend heavily on seed reserves during the first year, being even more important under shade. During the adult stage of *Quercus*, the relative growth rate decreases with plant size and in lower amount on temperature and precipitation variation. Oak radial growth is very constrained by spring to summer droughts. Oaks will modify their physiological response as a consequence of the progressive carbonization of the atmosphere, but the responses will be constrained by other co-limiting factors such as light, water, nutrients or temperature. The vulnerability of Iberian oaks to climate warming is illustrated by recent cases of drought-induced growth decline and canopy dieback. Drought and high spring-summer temperatures trigger a decrease in leaf production and transpiring area, and enhance leaf shedding. It is unclear how this loss in vigour associated to increased crown transparency and growth decline translate into increasing mortality rates in the long term, due to several resilience mechanisms (e.g., resprouting).

Forest densification in the Iberian Peninsula associated to land-use abandonment can lead to increased competition for water and wildfire risk exacerbating the negative impacts of drought on oak forests and decreasing their productivity making them more vulnerable to climate warming. Predicting long-term responses to future climate projections is uncertain partly due to scarce empirical studies made on the topic. Recently, the efforts of calibrating remote hyperspectral images at ecosystem level with tree crown and leaf behavior are promising to detect early-warning signals of forest dieback to anticipate managing strategies when symptoms are pre-visual and still reversible. The development of models based on remote sensing information can allow predicting the time evolution of the functioning and ecosystem services of Iberian *Quercus* forests.

ACKNOWLEDGMENTS

The seed of this collaborative chapter germinated during the UNIA Workshop at Baeza (Spain) in autumn 2013. We thank the Spanish Ministry of Science and Innovation for support of this work with the grants INTARSU (CGL2017-82254-R), DEGRAMED (CGL2017-85891-R), ESPECTRAMED (CGL2017-86161-R), ECO-MEDIT (CGL2014-53236-R), DIVERBOS (CGL2011-30285-C02), and RTI2018-096884-B-C31; to the Ansalusian Government for the grant FOR-CHANGE, and to the European FEDER funds. V. Acácio acknowledges the Portuguese Foundation for Science and Technology (FCT), fellowship number SFRH/BPD/80598/2011 and contract DL 57/2016/CP1382/CT0002. I.M.P.R. was funded by a "Ramón and Cajal" contract (RYC-2013-13937).

REFERENCES

Abrams, M. D. (1990). Adaptations and responses to drought in *Quercus* species of North America. *Tree Physiology* 7: 227-238.

Acácio, V., Dias, F. S., Catry, F. X., Rocha, M., Moreira, F. (2017). Landscape dynamics in Mediterranean oak forests under global change: understanding the role of anthropogenic and environmental drivers across forest types. *Global Change Biology* 23: 1199-1217.

Acácio, V., Holmgren, M., Jansen, P. A., Schrotter, O. (2007). Multiple recruitment limitation causes arrested succession in Mediterranean cork oak systems. *Ecosystems* 10: 1220-1230.

Acácio, V., Holmgren, M., Rego, F., Moreira, F., Mohren, G. M. J. (2009). Are drought and wildfires turning Mediterranean cork oak forests into persistent shrublands? *Agroforestry Systems* 76: 389-400.

Aguilar-Romero, R., Pineda-García, F., Paz, H., González-Rodriguez, A., Oyama, K. (2017). Differentiation in the water-use strategies among oak species from central Mexico. *Tree Physiology* 37: 915-925.

Ainsworth, E. A., Rogers, A. (2007) The response of photosynthesis and stomatal conductance to rising [CO_2]: mechanisms and environmental interactions. *Plant Cell and Environment* 30: 258–270.

Alberto, F. J., Aitken, S. N., Alía, R., González-Martínez, S. C., Hänninen, H., Kremer, A. et al. (2013). Potential for evolutionary responses to climate change–evidence from tree populations. *Global Change Biology* 19: 1645-1661.

Alla, A. Q., Camarero, J. J. (2012). Contrasting responses of radial growth and wood anatomy to climate in a Mediterranean ring-porous oak: implications for its future persistence or why the variance matters more than the mean. *European Journal of Forest Research* 131: 1537-1550.

Alla, A. Q., Camarero, J. J., Palacio. S, Montserrat-Martí, G. (2013). Revisiting the fate of buds: size and position drive bud mortality and bursting in two coexisting Mediterranean *Quercus* species with contrasting leaf habit. *Trees* 27: 1375–1386.

Allen, C. D. et al. (2010) A global overview of drought and heat-induced tree mortality reveals emerging climate change risks for forests. *Forest Ecology and Management* 259: 660–684.

Almeida, M. H., Merouani, H., Costa e Silva, F., Cortina, J., Trubat, R., Chirino, E., Vilagrosa, A., Khaldi, A., Stiti, B., El Alami, S. L., Vallejo, R. (2009). Germplasm selection and nursery techniques. In: J. Aronson, J. S. Pereira and J. G. Pausas (eds.), *Cork oak woodlands on the edge. Ecology, adaptive management, and restoration*, pp. 129-137. Island Press, Washington D.C.

Anderegg, W. R. L., Schwalm, C., Biondi, F., Camarero, J. J., Koch, G., Litvak, M., Ogle, K., Shaw, J. D., Shevliakova, E., Williams, A. P., Wolf, A., Ziaco, E., Pacala, S. (2015). Pervasive drought legacies in forest ecosystems and their implications for carbon cycle models. *Science* 349: 528-532.

Antúnez, I., Retamosa, E. C., Villar, R. (2001). Relative growth rate in phylogenetically related deciduous and evergreen woody species. *Oecologia* 128: 172-180.

APCOR (2019). *Cork. Information Bureau. Cortiça em números*. Associação Portuguesa de Cortiça.

Aranda, I., Cadahía, E., de Simón, B. F. (2020). Leaf ecophysiological and metabolic response in *Quercus pyrenaica* Willd seedlings to moderate drought under enriched CO_2 atmosphere. *Journal of Plant Physiology* 244: 153083.

Aranda, I., Castro, L., Alía, R., Pardos, J.A., Gil, L. (2005) Low temperature during winter elicits differential responses among populations of the Mediterranean evergreen cork oak (*Quercus suber*). *Tree Physiology* 25: 1085-90.

Aronson, J., Pereira, J. S., Pausas, J. G. (Eds.) (2012). *Cork oak woodlands on the edge: ecology, adaptive management, and restoration*. Island Press, Washington D. C.

Audet, P., Charest, C. (2007). Heavy metal phytoremediation from a meta-analytical perspective. *Environmental Pollution* 147: 231-237.

Augustine, D. J., McNaughton, S. J. (2006) Interactive effects of ungulate herbivores, soil fertility, and variable rainfall on ecosystem processes in a semi-arid savanna. *Ecosystems* 9: 1242–1256.

Baldocchi, D. D., Ma, S., Rambal, S., Misson, L., Ourcival, J. M., Limousin, J. M., et al. (2010). On the differential advantages of evergreenness and deciduousness in mediterranean oak woodlands: a flux perspective. *Ecological Applications* 20: 1583-1597.

Baraloto, C., Paine, C. E. T., Poorter, L., Beauchene, J., Bonal, D., Domenach, A. et al. (2010). Decoupled leaf and stem economics in rain forest trees. *Ecology Letters* 13: 1338–1347.

Barbeta, A., Ogaya, R., Peñuelas, J. (2013). Dampening effects of long-term experimental drought on growth and mortality rates of a Holm oak forest. *Global Change Biology* 19: 3133–3144.

Bates, B. C., Kundzewicz, Z. W., Wu, S., Palutikof, J. P. (2008). *Climate change and water*, IPCC Secretariat, Geneva.

Baudena, M., Santana, V. M., Baeza, M. J., Bautista, S., Eppinga M. B., Hemerik L., Mayor A. G., Rodriguez F. et al. (2020). Increased aridity drives post-fire recovery of Mediterranean forests towards open shrublands. *New Phytologist* 225: 1500-1515.

Beaufoy, G., Cooper, T. (2008). *Guidance Document to the Member States on the Application of the HNV Impact Indicator*. Report for the European Evaluation Network for Rural Development, Brussels.

Bergot, M., Cloppet, E., Pérarnaud, V., Déqué, M., Marçais, B., Desprez-Loustau, M. L. (2004). Simulation of potential range expansion of oak disease caused by *Phytophthora cinnamomi* under climate change. *Global Change Biology* 10: 1539-1552.

Boshier, D. H. (2000). Mating Systems. In: Young A., Boshier D. H., Boyle T. J. (eds.), *Forest Conservation Genetics: Principles and Practices*, pp. 63-69. CSIRO, Melbourne, Australia; CABI, Wallingford, UK.

Branco, M., Bragança, H., Sousa, E., Phillips A. J. (2014). Pests and diseases in Portuguese forestry: current and new threats. In: F. Reboredo (ed.), *Forest context and policies in Portugal. Present and future challenges*, pp. 117-154. World Forests 19. Springer International Publishing Switzerland.

Branco, M., Ramos, A. P. (2009). Coping with pests and diseases. In: J. Aronson, J. S. Pereira, J. G. Pausas (eds.), *Cork oak woodlands on the edge. Ecology, adaptive management, and restoration*, pp. 103-111. Island Press, Washington D.C., USA.

Brasier, C. M. (1996). *Phytophthora cinnamomi* and oak decline in southern Europe. Environmental constraints including climate change. *Annales des Sciences Forestières* 53: 347-358.

Brewer S., Cheddadi R., de Beaulieu J. L., Reille M. (2002). The spread of deciduous *Quercus* throughout Europe since the last glacial period. *Forest Ecology and Management* 156: 27–48.

Bugalho, M. N., Caldeira, M. C., Pereira, J. S., Aronson J., Pausas J. G. (2011). Mediterranean cork oak savannas require human use to sustain biodiversity and ecosystem services. *Frontiers in Ecology and the Environment* 9: 278–286.

Caldeira, M. C., Lecomte X., David, T. S., Pinto, J. G., Bugalho, M. N., Werner, C. (2015). Synergy of extreme drought and shrub invasion reduce ecosystem functioning and resilience in water-limited climates. *Scientific Reports* 5: 15110.

Camarero, J. J., Franquesa, M., Sangüesa-Barreda, G. (2015). Timing of drought triggers distinct growth responses in holm oak: implications to predict warming-induced forest defoliation and growth decline. *Forests* 6: 1576-1597.

Camarero, J. J., Lloret, F., Corcuera, L., Peñuelas, J., Gil-Pelegrín, E. (2004). Global change and forest decline. In: Valladares, F. (ed.), *Ecología del bosque mediterráneo en un mundo cambiante*, pp. 397-423. Ministerio de Medio Ambiente, Madrid (in Spanish).

Camarero, J. J., Sangüesa-Barreda, G., Vergarechea, M. (2016). Prior height, growth, and wood anatomy differently predispose to drought-induced dieback in two Mediterranean oak species. *Annals of Forest Science* 73: 341-351.

Camilo-Alves, C. S. P., Clara, M. I. E., Ribeiro, N. M. C. A. (2013). Decline of Mediterranean oak trees and its association with *Phytophthora cinnamomi*: a review. *European Journal of Forest Research* 132: 411-432.

Campelo, F., Nabais, C., García-González, I., Cherubini, P., Gutiérrez, E., Freitas, H. (2009) Dendrochronology of *Quercus ilex* L. and its potential use for climate reconstruction in the Mediterranean region. *Canadian Journal of Forest Research* 39: 2486–2493.

Cañellas, I., Montero, G. (2002). The influence of pruning on the yield of cork oak dehesa woodland in Extremadura (Spain). *Annals of Forest Science* 59: 753-760.

Carnicer, J., Coll, M, Ninyerola, M., Pons, X., Sánchez, G., Peñuelas, J. (2011). Widespread crown condition decline, food web disruption, and amplified tree mortality with increased climate change-type drought. *Proceedings of the National Academy of Sciences USA* 108: 1474–1478.

Carrión, J. S., Fernández, S., González-Sampériz, P., Gil-Romera, G., Badal, E., Carrión-Marco, Y., López-Merino, L., López-Sáez, J. A., Fierro, E., Burjachs, F. (2010). Expected trends and surprises in the Lateglacial and Holocene vegetation history of the Iberian Peninsula and Balearic Islands. *Review of Palaeobotany and Palynology* 162: 458–475.

Catry, F. X., Branco, M., Sousa, E., Caetano, J., Naves, P., Nóbrega, F. (2017). Presence and dynamics of ambrosia beetles and other xylophagous insects in a Mediterranean cork oak forest following fire. *Forest Ecology and Management* 404: 45-54.

Catry, F. X., Moreira, F., Pausas, J. G., Fernandes, P. M., Rego, F., Cardillo, E., Curt, T. (2012). Cork oak vulnerability to fire: the role of bark harvesting, tree characteristics and abiotic factors. *PLOS ONE* 7: e39810.

Catry, F. X., Pausas, J. G., Moreira, F., Fernandes, P. M., Rego, F. (2013). Post-fire response variability in Mediterranean Basin tree species in Portugal. *International Journal of Wildland Fire* 22: 919-932.

Cavender-Bares, J. (2005) Impacts of freezing on long-distance transport in woody plants. In: Hoolbrook, N. M, Zwieniecki, M. (eds.), *Vascular transport in plants*, pp: 401-424. Elsevier, Amsterdam.

Cavender-Bares, J. (2016). Diversity, distribution and ecosystem services of the North American oaks. *International Oaks* 27: 37-48.

Cavender-Bares, J. (2019). Diversification, adaptation, and community assembly of the American oaks (*Quercus*), a model clade for integrating ecology and evolution. *New Phytologist* 221: 669-692.

Cavender-Bares, J., Ackerly, D. D., Hobbie, S. E., Townsend, P. A. (2016). Evolutionary legacy effects on ecosystems: biogeographic origins, plant traits, and implications for management in the era of Global Change. *Annual Review of Ecology, Evolution, and Systematics* 47: 433–462.

Cavender-Bares, J., Gonzalez-Rodriguez, A., Pahlich, A., Koehler, K., Deacon, N. (2011). Phylogeography and climatic niche evolution in live oaks (*Quercus* series Virentes) from the tropics to the temperate zone. *Journal of Biogeography* 38: 962–981.

Cavender-Bares, J., Kitajima, K., Bazzaz, F. A. (2004) Multiple trait associations in relation to habitat differentiation among 17 Floridian oak species. *Ecological Monographs* 74: 635–662.

Cavender-Bares, J., Ramírez-Valiente, J. A. (2017). Physiological evidence from common garden experiments for local adaptation and adaptive plasticity to climate in American live oaks (*Quercus* Section Virentes): implications for conservation under global change. In: Gil-Pelegrín, E.,

Peguero-Pina, J., Sancho-Knapik, D. (eds.) *Oaks physiological ecology. Exploring the functional diversity of genus Quercus L.*, Tree Physiology, vol 7. pp. 107-135. Springer, Cham, Switzerland.

Chae, H. M, Lee, S. H, Cha, S, Shin, J. H, Choi, M. J., Shim, J. K. (2016). Effects of elevated CO_2 on aboveground growth in seedlings of four dominant *Quercus* species. *Applied Ecology and Environmental Research* 14: 597-611.

Cheaib, A., Badeau, V., Boe, J., Chuine, I., Delire, C., Dufrêne, E., François, C., Gritti, E. S., Legay, M., Pagé, C., Thuiller, W., Viovy, N., Leadley, P. (2012). Climate change impacts on tree ranges: model intercomparison facilitates understanding and quantification of uncertainty. *Ecology Letters* 15: 533-544.

Clark, J. S., Beckage, B., Camill, P., Cleveland, B., HilleRisLambers, J., Lichter, J., Mclachlan, J., Mohan, J., Wyckoff, P. (1999). Interpreting recruitment limitation in forests. *American Journal of Botany*, 86: 1-16.

Coello, J., Cortina, J., Valdecantos, A., Varela, E. (2015). Forest landscape restoration experiences in southern Europe: sustainable techniques for enhancing early tree performance. *Unasylva* 66: 82-90.

Corcobado, T., Cubera, E., Juárez, E., Moreno, G., Solla, A. (2014). Drought events determine performance of *Quercus ilex* seedlings and increase their susceptibility to *Phytophthora cinnamomi*. *Agricultural and Forest Meteorology* 192-193: 1-8.

Corcuera, L., Camarero, J. J., Gil-Pelegrín, E. (2004a). Effects of a severe drought on growth and wood-anatomical properties of *Quercus faginea*. *IAWA Journal* 25: 185-204.

Corcuera, L., Camarero, J. J., Gil-Pelegrín, E. (2004b). Effects of a severe drought on *Quercus ilex* radial growth and xylem anatomy. *Trees* 18: 83-92.

Corcuera, L., Camarero, J. J., Sisó, S., Gil-Pelegrín, E. (2006). Radial-growth and wood anatomical changes in overaged *Quercus pyrenaica* coppice stands: functional responses in a new Mediterranean landscape. *Trees-Structure and Function* 20: 91–98.

Corral, R., Pérez-García, F., Pita, J. M. (1990). Some aspects of seed germination in four species of *Cistus* L. *Seed Science and Technology* 18: 321-325.

Correia, R. A., Bugalho, M. N., Franco, A. M. A., Palmeirim, J. M. (2018). Contribution of spatially explicit models to climate change adaptation and mitigation plans for a priority forest habitat. *Mitigation and Adaptation Strategies for Global Change* 23: 371-386.

Cortina, J., Pérez-Devesa, M., Vilagrosa, A., Abourouh, M., Messaoudène, M., Berrahmouni, N., Sousa, L. N., Almeida, M. H., Khaldi, A. (2009). Field techniques to improve cork oak establishment. In: Aronson, J., Pereira, J. S., Pausas, J. G. (eds.), *Cork oak woodlands on the edge. Ecology, adaptive management, and restoration*, pp. 141-149. Island Press, Washington D.C., USA.

Costa, A., Madeira, M., Santos, J. L. (2014). Is cork oak (*Quercus suber* L.) woodland loss driven by eucalyptus plantation? A case study in southwestern Portugal. *iForest - Biogeosciences and Forestry* 7: 193-203.

Costa A., Pereira, H. (2007). Montados e sobreirais: uma espécie, duas perspectivas [Mounted and cork oak trees: one species, two perspectives]. In: Silva JS (eds.) *Os Montados - Muito para além das árvores, Colecção Árvores e Florestas de Portugal* [*The Montados - Far beyond the trees, the Portuguese Trees and Forests Collection*], Vol. 3, pp.17-37. Jornal Público/ Fundação Luso-Americana para o Desenvolvimento/ Liga para a Protecção da Natureza, Lisboa.

Costa, A., Pereira, H., Madeira, M. (2009). Landscape dynamics in endangered cork oak woodlands in Southwestern Portugal (1958–2005). *Agroforestry Systems* 77: 83-96.

Cubera, E., Moreno, G. (2007). Effect of land-use on soil water dynamic in dehesas of Central-Western Spain. *Catena* 71: 298-308.

Curt, T., Pausas, J. G. (2010). Are changes in fire regime threatening cork oak-shrubland mosaics? *IUFRO Landscape Ecology International Conference*, Bragança, Portugal, September 21-27, 5pp.

Davis, M. A., Reich, P., Knoll, M. J. B., Dooley, L., Hundtoft, M., Attleson, I. (2007). Elevated atmospheric CO_2: a nurse plant substitute for oak seedlings establishing in old fields. *Global Change Biology* 13: 2308–2316.

De la Cruz A. C., Gil P. M., Fernández-Cancio A., Minaya M., Navarro-Cerrillo R. M., Sánchez-Salguero R., Grau J. M. (2014). Defoliation triggered by climate induced effects in Spanish ICP Forests monitoring plots. *Forest Ecology and Management* 331: 245-255.

de la Riva E. G., Olmo M., Poorter H., Ubera J. L., Villar R. (2016a). Leaf Mass per Area (LMA) and its relationship with leaf structure and anatomy in 34 Mediterranean woody species along a water availability gradient. *Plos One* 11(2): e0148788.

de la Riva E. G., Pérez-Ramos I. M., Navarro-Fernández C. M., Olmo M., Marañón T. Villar R. (2014). Functional traits in the *Quercus* genus: acquisitive versus conservative strategies in the use of resources. *Ecosistemas* 23(2): 82-89 (in Spanish).

de la Riva E. G., Pérez-Ramos I. M., Tosto A, Navarro-Fernández C. M., Olmo M., Marañón T., Villar R. (2016c). Disentangling the relative importance of species occurrence, abundance and intraspecific variability in community assembly: a trait-based approach at the whole-plant level in Mediterranean forests. *Oikos* 125: 354–363.

de la Riva E. G., Tosto A., Perez-Ramos I. M., Navarro-Fernandez C. M., Olmo M., Anten N. P. R., Marañón T., Villar R. (2016b). A plant economics spectrum in Mediterranean forests along environmental gradients: is there coordination among leaf, stem and root traits? *Journal of Vegetation Science* 27: 187-199.

Denk, T., Grimm, G. W. (2010). The oaks of western Eurasia: traditional classifications and evidence from two nuclear markers. *Taxon* 59: 351-366.

Desprez-Loustau, M.L., Marçais, B., Nageleisen, L.M., Piou, D., Vannini, A. (2006). Interactive effects of drought and pathogens in forest trees. *Annals of Forest Science* 63: 597-612.

Díaz, S.; Lavorel, S.; Mcintyre, S.; Falczuk, V.; Casanoves, F.; Milchunas, Dg.; Skarpe, C.; Rusch, G.; Sternberg, M.; Noy-Meir, I.; Landsberg, J.;

Zhang, W.; Clark, H., Campbell, Bd. (2007). Plant trait responses to grazing –a global synthesis. *Global Change Biology* 13: 313-341.

Dobrowski, S. Z., Pushnik, J. C., Zarco-Tejada, P. J., Ustin, S. L. (2005). Simple reflectance indices track heat and water stress-induced changes in steady-state chlorophyll fluorescence at the canopy scale. *Remote Sensing of Environment* 97: 403-414.

Domínguez, M.T., F. Madrid, T. Marañón, J. M. Murillo. (2009). Cadmium availability in soil and retention in oak roots: Potential for phytostabilization. *Chemosphere* 76 (4): 480–86.

Domínguez, M. T., I. M. Pérez-Ramos, J. M. Murillo, T. Marañón (2015). Facilitating the afforestation of Mediterranean polluted soils by nurse shrubs. *Journal of Environmental Management* 161: 276–86.

Domínguez, M. T., T. Marañón, J. M. Murillo, R. Schulin, B. H. Robinson. (2010a). Nutritional status of Mediterranean trees growing in a contaminated and remediated area. *Water Air and Soil Pollution* 205 (1–4): 305–21.

Domínguez, M. T., T. Marañón, J. M. Murillo, R. Schulin, B. H. Robinson. (2008). Trace element accumulation in woody plants of the Guadiamar Valley, (SW) Spain: A large-scale phytomanagement case study. *Environmental Pollution* 152 (1): 50–59.

Domínguez, M. T., P. Madejón, T. Marañón, J. M. Murillo. (2010b). Afforestation of a trace-element polluted area in SW Spain: Woody plant performance and trace element accumulation. *European Journal of Forest Research* 129 (1): 47–59.

Drava, G., D. Brignole, P. Giordani, V. Minganti. (2017). The Bark of the Branches of Holm Oak (*Quercus Ilex* L.) for a Retrospective Study of Trace Elements in the Atmosphere. *Environmental Research* 154: 291–95.

Escudero A, Mediavilla S, Olmo M, Villar R, Merino J. (2017). Coexistence of deciduous and evergreen oak species in Mediterranean environments: costs associated with the leaf traits of both habits. In: Gil-Pelegrín E, Peguero-Pina JJ, Sancho-Knapik D (eds.), *Oaks Physiological Ecology. Exploring the Functional Diversity of Genus Quercus L.* pp. 195-237. Tree Physiology, vol 7. Springer, Cham, Switzerland.

Espelta, J. M., Cortés, P., Molowny-Horas, R., Sánchez-Humanes, B., Retana, J. (2008). Masting mediated by summer drought reduces acorn predation in Mediterranean oak forests. *Ecology* 89: 805-817.

European Commission (2015). *Natura 2000 and Forests. Technical Report.* Luxembourg: Publications Office of the European Union.

Evans G. C. (1972). *The quantitative analysis of plant growth.* Blackwell Scientific, Oxford.

Evans, J. A., Foster, J. G. (2011). Metaknowledge. *Science*, 331: 721-725.

Fernández-Habas J, Fernández-Rebollo P, Rivas M, García A, Abellanas B. (2019). Spatio-temporal analysis of oak decline process in open woodlands: A case study in SW Spain. *Journal of Environmental Management* 248: 109308.

Fernández-Martínez, M., Pearse, I., Sardans, J., Sayol, F., Koenig, W. D., LaMontagne, J. M., Bogdziewicz, M., Collalti, A., Hacket-Pain, A., Vacchiano, G., Espelta, J. M., Peñuelas, J. Janssens, I. A. (2019). Nutrient scarcity as a selective pressure for mast seeding. *Nature Plants* 5: 1222–1228.

Fernández-Rebollo P., Carbonero M. D., García A., Leal J. R. (2014). Dehesas de encinas: el árbol que modeló el sistema y el sistema que moldeó al árbol [Holm oak pastures: the tree that shaped the system and the system that shaped the tree]. *Ambienta* 108: 94-105.

Fernández-Rebollo P., Hidalgo M. T., Leal J. R., Carbonero M. D., Fernández-Habas J., Andicoberri S., Jiménez J. M., Pacheco J. M. (2017). Variación morfológica y bromatológica de la bellota de encina dentro del árbol [Morphological and bromatological variation of the oak acorn inside the tree]. In: Bartolomé et al. (eds.). *Renaturalización vs ruralización.* 56ª Reunión Científica de la SEEP. Barcelona. pp: 330-336.

Finch-Savage, W. E. (1992). Embryo water status and survival in the recalcitrant species *Quercus robur* L: Evidence for a critical moisture content. *Journal of Experimental Botany* 43: 663-669.

Forner, A., Valladares, F., Aranda, I. (2018). Mediterranean trees coping with severe drought: Avoidance might not be safe. *Environmental and Experimental Botany* 155: 529-540.

Fortunel, C., Fine, P. V., Baraloto, C. (2012). Leaf, stem and root tissue strategies across 758 Neotropical tree species. *Functional Ecology* 26(5): 1153-1161.

Frank, D. C., Poulter, B., Saurer, M., et al., (2015). Water-use efficiency and transpiration across European forests during the Anthropocene. *Nature Climate Change* 5: 579–583.

Freschet, G. T., Cornelissen, J. H. C., van Logtestijn, R. S. P., Aerts, R. (2010). Evidence of the "plant economics spectrum" in a subarctic flora. *Journal of Ecology* 98: 362–373.

Garbelotto, M., Pautasso, M. (2012). Impacts of exotic forest pathogens on Mediterranean ecosystems: four case studies. *European Journal of Plant Pathology* 133: 101-116.

Garbisu, C., J. H. Allica, O. Barrutia, I. Alkorta, J. M. Becerril. (2002). Phytoremediation: A Technology Using Green Plants to Remove Contaminants from Polluted Areas. *Reviews on Environmental Health* 17(3):173–88.

García-Moreno, A. M., Carbonero-Muñoz, M. D., Serrano-Moral, M., Fernández-Rebollo, P., (2014). Grazing affects shoot growth, nutrient and water status of *Quercus ilex* L. in Mediterranean open woodlands. *Annals of Forest Science* 71: 917–926.

García-Plazaola J. I., Hernández A., Fernández-Marín B., Esteban R., Peguero-Pina J. J., Verhoeven A., Cavender-Bares J. (2017) Photoprotective mechanisms in the genus *Quercus* in response to winter cold and summer drought. In E. Gil-Pelegrín et al. (eds.), *Oaks Physiological Ecology. Exploring the Functional Diversity of Genus Quercus L.*, Tree Physiology 7, pp 361-392, Springer, Cham, Switzerland.

Gardi, C., S. Jeffery, A. Saltelli. (2013). An Estimate of Potential Threats Levels to Soil Biodiversity in EU. *Global Change Biology* 19 (5): 1538–48.

Gea-Izquierdo G., Cañellas I. (2014) Local climate forces instability in long-term productivity of a Mediterranean oak along climatic gradients. *Ecosystems* 17: 228-241.

Gea-Izquierdo G., Fonti P., Cherubini P., Martín-Benito D., Chaar H., Cañellas I. (2012). Xylem hydraulic adjustment and growth response of *Quercus canariensis* Willd. to climatic variability. *Tree Physiology* 32: 401-413.

Gil-Martínez, M., Á. López-García, M. T. Domínguez, C. M. Navarro-Fernández, R. Kjøller, M. Tibbett, T. Marañón. (2018). Ectomycorrhizal Fungal Communities and Their Functional Traits Mediate Plant–Soil Interactions in Trace Element Contaminated Soils. *Frontiers in Plant Science* 9: 1682.

Gil-Pelegrín E., Saz M. A., Cuadrat J. M., Peguero-Pina J. J., Sancho-Knapik D. (2017). Oaks under Mediterranean-type climates: functional response to summer aridity. In: E. Gil-Pelegrín et al. (eds.), *Oaks Physiological Ecology. Exploring the Functional Diversity of Genus Quercus L.*, Tree Physiology 7, pp: 137-194, Springer, Cham, Switzerland.

Godinho, S., Guiomar, N., Machado, R., Santos, P., Sá-Sousa, P., Fernandes, J. P., Neves, N., Pinto-Correia, T. (2016). Assessment of environment, land management, and spatial variables on recent changes in montado land cover in southern Portugal. *Agroforestry Systems* 90: 177-192.

Gómez, J. M. (2004). Bigger is not always better: conflicting selective pressures on seed size in *Quercus ilex*. *Evolution* 58: 71-80.

Gómez-Aparicio L, García-Valdés R, Ruíz-Benito P, Zavala MA. (2011). Disentangling the relative importance of climate, size and competition on tree growth in Iberian forests: implications for forest management under global change. *Global Change Biology* 17: 2400–2414.

Gómez-Aparicio, L., Ibáñez, B., Serrano, M. S., De Vita, P., Ávila, J. M., Pérez-Ramos, I. M., García, L. V., Sánchez, M. E., Marañón, T. (2012). Spatial patterns of soil pathogens in declining Mediterranean forests: implications for species regeneration. *New Phytologist* 194: 1014–1024.

Gómez-Aparicio, L., Pérez-Ramos, I. M., Mendoza, I, Matias, L., Quero, J. L., Castro, J., Zamora, R., Marañón, T. (2008). Oak seedling survival and growth along resource gradients in Mediterranean forests: implications for regeneration in current and future environmental scenarios. *Oikos* 117: 1683-1699.

González-Rodríguez V., Barrio I. C., Villar, R. (2012). Within-population variability influences early seedling establishment in four Mediterranean oaks. *Acta Oecologica* 41: 82- 89.

González-Rodríguez V., Navarro Cerrillo R., Villar R. (2010). Maternal influences on seed mass effect and initial seedling growth in four *Quercus* species. *Acta Oecologica* 37: 1-9.

González-Rodríguez V., Villar R. (2012). Post-dispersal seed removal in four Mediterranean oaks: species and microhabitat selection differ depending on large herbivore activity. *Ecological Research* 27: 587-594.

Gratani L. (1996). Leaf and shoot growth dynamics of *Quercus ilex* L. *Acta Oecologica* 17: 17–27.

Gratani, L., Meneghini, M., Pesoli, P., Crescente, M. F. (2003). Structural and functional plasticity of *Quercus ilex* seedlings of different provenances in Italy. *Trees*, 17: 515-521.

Gugger P. F., Penaloza-Ramirez J. M., Wright J. W., Sork V. L. (2017) Whole-transcriptome response to water stress in a California endemic oak, *Quercus lobata*. *Tree Physiology* 37: 632–644.

Gutiérrez, E., Campelo, F., Camarero, J. J., Ribas, M., Muntán, E., Nabais, C., Freitas, H. (2011). Climate controls act at different scales on the seasonal pattern of *Quercus ilex* L. stem radial increments in NE Spain. *Trees* 25: 637-646.

Hartmann, H., Moura, C. F., Anderegg, W. R. L., Ruehr, N. K., Salmon, Y., Allen, C. D., Arndt, S. K., Breshears, D. D., Davi, H., Galbraith, D. et al. (2018). Research frontiers for improving our understanding of drought-induced tree and forest mortality. *New Phytologist* 218: 15–28.

Helmisaari, H. S., J. Derome, H. Fritze, T. Nieminen, K. Palmgren, M. Salemaa, and I. Vanha-Majamaa. (1995). Copper in Scots Pine Forests around a Heavy-Metal Smelter in South-Western Finland. *Water, Air, and Soil Pollution* 85: 1727–32.

Hernández-Clemente, R., Navarro-Cerrillo, R. M., Zarco-Tejada, P. J., (2012). Carotenoid content estimation in a heterogeneous conifer forest using narrow-band indices and PROSPECT + DART simulations. *Remote Sensing of Environment* 127: 298–315.

Hernández-Clemente, R., Navarro-Cerrillo, R. M., Suarez, L., Morales, F., & Zarco-Tejada, P. J., (2011). Assessing structural effects on PRI for stress detection in conifer forests. *Remote Sensing of Environment* 115: 2360–2375.

Hernández-Clemente, R., North, P. R. J., Hornero, A., Zarco-Tejada, P. J. (2017). Assessing the effects of forest health on sun-induced chlorophyll fluorescence using the FluorFLIGHT 3-D radiative transfer model to account for forest structure. *Remote Sensing of Environment* 193: 165-179.

Hipp, A. L., Manos, P. S., Hahn, M., Avishai, M., Bodénès, C., Cavender-Bares, J. et al. (2019). The genomic landscape of the global oak phylogeny. *New Phytologist* (in press). doi:10.1111/nph.16162.

Homet, P., González, M., Matías, L. Godoy, Ó., Pérez-Ramos, I. M., García, L. V. Gómez-Aparicio, L. (2019). Exploring interactive effects of climate change and exotic pathogens on *Quercus suber* performance: Damage caused by *Phytophthora cinnamomi* varies across contrasting scenarios of soil moisture. *Agricultural and Forest Meteorology* 276–277: 1076052.

Hulme, P. E., (1997). Post-dispersal seed predation and the establishment of vertebrate dispersed plants in Mediterranean scrublands. *Oecologia* 111: 91–98.

Hunt R. (1982). *Plant growth curves: a functional approach to plant growth analysis*. Edward Arnold, London.

Ibáñez, B., Gómez-Aparicio, L., Stoll, P., Ávila, J. M., Pérez-Ramos, I. M., Marañón, T. (2015). A neighborhood analysis of the consequences of *Quercus suber* decline for regeneration dynamics in Mediterranean forests. *Plos One* 10(2): e0117827.

ICNF. (2015). *6.º Inventário Florestal Nacional – IFN6. Relatório Final* [*6th National Forest Inventory - IFN6. Final report*]. Lisboa: Instituto da Conservação da Natureza e das Florestas.

Infante, J. M., Rambal, S. et al. (1997). Modeling transpiration in holm-oak savannah: scaling up from the leaf to the tree scale. *Agricultural and Forest Meteorology* 87: 273–289.

Ingram, L. J., Stahl, P. D., Schuman, G. E., et al., (2008). Grazing impacts on soil carbon and microbial communities in a mixed-grass ecosystem. *Soil Science Society of America Journal* 72: 939-948.

Jackson, R. D., Idso, S. B., Reginato, R. J., Pinter, P. J. (1981). Canopy temperature as a crop water stress indicator. *Water Resources Research* 17: 1133-1138.

Jamnická, G., K. Bučinová, I. Havranová, A. Urban. (2007). Current State of Mineral Nutrition and Risk Elements in a Beech Ecosystem Situated near the Aluminium Smelter in Žiar Nad Hronom, Central Slovakia. *Forest Ecology and Management* 248: 26–35.

Jiménez P., López de Heredia U., Collada C., Lorenzo Z., Gil L. (2004). High variability of chloroplast DNA in three Mediterranean evergreen oaks indicates complex evolutionary history. *Heredity* 93:510–515.

Jones, A., et al. (2012). *The state of soil in Europe* - A contribution of the JRC to the European Environment Agency's environment state and outlook report – SOER 2010. Publication Office of the European Union.

Jordano, P., Herrera, C. M., (1995). Shuffling the offspring: uncoupling and spatial discordance of multiple stages in vertebrate seed dispersal. *Ecoscience* 2: 230–237.

Jump A., Marchant R., Peñuelas, J. (2009). Environmental change and the option value of genetic diversity. *Trends in Plant Science* 14: 51–58.

Juntila, S., Vastaranta, M., Hämäläinen, J., Latva-Käyrä, P., Hannu, H., Navarro Cerrillo, R. M., Hernandez-Clemente, R., (2016). Effect of forest structure and health on the relative surface temperature captured by airborne thermal imagery – Case study in Norway Spruce-dominated stands in Southern Finland. *Scandinavian Journal of Forestry Research* 32: 154–165.

Kelly, D. Sork, V. L. (2002). Mast-seeding in perennial plants: why, how, where? *Annual Review of Ecology and Systematics* 33: 427–447.

Kembel, S. W., Cahill, J. F. (2011). Independent evolution of leaf and root traits within and among temperate grassland plant communities. *PloS One* 6(6): e19992.

Killi D., Bussotti F., Gottardini E., Pollastrini M., Mori J., Tani C., Papini A., Ferrini F., Fini A. (2018) Photosynthetic and morphological responses of oak species to temperature and [CO_2] increased to levels predicted for 2050. *Urban Forestry and Urban Greening* 31: 26-37.

King, D. A. (2004). The scientific impact of nations. *Nature* 430: 311-316.

Kitao, M., Hida T., Eguchi N., Tobita H., Utsugi H., Uemura A., Kitaoka S., Koike T. (2016) Light compensation points in shade-grown seedlings of deciduous broadleaf tree species with different successional traits raised under elevated CO_2. *Plant Biology* 18 (Suppl. 1): 22–27. Körner, C., Asshoff, R., Bignucolo, O., Hättenschwiler, S., Keel, S. G., Peláez-Riedl, S., Pepin, S., Siegwolf, R.T.W., Zotz, G. (2005). Carbon flux and growth in mature deciduous forest trees exposed to elevated CO_2. *Science* 309: 1360-1362.

Kremer, A. (2013). Evolutionary responses of European oaks to climate change. *International Oaks* 24: 11-20.

Kremer, A. Hipp, A. L. (2019). Oaks: an evolutionary success story. *New Phytologist* (in press). doi:10.1111/nph.16274.

Larsen, P., von Ins, M. (2010). The rate of growth in scientific publication and the decline in coverage provided by Science Citation Index. *Scientometrics* 84: 575-603.

Larson-Johnson, K. (2016). Phylogenetic investigation of the complex evolutionary history of dispersal mode and diversification rates across living and fossil Fagales. *New Phytologist* 209: 418-435.

Lempereur M., Martin-StPaul N. K., Damesin C., Joffre R., Ourcival J. M., Rocheteau A., Rambal S. (2015). Growth duration is a better predictor of stem increment than carbon supply in a Mediterranean oak forest: implications for assessing forest productivity under climate change. *New Phytologist* 207: 579–590.

Leroy, T., Plomion, C., Kremer, A. (2019). Oak symbolism in the light of genomics. *New Phytologist* (in press). doi:10.1111/nph.15987.

Leuzinger, S., Körner, C. (2007). Water savings in mature deciduous forest trees under elevated CO_2 *Global Change Biology* 13: 2498–2508.

Limousin J. M., Longepierre D., Huc R., Rambal S. (2010) Change in hydraulic traits of Mediterranean *Quercus ilex* subjected to long-term throughfall exclusion. *Tree Physiology* 30: 1026–1036.

Limousin, J. M., S. Rambal, J. M. Ourcival, J. Rodríguez-Calcerrada, I. M. Pérez-Ramos, R. Rodríguez-Cortina, L. Misson, R. Joffre. (2012). Morphological and phenological shoot plasticity in a Mediterranean evergreen oak facing long-term increased drought. *Oecologia* 169: 565–577.

Liu, G., Freschet, G. T., Pan, X., Cornelissen, J. H. C., Li, Y. Dong, M. (2010). Coordinated variation in leaf and root traits across multiple spatial scales in Chinese semi-arid and arid ecosystems. *New Phytologist* 188: 543–553.

Lloret F., Casanovas C., Peñuelas J. (1999). Seedling survival of Mediterranean shrubland species in relation to root:shoot ratio, seed size and water and nitrogen use. *Functional Ecology* 13: 210–216.

Lloret, F., and Siscart, D. (1995). Los efectos demográficos de la sequía en poblaciones de encina [The demographic effects of drought on oak populations]. *Cuadernos de la Sociedad Española de Ciencias Forestales* 2: 77–81.

Lloret, F., Escudero, A., Iriondo, J. M., Martínez-Villalta, J., Valladares, F. (2015). Mecanismos de estabilización y resiliencia de la vegetación frente a eventos climáticos extremos [Vegetation stabilization and resilience mechanisms against extreme climatic events]. In: A. Herrero, MA Zavala (eds.) *Los bosques y la Biodiversidad frente al cambio climático: impactos, vulnerabilidad y adaptación en España,* pp. 369-375. Ministerio de Agricultura, Alimentación y Medio Ambiente, Madrid.

Lloret, F., Peñuelas, J., Estiarte, M. (2004). Experimental evidence of reduced diversity of seedlings due to climate modification in a Mediterranean-type community. *Global Change Biology* 10: 248–258.

López de Heredia U., Carrión J. S., Jiménez P., Collada C., Gil L. (2007). Molecular and palaeobotanical evidence for multiple glaciar refugia for evergreen oaks on the Iberian Peninsula. *Journal of Biogeography* 34: 1505–1517.

López de Heredia U., Jiménez P., Díaz-Fernández P., Gil L. (2005). The Balearic Islands: a reservoir of cpDNA genetic variation for evergreen oaks. *Journal of Biogeography* 32: 939–949.

López-García, Á., M. Gil-Martínez, C. M. Navarro-Fernández, R. Kjøller, C. Azcón-Aguilar, M. T. Domínguez, T. Marañón. (2018). Functional Diversity of Ectomycorrhizal Fungal Communities Is Reduced by Trace Element Contamination. *Soil Biology and Biochemistry* 121: 202–11.

López-Sánchez, A., Perea, R., Dirzo, R., Roig, S. (2016). Livestock vs. wild ungulate management in the conservation of Mediterranean dehesas: Implications for oak regeneration. *Forest Ecology and Management* 362: 99–106.

Lorenzo Z., Burgarella C., López de Heredia U., Lumaret R., Petit R. J., Soto A., Gil L. (2009). Relevance of genetics for conservation policies: the case of Minorcan cork oaks. *Annals of Botany* 104: 1069–1076.

Lukac M., Calfapietra C., Lagomarsino A., Loreto F. (2010). Global climate change and tree nutrition: effect of elevated CO_2 and temperature. *Tree Physiology* 30:1209–1220.Madejón, P., Domínguez, M. T., Madejón, E., Cabrera, F., Marañón, T., Murillo, J. M. (2018). Soil-plant relationships and contamination by trace elements: a review of twenty years of experimentation and monitoring after the Aznalcóllar (SW Spain) mine accident. *Science of the Total Environment* 625: 50-63.

Madejón, P., Marañón, T., Murillo, J.M. (2006). Biomonitoring of Trace Elements in the Leaves and Fruits of Wild Olive and Holm Oak Trees. *Science of The Total Environment* 355 (1): 187–203.

Main, R., Cho, M. A., Mathieu, R., O'Kennedy, M. M., Ramoelo, A., Koch, S. (2011). An investigation into robust spectral indices for leaf chlorophyll estimation. *ISPRS Journal of Photogrammetry and Remote Sensing* 66(6): 751-761.

Maisto, G., A. Alfani, D. Baldantoni, A. De Marco, A. Virzo De Santo. (2004). Trace Metals in the Soil and in Quercus Ilex L. Leaves at Anthropic and Remote Sites of the Campania Region of Italy. *Geoderma* 122 (2): 269–79.

Maňkovská, B., B. Godzik, O. Badea, Y. Shparyk, P. Moravčík. (2004). Chemical and Morphological Characteristics of Key Tree Species of the Carpathian Mountains. *Environmental Pollution* 130 (1): 41–54.

Manos, P. S. Stanford, A. M. (2001). The historical biogeography of Fagaceae: tracking the tertiary history of temperate and subtropical forests of the Northern Hemisphere. *International Journal of Plant Sience* 162: S77-S93.

Mansourian S., D. Vallauri. (2012). *Lessons Learnt from WWF's Worldwide Field Initiatives Aiming at Restoring Forest Landscapes*. Marseille: WWF France.

Marañón, T., Ibáñez, B., Anaya-Romero, M., Muñoz-Rojas, M., Pérez-Ramos, I. M. (2012). Oak trees and woodlands providing ecosystem services in Southern Spain. In: I. D. Rotherham, C. Handley, M. Agnoletti, T. Samojlik (eds.). *Trees beyond the wood.* pp. 291-299, Wildtrack Publishing, Sheffield.

Marañón, T., Padilla-Díaz, C. M., Pérez-Ramos, I. M., Villar, R. (2014). Research trends on ecology and management of *Quercus* species. *Ecosistemas* 23: 124-129 (in Spanish).

Marques, S., Borges, J. G., Garcia-Gonzalo, J., Moreira, F., Carreiras, J. M. B., Oliveira, M. M., Cantarinha, A., Botequim B., Pereira, J. M. C. (2011). Characterization of wildfires in Portugal. *European Journal of Forest Research* 130: 775-784.

Martinez-Frias, J. (1997). Mine Waste Pollutes Mediterranean. *Nature* 388 (6638): 120.

Matías, L., Zamora, R., Castro, J. (2011). Repercussions of simulated climate change on the diversity of woody-recruit bank in a Mediterranean-type ecosystem. *Ecosystems* 14: 672–682.

Matías, L., Zamora, R., Castro, J. (2012). Sporadic rainy events are more critical than increasing of drought intensity for woody species recruitment in a Mediterranean community. *Oecologia* 169: 833–844.

McDowell, N. G., Pockman, W. T., Allen, C. D., Breshears, D. D., Cobb, N., Kolb, T., Plaut, J., Sperry, J., West, A., Williams, D. G. Yepez, E. A. (2008). Mechanisms of plant survival and mortality during drought: why do some plants survive while others succumb to drought? *New Phytologist* 178: 719-739.

Mediavilla, S., Santiago, H., Escudero, A. (2002). Stomatal and mesophyll limitations to photosynthesis in one evergreen and one deciduous Mediterranean oak species. *Photosynthetica* 40: 553–559.

Mendes, A. M. S. C., Graça, J. A. R. (2009). Cork bottle stoppers and other cork products. In: Aronson, J., Pereira, J. S., Pausas, J. G. (eds.), *Cork oak woodlands on the edge. Ecology, adaptive management, and restoration*, pp. 59-69. Island Press, Washington D.C.

Metcalf J. C., Rose K. E., Rees M. (2003). Evolutionary demography of monocarpic perennials. *Trends in Ecology and Evolution* 18: 471–480.

Misson L., Degueldre D., Collin C., Rodriguez R., Rocheteau A., Ourcival J. M., Rambal S. (2011) Phenological responses to extreme droughts in a Mediterranean forest. *Global Change Biology* 17: 1036–1048.

Mohan J. E., Clark J. S., Schlesinger W. H. (2007) Long-term CO_2 enrichment of a forest ecosystem: implications for forest regeneration and succession. *Ecological Applications* 17: 1198–1212.

Montero G, Ruiz-Peinado R, Muñoz M. (2005). *Producción de biomasa y fijación de CO_2 por los bosques españoles* [*Biomass production and CO2 fixation by Spanish forests*]. Instituto Nacional de Investigación y Tecnología Agraria y Alimentaria.

Montserrat-Martí, G., Camarero, J. J., Palacio, S., Pérez-Rontomé, C., Milla, R., Albuixech, J. and Maestro, M. (2009). Summer-drought constrains the phenology and growth of two co-existing Mediterranean oaks with contrasting leaf habit: implications for their persistence and reproduction. *Trees-Structure and Function* 23: 787-799.

Moran, M.S., Clarke, T.R, Inoue, Y., Vidal, A. (1994). Estimating crop water deficit using the relation between surface-air temperature and spectral vegetation index. *Remote Sensing of Environment* 49: 246-263.

Moreira, A.C., Martins, M.S. (2005). Influence of site factors on the impact of *Phytophthora cinnamomi* in cork oak stands in Portugal. *Forest Pathology* 35: 145-162.

Moreira, F., Viedma, O., Arianoutsou, M., Curt, T., Koutsias, N., Rigolot, E., Barbati, A., Corona, P., Vaz, P., Xanthopoulos, G., Mouillot, F., Bilgili, E.. (2011). Landscape and wildfire interactions in southern Europe: implications for landscape management. *Journal of Environmental Management* 92: 2389-2402.

Moreno G., Pulido F. (2009) The functioning, management, and persistence of dehesas. In: Rigueiro-Rodríguez A, Mcadam J, Mosquera-Losada

MR (eds) *Agroforestry in Europe: Current Status and Future Prospects*, pp. 127-160, Springer, Berlin.

Naidu S. L., DeLucia E. H. (1998) Physiological and morphological acclimation of shade-grown tree seedlings to late-season canopy gap formation. *Plant Ecology* 138: 27-40.

Naidu, R., G. S. R. Krishnamurti, N. S. Bolan, W. Wenzel, M. Megharaj. (2001). Heavy metal interaction in soils and implications for soil microbial biodiversity. In: M. N. V. Prasad (ed.), *Metals in the environment, Analysis by biodiversity*, pp. 401– 431. New York: Marcel Dekker.

Nicola, F. De, V. Spagnuolo, D. Baldantoni, L. Sessa, A. Alfani, R. Bargagli, F. Monaci, S. Terracciano, S. Giordano. (2013). Improved Biomonitoring of Airborne Contaminants by Combined Use of Holm Oak Leaves and Epiphytic Moss. *Chemosphere* 92: 1224–30.

Nieminen, T., H.-S. Helmisaari. (1996). Nutrient retranslocation in the foliage of *Pinus sylvestris* L. growing along a heavy metal pollution gradient. *Tree Physiology* 16: 825–31.

Niinemets, Ü. (2015). Is there a species spectrum within the world-wide leaf economics spectrum? Major variations in leaf functional traits in the Mediterranean sclerophyll *Quercus ilex*. *New Phytologist* 205: 79-96.

Niinemets, U., Tenhunen J. D., Beyschlag, W. (2004). Spatial and age-dependent modifications of photosynthetic capacity in four Mediterranean oak species. *Functional Plant Biology* 31: 1179–1193.

Norby R. (1996) Oaks in a high-CO_2 world. *Annales des Sciences Forestières*. 53(2-3): 413-429.

Norby R. J., O'Neill E. G. (1989) Growth dynamics and water use of seedlings of *Quercus alba* L. in CO_2- enriched atmospheres. *New Phytologist* 111: 491–500.

Ogaya R., Peñuelas J. (2006) Contrasting foliar responses to drought in *Quercus ilex* and *Phillyrea latifolia*. *Physiologia Plantarum* 50: 373–382.

Ogaya R., Peñuelas J. (2007) Tree growth, mortality, and aboveground biomass accumulation in a holm oak forest under a five-year experimental field drought. *Plant Ecology* 189: 291–299.

Olalde M., Herrán A., Espinel S., Goicoechea P. G. (2002). White oaks phylogeography in the Iberian Peninsula. *Forest Ecology and Management* 156: 89–102.

Olsson, P., Kantola, T., Lyytikäinen-Saarenmaa, P., Jönsson, A. M., Eklundh, L. (2016). Development of a method for monitoring of insect induced forest defoliation – limitation of MODIS data in Fennoscandian forest landscapes. *Silva Fennica* 50 (2) 1495.

Ortego J., Bonal R., Muñoz A. (2010). Genetic consequences of habitat fragmentation in long-lived tree species: the case of the Mediterranean holm oak (*Quercus ilex*, L.). *Journal of Heredity* 101: 717–26.

Ortego J., Bonal R., Muñoz A., Aparicio J. M. (2014). Extensive pollen immigration and no evidence of disrupted mating patterns or reproduction in a highly fragmented holm oak stand. *Journal of Plant Ecology* 7: 384–395.

Ovando, Paola, Campos, Pablo, Oviedo, José L., Montero, Gregorio. (2009). Cost-benefit analysis of cork oak woodland afforestation and facilitated natural regeneration in Spain. In: Aronson, J., Pereira, J. S., Pausas, J. G. (eds.), *Cork oak woodlands on the edge. Ecology, adaptive management, and restoration*, pp. 177-188. Island Press, Washington D.C.

Panda, S., Amatya, D. M., Hoogenboom, G. (2014). Stomatal conductance, canopy temperature, and leaf area index estimation using remote sensing and OBIA techniques. *Journal of Spatial Hydrology* 12(1): 24 p.

Parker, J. N., Allesina, S., Lortie, C. J. (2013). Characterizing a scientific elite (B): publication and citation patterns of the most highly cited scientists in environmental science and ecology. *Scientometrics* 94: 469-480.

Paulo, J. A., Faias, S. P., Ventura-Giroux, C., Tomé, M. (2015). Estimation of stand crown cover using a generalized crown diameter model: application for the analysis of Portuguese cork oak stands stocking evolution. *iForest* 9: 437-444.

Pausas, J. G., Bladé, C., Valdecantos, A., Seva, J. P., Fuentes, D., Alloza, J. A. et al. (2004). Pines and oaks in the restoration of Mediterranean landscapes of Spain: new perspectives for an old practice—a review. *Plant Ecology* 171: 209-220.

Pausas, J. G., Marañón, T., Caldeira, M., Pons, J. (2009b). Natural regeneration. In: Aronson, J., Pereira, J. S., Pausas, J. G. (eds.), *Cork oak woodlands on the edge. Ecology, adaptive management, and restoration*, pp. 115-124. Island Press, Washington D.C.

Pausas, Juli G., Pereira, João S., Aronson James. (2009a). The tree. In: Aronson, J., Pereira, J. S., Pausas, J. G. (eds.), *Cork oak woodlands on the edge. Ecology, adaptive management, and restoration*, pp. 11-21. Island Press, Washington D.C.

Pautasso M. (2009). Geographical genetics and the conservation of forest trees. *Perspectives in Plant Ecology, Evolution and Systematics* 11: 157–189.

Pautasso, M., Dehnen-Schmutz, K., Holdenrieder, O., Pietravalle, S., Salama, N., Jeger, M. J., Lange, E., Hehl-Lange, S. (2010). Plant health and global change - some implications for landscape management. *Biological Reviews* 85: 729-755.

Peñuelas, J., F. Lloret, R. Montoya. (2001). Severe drought effects on Mediterranean woody flora in Spain. *Forest Science* 47: 214–218.

Peñuelas, J., Sardans, J., Filella, I., Estiarte, M., Llusià, J., Ogaya, R., et al. (2018). Assessment of the impacts of climate change on Mediterranean terrestrial ecosystems based on data from field experiments and long-term monitored field gradients in Catalonia. *Environmental and Experimental Botany* 152: 49-59.

Pereira, J. S., Correia, A. V., Correia, A. P., Branco, M., Bugalho, M., Caldeira, M. C., Cruz, C. S., de Freitas, H., Oliveira, Â. C., Pereira, J. M. C., Reis, R. M., Vasconcelos, M. J. (2002). Forests and biodiversity. In: F. D. Santos, K. Forbes, R. Moita (eds.), *Climate change in Portugal. Scenarios, impacts and adaptation measures*, pp. 363-413. Lisboa: Gradiva.

Pérez de Mora, A., E. Madejón, P. Burgos, P. Madejón, M. T. Domínguez, F. Madrid, T. Marañón, J. M. Murillo, F. Cabrera. (2009). Accumulation, Transfer and Remediation of Cd in Soils Affected by the Aznalcóllar Mine Spill (SW Spain): A Decade of Experience (1998-

2008). In: R. G. Parvau (ed.), *Cadmium in the Environment*, pp. 367–388. New York: Nova Science Publishers.

Pérez-Ramos, I. M. 2014. The miracle of regenerating in *Quercus* species. How will be oak forests in future? *Ecosistemas* 23: 13-17 (in Spanish).

Pérez-Ramos, I. M., Gomez-Aparicio L., Villar R., García L. V., Marañón T. (2010a). Seedling growth and morphology of three oak species along field resource gradients and seed mass variation: a seedling age-dependent response. *Journal of Vegetation Science* 21: 419–437.

Pérez-Ramos, I. M., Marañón, T. (2012). Community-level seedling dynamics in Mediterranean forests: uncoupling between the canopy and the seedling layers. *Journal of Vegetation Science* 23:526-540.

Pérez-Ramos, I. M., Ourcival, J. M., Limousin, J. M., Rambal, S. (2010b). Mast seeding under increasing drought: results from a long-term data set and from a rainfall exclusion experiment. *Ecology* 91: 3057-3068.

Pérez-Ramos, I. M., Padilla-Díaz, C. M., Koenig, W. D. Marañón, T. (2015). Environmental drivers of mast-seeding in Mediterranean oak species: does leaf habit matter? *Journal of Ecology* 103: 691–700.

Pérez-Ramos, I. M., Rodríguez-Calcerrada, J., Ourcival, J. M., Rambal, S. (2013b). *Quercus ilex* recruitment in a drier world: A multi-stage demographic approach. *Perspectives in Plant Ecology, Evolution and Systematics* 15:106-117.

Pérez-Ramos, I. M., Urbieta, T. I., Zavala, M. A., Marañón, T. (2012). Ontogenetic demographic conflicts and rank reversals in two Mediterranean oak species: implications for coexistence. *Journal of Ecology* 100:267-277.

Pérez-Ramos, I. M., Verdú, J. R., Numa, C., Marañón, T., Lobo, J. M. (2013a). The comparative effectiveness of rodents and dung beetles as local seed dispersers in Mediterranean oak forests. *PLoS One* 8(10):e77197.

Petit, R. J., Brewer, S., Bordács, S., Burg, K., Cheddadi, R., Coart, E. et al. (2002). Identification of refugia and post-glacial colonisation routes of European white oaks based on chloroplast DNA and fossil pollen evidence. *Forest Ecology and Management* 156: 49-74.

Petit R. J., Csaikl U. M., Bordacs S., Burg K., Coart E., Cottrell J., van Dam B., Deans J. D., Dumolin-Lapegue S., Fineschi S., Finkelday R., Gillies A., Glaz I., Goicoechea P. G., Jensen J. S., König A., Lowe A. J., Madsen S. F., Mátyás G., Munro R. C., Pemonge M. H., Popescu F., Slade D., Tabbener H., Taurchini D., de Vries S. M. G., Ziegenhagen B., Kremer A. (2002). Chloroplast DNA variation in European white oaks – Phylogeography and patterns of diversity based on data from over 2600 populations. *Forest Ecology and Management* 156: 5–26.

Petit R. J., Duminil J., Fineschi S., Hampe A., Salvini D., Vendramin G. G. (2005). Comparative organization of chloroplast, mitochondrial and nuclear diversity in plant populations. *Molecular Ecology* 14: 689–701.

Petit R. J., Excoffier L. (2009). Gene flow and species delimitation. *Trends in Ecology and Evolution* 24: 386–393.

Petit R. J., Hampe A. (2006). Some evolutionary consequences of being a tree. *Annual Review of Ecology, Evolution and Systematics* 37: 187-214.

Pinto-Correia, T., Godinho, S. (2013). Changing agriculture – changing landscapes: what is going on in the high valued montado landscapes of southern Portugal. In: D. Ortiz-Miranda, A. Moragues-Faus, E. Arnalte-Alegre (eds.), *Agriculture in Mediterranean Europe: Between Old and New Paradigms*, pp. 75-90. Emerald Group Publishing Limited, Bingley, UK.

Pinto-Correia, T., J., Mascarenhas. (1999). Contributing to the extensification/intensification debate: new trends in the Portuguese Montado. *Landscape and Urban Planning* 46: 125-131.

Pinto-Correia, T., Ribeiro, N., Sá-Sousa, P. (2011). Introducing the montado, the cork and holm oak agroforestry system of Southern Portugal. *Agroforestry Systems* 82: 99-104.

Polle, A., McKee, I., Blaschke, L. (2001). Altered physiological and growth responses to elevated [CO_2] in offspring from holm oak (*Quercus ilex* L.) mother trees with lifetime exposure to naturally elevated [CO2]. *Plant, Cell and Environment* 24: 1075-1083.

Poorter H., Niklas K. J., Reich P. B., Oleksyn J., Poot P., Mommer L. (2012). Biomass allocation to leaves, stems and roots: meta-analyses of interspecific variation and environmental control. *New Phytologist* 193:30–50.

Poorter H., Remkes C. (1990). Leaf area ratio and net assimilation rate of 24 wild species differing in relative growth rate. *Oecologia* 83: 553–559.

Porqueddu C., Ates S., Louhaichi M., Kyriazopoulost A. P., Moreno G., del Pozo A., Ovalle C. Ewing M. A., Nichols P. G. H. (2015). Graslands in Old World and New World Mediterranean-climate zones: past trends, current status and future research priorities. *Grass and Forage Science* 71: 1-35.

Poyatos, R., Aguadé, D., Galiano, L., Mencuccini, M. Martíınez-Vilalta, J. (2013). Drought-induced defoliation and long periods of near-zero gas exchange play a key role in accentuating metabolic decline of Scots pine. *New Phytologist* 200: 388–401.

Puerta-Piñero, C., Gómez, J. M., Schupp, E. W. (2010). Spatial patterns of acorn dispersal by rodents: do acorn crop size and ungulate presence matter? *Oikos* 119: 179–187.

Pulido, F. J. (2002). Biología reproductiva y conservación: el caso de la regeneración de bosques templados y subtropicales de robles (*Quercus* spp.). *Revista Chilena de Historia Natural* 75: 5-15.

Pulido, F. J., Díaz, M. (2002). Dinámica de la regeneración natural del arbolado de encina y alcornoque [Dynamics of the natural regeneration of oak and cork oak Woodland]. In: F. Pulido, P. Campos, G. Montero (eds.), *La gestión forestal de las dehesas. Historia, ecología, selvicultura y economia*, pp. 39–62. Mérida: Instituto del Corcho, la Madera y el Carbón Vegetal (ICMC). Junta de Extremadura.

Pulido, F. J., Díaz, M. (2005). Regeneration of a Mediterranean oak: a wholecycle approach. *Écoscience* 12: 92–102.

Pulido, F. J., García, E., Obrador, J. J., Moreno, G. (2010). Multiple pathways for tree regeneration in anthropogenic savannas: incorporating biotic and abiotic drivers into management schemes. *Journal of Applied Ecology* 47: 1272-1281.

Quero, J. L., Villar R., Marañón T., Zamora R. (2006). Interactions of drought and shade effects on seedlings of four Quercus species:

physiological and structural leaf responses. *New Phytologist* 170: 819–834.

Quero J. L., Villar R., Marañón T., Zamora R., Poorter L. (2007). Seed mass effects in four mediterranean *Quercus* species (Fagaceae) growing in contrasting light environments. *American Journal of Botany* 94: 1795-1803.

Quero, J. L., Villar, R., Marañón, T., Zamora, R., Vega, D., Sack, L. (2008). Relating leaf photosynthetic rate to whole-plant growth: drought and shade effects on seedlings of four *Quercus* species. *Functional Plant Biology* 35: 725–737.

Ramírez-Valiente J. A., Cavender-Bares J. (2017) Evolutionary trade-offs between drought resistance mechanisms across a precipitation gradient in a seasonally dry tropical oak (*Quercus oleoides*). *Tree Physiology* 37: 889–901.

Ramírez-Valiente J. A., Deacon N. J., Etterson J., Center A., Sparks J. P., Sparks K. L., Longwell T., Pilz G., Cavender-Bares J. (2018). Natural selection and neutral evolutionary processes contribute to genetic divergence in leaf traits across a precipitation gradient in the tropical oak *Quercus oleoides*. *Molecular Ecology* 27: 2176–2192.

Ramírez-Valiente J. A., Koehler K., Cavender-Bares J. (2015) Climatic origins predict variation in photoprotective leaf pigments in response to drought and low temperatures in live oaks (*Quercus* series Virentes). *Tree Physiology* 35: 521–534.

Ramirez-Valiente J. A., López R., Hipp A. L., Aranda I. (2020). Correlated evolution of morphology, gas exchange, growth rates and hydraulics as a response to precipitation and temperature regimes in oaks (*Quercus*). *New Phytologist* (in press), doi:10.1111/nph.16320.

Ramírez-Valiente, J. A., Etterson, J. R., Deacon, N. J., Cavender-Bares, J. (2019). Evolutionary potential varies across populations and traits in the neotropical oak *Quercus oleoides*. *Tree Physiology*, 39: 427-439.

Ramírez-Valiente, J.A., Lorenzo, Z., Soto, A., Valladares, F., Gil, L., Aranda, I. (2009). Elucidating the role of genetic drift and natural selection in cork oak differentiation regarding drought tolerance *Molecular Ecology* 18: 3803-3815.

Ramírez-Valiente, J. A., Sánchez-Gómez, D., Aranda, I., Valladares, F. (2010). Phenotypic plasticity and local adaptation in leaf ecophysiological traits of 13 contrasting cork oak populations under different water availabilities. *Tree Physiology* 30: 618-627.

Ramírez-Valiente, J. A., Valladares, F., Huertas, A. D., Granados, S., Aranda, I. (2011). Factors affecting cork oak growth under dry conditions: local adaptation and contrasting additive genetic variance within populations. *Tree Genetics and Genomes* 7: 285-295.

Ramírez-Valiente, J. A., Valladares, F., Sánchez-Gómez, D., Delgado, A., Aranda, I. (2014). Population variation and natural selection on leaf traits in cork oak throughout its distribution range. *Acta Oecologica* 58: 49-56.

Reed D. H., Frankham R. (2003). Correlation between fitness and genetic diversity. *Conservation Biology* 17: 230–237.

Rees M., Osborne C. P., Woodward F. I., Hulme S. P., Turnbull L. A., Taylor S. H. (2010). Partitioning the components of relative growth rate: How important is plant size variation? *American Naturalist* 176: E152–E161.

Regato, P., Berrahmouni, N. (2005). Using nontimber forest products for restoring environmental, social, and economic functions. In: S. Mansourian, D. Vallauri, N. Dudley (eds.), *Forest restoration in landscapes. Beyond planting trees*, pp. 215-222. Springer, New York.

Reich, P. B. (2014). The world-wide 'fast–slow'plant economics spectrum: a traits manifesto. *Journal of Ecology* 102: 275-301.

Reich, P. B., Walters, M. B., Ellsworth, D. S. (1997). From tropics to tundra: global convergence in plant functioning. *Proceedings of the National Academy of Sciences* 94: 13730-13734.

Reimann, C., F. Koller, G. Kashulina, H. Niskavaara, P. Englmaier. (2001). Influence of Extreme Pollution on the Inorganic Chemical Composition of Some Plants. *Environmental Pollution* 115: 239–52.

Rey, P., Alcántara, J. M. (2000). Recruitment dynamics of a fleshy-fruited plant (*Olea europaea*): connecting patterns of seed dispersal to seedling establishment. *Journal of Ecology* 88: 622-633.

Ribeiro, N. A., P. Surový. (2008). *Inventário nacional de mortalidade de sobreiro na fotografia aérea digital de 2004/2006* [*National inventory of cork oak mortality in 2004/2006 digital aerial photography*]. Évora: Instituto de Ciência Agrárias Mediterrânicas, Universidade de Évora.

Robert E. M. R., Mencuccini M., Martínez-Vilalta J. (2017). The anatomy and Functioning of the xylem in oaks. In: E. Gil-Pelegrín et al. (eds.), *Oaks Physiological Ecology. Exploring the Functional Diversity of Genus Quercus L.*, Tree Physiology 7, pp. 261-302, Springer, Cham, Switzerland.

Rodríguez-Calcerrada J., López R., Salomón R., Gordaliza G. G., Valbuena-Carabaña M., Oleksyn J., Gil L. (2015). Stem CO_2 efflux in six co-occurring tree species: underlying factors and ecological implications. *Plant Cell and Environment* 38: 1104-1115.

Rodríguez-Calcerrada. J., Pérez-Ramos, I., Ourcival, J. M., Limousin, J. M., Joffre, R., Rambal, S. (2011). Is selective thinning an adequate practice for adapting *Quercus ilex* coppices to climate change? *Annals of Forest Science* 68: 575-585.

Rodríguez-Calcerrada J., Reich P. B., Rosenqvist E., Pardos J. A., Cano F. J., Aranda I. (2008) Leaf physiological versus morphological acclimation to high-light exposure at different stages of foliar development in oak. *Tree Physiology* 28: 761-771.

Rodríguez-Calcerrada J., Sancho-Knapik D., Martin-StPaul N. K., Limousin J. M., McDowell N. G., Gil-Pelegrín E. (2017) Drought-induced oak decline – factors involved, physiological dysfunctions, and potential attenuation by forestry practices. In: Gil-Pelegrín et al. (eds.), *Oaks physiological ecology. Exploring the functional diversity of genus Quercus L.* Tree Physiology 7, pp: 419-451, Springer, Cham, Switzerland.

Rodríguez-Molina, M. C., Torres-Vila, L. M., Blanco-Santos, A, Palo Núñez, E. J., Torres-Álvarez E. (2002). Viability of holm and cork oak seedlings from acorns sown in soils naturally infected with *Phytophthora cinnamomi. Forest Pathology* 32: 365-372.Rolo, V., López-Díaz, M. L., Moreno, G. (2012). Shrubs affect soil nutrients availability with contrasting consequences for pasture understory and tree overstory production and nutrient status in Mediterranean grazed open woodlands. *Nutrient Cycling in Agroecosystems 93*(1): 89-102.

Rolo, V. Moreno, G. (2011). Shrub species affect distinctively the functioning of scattered *Quercus ilex* trees in Mediterranean open woodlands. *Forest Ecology and Management* 261: 1750-1759.

Romanyà, J., Vallejo, R. (2004). Productivity of *Pinus radiata* Plantations in Spain in Response to Climate and Soil. *Forest Ecology and Management* 195: 177–89.

Rosário, I. T., Rebelo, R., Caser, U., Vasconcelos, L., Santos-Reis M. (2019). Valuation of ecosystem services by stakeholders operating at different levels: insights from the Portuguese cultural montado landscape. *Regional Environmental Change* 19: 2173-2185.

Roubtsova, T. V. Bostock, R. M. (2009) Episodic Abiotic Stress as a Potential Contributing Factor to Onset and Severity of Disease Caused by *Phytophthora ramorum* in Rhododendron and Viburnum. *Plant Disease* 93: 912-918.

Ruiz-Labourdette, D., Schmitz, M. F., Pineda, F. D. (2013). Changes in tree species composition in Mediterranean mountains under climate change: indicators for conservation planning. *Ecological Indicators* 24: 310-323.

Ruíz-Robleto J, Villar R. (2005). Relative growth rate and biomass allocation in ten woody species with different leaf longevity using phyllogenetic independent contrasts (PICs). *Plant Biology* 7: 484-494.

Salomón, R., Rodríguez-Calcerrada, J., Zafra, E., Morales-Molino, C., Rodríguez-García, A., González-Doncel et al. (2016). Unearthing the roots of degradation of *Quercus pyrenaica* coppices: a root-to-shoot imbalance caused by historical management? *Forest Ecology and Management* 363: 200-211.

Sampaio, C., Clara, M., Almeida-Ribeiro, N. (2013). Decline of Mediterranean oak tres and its association with *Phythopthora cinnamomi*: a review. *European Journal of Forest Research* 132: 411-432.

Sánchez M. E., Caetano, P., Romero, M. A., Navarro, R. M. Trapero, A. (2006). *Phytophthora* root rot as the main factor of oak decline in southern Spain. In: Brasier C et al. (eds.), *Progress in Research on Phytophthora Diseases of Forest Trees*, pp. 149-154. Forest Research, Farnham, UK.

Sánchez-Salguero, R., Camarero, J. J., Grau, J. M., De la Cruz, A., Gil, P. M., Minaya, M., Fernández-Cancio, A. (2017). Analysing atmospheric processes and climatic drivers of tree defoliation to determine forest vulnerability to climate warming. *Forests* 8: 13.

Sardans, J., Peñuelas, J. (2005). Drought decreases soil enzyme activity in a Mediterranean *Quercus ilex* L. forest. *Soil Biology and Biochemistry* 37: 455-461.

Sardans, J., Rodà, F., Peñuelas, J. (2004). Phosphorus limitation and competitive capacities of *Pinus halepensis* and *Quercus ilex* subsp. *rotundifolia* on different soils. *Plant Ecology* 174: 307.

Savolainen, O., Pyhäjärvi, T., Knürr, T. (2007). Gene flow and local adaptation in trees. *Annual Review of Ecology, Evolution and Systematics* 38: 595-619.

Scarascia-Mugnozza G., De Angelis P., Matteucci G., Valentini R. (1996). Long-term exposure to elevated [CO_2] in a natural *Quercus ilex* L. community: net photosynthesis and photochemical efficiency of PSII at different levels of water stress. *Plant, Cell and Environment* 19: 643-654.

Schupp, E. W. (1995). Seed seedling conflicts, habitat choice, and patterns of plant recruitment. *American Journal of Botany* 82: 399-409.

Schupp, E. W., Fuentes, M., (1995). Spatial patterns of seed dispersal and the unification of plant population ecology. *Ecoscience* 2: 267–275.

Shi C., Kitao M., Agathokleous E., Watanabe M., Tobita H., Yazaki K., Kitaoka S., Koike T. (2016) Foliar chemical composition of two oak species grown in a free-air enrichment system with elevated O_3 and CO_2. *Journal of Agricultural Meteorology* 72: 50–58.

Soto A., Lorenzo Z., Gil L. (2007). Differences in fine-scale genetic structure and dispersal in *Quercus ilex* L. and *Q. suber* L.: consequences for regeneration of Mediterranean open woods. *Heredity* 99: 601–607.

Sousa, E. M. R., Santos, M. N. S., Varela, M. C., Henriques, J. (2007). *Perda de vigor dos montados de sobro e azinho: análise da situação e perspetivas* [*Loss of vigor of cork oak and holm oak: analysis of the situation and perspectives*]. Lisboa: Ministério da Agricultura, do

Desenvolvimento Rural e das Pescas. Direcção-Geral dos Recursos Florestais. Instituto Nacional dos Recursos Biológicos.

Sulla-Menashe, D., Friedl, M. A., Woodcock, C. E. (2016). Sources of bias and variability in long-term Landsat time series over Canadian boreal forests. *Remote Sensing of Environment* 177: 206–219.

Sullivan, A.R., Lind, J.F., McCleary, T.S., Romero-Severson, J., Gailing, O. (2013). Development and characterization of genomic and gene based microsatellite markers in North American red oak species. *Plant Molecular Biology Reporter* 31: 231–239.Tang, C., Liu, D., Wei, H., Wang, Y., Dai, C., Wu, P., Zhu, W, Rao, R. (2016). The response of the temperature of cold-point mesopause to solar activity based on SABER data set. *Journal of Geophysical Research* 121:7245-7255.

Tárrega R., Calvo L., Taboada A., García-Tejero S., Marcos E. (2009). Abandonment and management in Spanish dehesa systems: Effects on soil features and plant species richness and composition. *Forest Ecology and Management* 257: 731–738.

Teshera-Levye J., Miles B., Terwilliger V., Lovelock C. E., Cavender-Bares J. (2018) Drivers of habitat partitioning among three Quercus species along a hydrologic gradient. *Tree Physiology* 40: 142-157.

Tessier, L., Nola, P., Serre-Bachet, F. (1994). Deciduous *Quercus* in the Mediterranean region – tree-ring/climate relationships. *New Phytologist* 126: 355–367.

Thenkabail, P. S., Lyon, J. G. (Eds.). (2016). *Hyperspectral remote sensing of vegetation*. CRC Press.

Tiberi, R., Branco, M., Bracalini, M., Croci, F., Panzavolta, T. (2016). Cork oak pests: a review of insect damage and management. *Annals of Forest Science* 73: 219-232.

Tognetti R., Johnson J. D., Michelozzi M., Raschi A. (1998). Response of foliar metabolism in mature trees of *Quercus pubescens* and *Quercus ilex* to long-term elevated CO_2. *Environmental and Experimental Botany* 39: 233–245.

Urbieta, T. I., Pérez-Ramos, I. M., Zavala, M. A., Marañón, T., Kobe, R. (2008). Soil water heterogeneity and emergence time control seedling establishment in three coexisting oaks species. *Canadian Journal of Forest Research* 38: 2382-2393.

Valbuena-Carabaña M., Gil L. (2017). Centenary coppicing maintains high levels of genetic diversity in a root resprouting oak (*Quercus pyrenaica* Willd.). *Tree Genetics and Genomes* 13: 1–14.

Valbuena-Carabaña M., González-Martínez S. C., Sork V. L., Collada C., Soto A., Goicoechea P. G., Gil L. (2005). Gene flow and hybridisation in a mixed oak forest [*Quercus pyrenaica* Willd. and *Q. petraea* (Matts.) Liebl.] in central Spain. *Heredity* 95: 457–465.

Valladares, F. (ed.) (2004). *Ecology of the Mediterranean forest in a changing world*. Ministerio de Medio Ambiente, Madrid (in Spanish).

Valladares, F., Benavides, R., Rabasa, S. G., Díaz, M., Pausas, J. G., Paula, S., Simonson, W. D. (2014). Global change and Mediterranean forests: current impacts and potential responses. In: D. A. Coomes, D. F. R. P. Burslem, W. D. Simonson (eds.), *Forests and global change*, pp. 47-75, Cambridge University Press.

Valladares, F., Gianoli, E., Gómez, J. M. (2007) Ecological limits to plant phenotypic plasticity. *New Phytologist* 176: 749-763.

Valladares, F., Martinez-Ferri, E., Balaguer, L., Perez-Corona, E., Manrique, E. (2000). Low leaf-level response to light and nutrients in Mediterranean evergreen oaks: a conservative resource-use strategy? *New Phytologist* 148: 79-91.

Vallejo, R. (2005). Restoring Mediterranean forests. In: S. Mansourian, D. Vallauri, N. Dudley (eds.), *Forest restoration in landscapes. Beyond planting trees*, pp. 313-319. Springer, New York.

Villar R., Lopez-Iglesias B., Ruiz-Benito P., De la Riva E. G., Zavala M. A. (2014). Crecimiento de plántulas y árboles de seis especies de *Quercus* [Seedling and tree growth of six *Quercus* species]. *Ecosistemas* 23: 64-72.

Villar R., Ruiz-Benito P., de la Riva E. G., Poorter H., Cornelissen J. H. C., Quero J. L. (2017). Growth and growth-related traits for a range of *Quercus* species grown as seedlings under controlled conditions and for adult plants from the field. In: Gil-Pelegrín et al. (eds.), *Oaks physiological ecology. Exploring the functional diversity of genus*

Quercus L. Tree Physiology 7, pp. 393-418, Springer, Cham, Switzerland.

Villar-Salvador P., Castro-Diez P., Pérez-Rontomé C., Montserrat-Marti G. (1997). Stem xylem features in three Quercus (Fagaceae) species along a climatic gradient in NE Spain. *Trees* 12: 90–96.

Voelker S. L., Muzika R. M., Guyette R. P., Stambaugh M. C. (2006) Historical CO_2 growth enhancement declines with age in *Quercus* and *Pinus*. *Ecological Monographs* 76: 549–564.

Wright I. J., Reich P. B., Westoby M., Ackerly D. D., Baruch Z., Bongers F., Cavender-Bares J., Chapin F. S., Cornelissen J. H. C., Diemer M., Flexas J., Garnier E., Groom P. K., Gulias J., Hikosaka K., Lamont B. B., Lee T., Lee W., Lusk C., Midgley J. J., Navas M. L., Niinemets Ü., Oleksyn J., Osada N., Poorter H., Poot P., Prior L., Pyankov V. I., Roumet C., Thomas S. C., Tjoelker M. G., Veneklaas E. J., Villar R. (2004). The world-wide leaf economics spectrum. *Nature* 428: 821-827.

Zarco-Tejada, P. J., González-Dugo, V., Williams, L. E., Suárez, L., Berni, J. A., Goldhamer, D., Fereres, E. (2013). A PRI-based water stress index combining structural and chlorophyll effects: Assessment using diurnal narrow-band airborne imagery and the CWSI thermal index. *Remote Sensing of Environment* 138: 38-50.

Zavala, M. A., Bravo de la Parra, R. (2005). A mechanistic model of tree competition and facilitation for Mediterranean forests: Scaling from leaf physiology to stand dynamics. *Ecological Modelling* 188: 76–92.

In: *Quercus*: Classification, Ecology and Uses ISBN: 978-1-53618-026-8
Editor: Bent J. Steffensen

Chapter 2

QUERCUS SPP. NUTS FOR HUMAN CONSUMPTION

***Luís M. G. Castro*[1,2], *Rita Sardão*[1,2], *Elisabete M. C. Alexandre*[2], *Jorge A. Saraiva*[2] *and Manuela Pintado*[1,*]**

[1]Universidade Católica Portuguesa, CBQF - Centro de Biotecnologia e Química Fina – Laboratório Associado, Escola Superior de Biotecnologia, Porto, Portugal
[2]LAQV-REQUIMTE, Departamento de Química, Universidade de Aveiro, Aveiro, Portugal

ABSTRACT

Acorns from the *Quercus* spp. trees can be found in copious amounts in the tropical and temperate regions and have a significant role in gastronomy, folk medicine and animal feed in these regions. Erstwhile, these nuts were an excellent food source during scarcity times, but nowadays acorns are largely perceived as animal feed and are far from being as widely used as other common nuts. Despite their large availability,

[*] Corresponding Author's Email: mpintado@porto.ucp.pt.

only reduced quantities of acorns are curretly being used for human nutrition, so a new management approaches and usage of the produced wordwilde acorns would foment and contribute positively to the planet's sustainability. These nuts have an excelent nutritional profile and are a good source of bioactive compounds with numerous health benefits and relevant biological activities, which can contribute to health maintenance and well-being. In other to take advantage of these benefits, various technologies have been used to extract several compounds from different acorn species and several acorn-based functional products are under development, among which some are already found in today's market. The incorporation of acorn in food matrices has shown to improve food properties and nutritional profiles without affecting negatively the sensory characteristics and consumers' perception of these new foods. A brief description of the acorn history, production, and sustainability was done in this chapter. The core of this chapter involves the acorn transformation and use for human consumption. Thus, an extensive revision about the study and the use of acorn oils, flours, extracts, and beverages was done. The nutritional composition and the bioactive compounds naturally present in acorn were described, as well as the effect of their bioactivities on human health. Some green processing and extracting technologies were also discussed. Additionally, a section about tannins and acorn toxicity was also presented and discussed.

Keywords: Quercus, acorn, production, consumption, nutricional composition, bioactivities, processing and extracting technologies

INTRODUCTION

The acorns have a huge potential as an alternative functional food, specifically considering their high nutritional value in carbohydrates (70 to 90%), of which aproximately 50% is starch, lipids (high unsaturated/saturated fatty acid racio) and richness in bioactive phytochemicals with important biological action (antioxidant compounds, namely phenolic compounds and sterols), contributing to the consumer well-being due to its anti-mutagenic, hypoglycemic, anti-inflammatory and hepatoprotective properties (León-Camacho et al. 2004; Custódio et al. 2015; Rakić et al. 2006). Furthermore, they promote the use of a natural product in accordance with the principles of sustainability, since much of their production is left behind in the fields. The consumption as nuts (resembling chestnuts) or flour (high starch contents) are the most common form of acorn consumption in prehistoric times. New products made with acorn are being developed such as liquors (exemple of *Licor de bolota* in Portugal), beer, coffee and other beverages like *Eichelkaffee* and *Raccahout*, with benefical effects to human health. Other products such as caramels, honey, bread, biscuits and yogurts are also being developed by replacing/adding acorn flour into the matrixes without affecting negatively the consumers preception and improving its nutritional value and properties (García-Gómez et al. 2017; Sanchez-Bel et al. 2015; Dabija et al. 2019; J. Korus et al. 2015; Silva et al. 2016; Ajo 2018; Parsaei et al. 2018). However, and despite their availability and nutritional importance, acorns are currently far from being as widely used as other common nuts. On the other hand, the reincorporation of acorn in human nutrition has shown promising results. Acorn extracts can be a natural alternative to the synthetic food antioxidants and have a positive effect on the chicken meat characteristics and on the food processing (Özünlüet et al. 2018; Ferreira et al. 2017). These extracts can be used to prepare functional drinks in health maintenance and prevention of illnesses (Sanchez-Bel et al. 2015; Coelho et al. 2018). The usage of acorn flour in food products has a positive effect on food properties and nutritional profile without affecting the sensory characteristics and consumers perception (Dabija et al. 2019; Ajo 2018; Pasqualone et al. 2019; Parsaei et al. 2018; Korus et al. 2017; Korus et al. 2015).

Various technologies have been used to extract various materials from different acorn *Quercus* spp. The classical techniques are dependent on solvent extraction capacity, polar nature of the interest compounds, agitation, and temperature usage. Solvent toxicity and its impact on environment and health must also be considered. However, these techniques present several disadvantages such as the long extraction times, high temperature, the usage of large quantities of organic solvents, low extraction yields, degradation of compounds sensible to temperature, among others (Alexandre et al. 2017; Azmir et al. 2013). The use of emergent extraction technologies arises from the need to surpass such difficulties. These are greener and environmental-friendlier technologies since they use fewer amounts of organic solvents, which can reduce the extraction time and allows them to obtain higher extraction yields with superior quality (Mena-García et al. 2019; Armenta et al. 2019; Zhang et al. 2018). Ultrasounds, microwave, and pressurized fluid technologies have been used to extract phenolic compounds. The first technique relies on the usage of sound waves and cavitation phenomen to extract compounds, while microwave extraction relies on the interaction between electromagnetic waves and the material, which leads to magnetic and dielectric losses. Pressurized fluid technology is based on the usage of high pressures and temperatures to maintain the extraction of solvents liquid above its boiling point. Supercritic carbon dioxide technique has been used to extract oils from acorns and relies on the usage of carbon dioxide (extraction solvent) in its supercritical state where it has both liquid and gas properties.

History, Production, and Sustainability

Quercus spp. (family *Fagaceae*) represent an important group of evergreen or deciduous trees from temperate and tropical climatic areas. These trees spread all over the northern hemisphere in Asia, North America, Europe, and North Africa (Axelrod, 1983). Mexico is the country with the largest number of species in the world, and one of the most common in the country (Valencia-A. 2004). Since different species of *Quercus* exist worldwide, roughly 500 species, acorns come in a variety of shapes, sizes, colours and bitterness (Bakiş et al. 2014; García-Gómez et al. 2017; Costas 2013; Rodríguez-Estevez et al. 2012; Aldrich et al. 2011). Acorns are nuts with mean width values around 1.5 and 3.0 cm, while usually weighing in the range from 1.2 to 6.5 g (Gea-Izquierdo et al. 2006). Its morphological characteristics are illustrated and described in Figure 1.

The *Quercus* and their nuts are considered sacred, symbols of strength and endurance, with a prominent place in human culture and history (Johnson et al. 2001). The trees are a keystone taxon from both ecological and economic perspectives, forming an important source of food and habitat for wildlife and wood for human use (Aldrich et al. 2011), in addition to providing carbon sequestration and protection against erosion and desertification (Allard et al. 2013). *Q. suber* has been traditionally used also to produce high-quality cork (León-Camacho et al. 2004). In times of scarcity, the *Quercus* acorns served as fundamental nuts for human consumption and to support the local economy (Correia et al. 2013). Besides that, the use of acorn has a worldwide long history in folk medicine being used as astringents, anti-diarrhoeal, anti-ulcer and antidotes (Papoti et al. 2018).

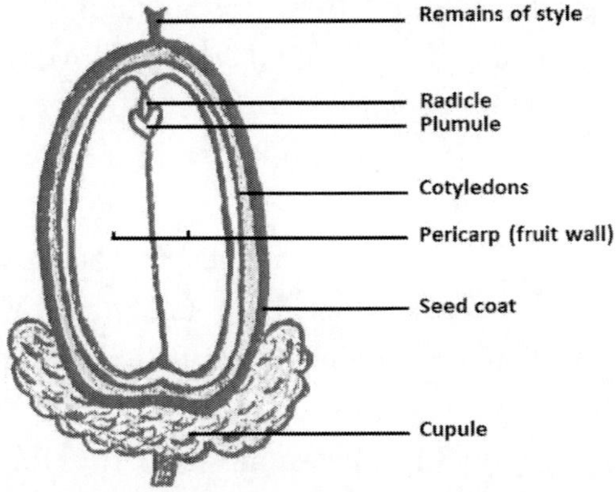

Figure 1. Schematic representation of a *Quercus* acorn and legend of morphological characters in longitudinal cross-section.

Despite their availability, acorns are now far from being as widely used as other common nuts and end up being used essentially to feed domestic fauna (Rodríguez-Estévez et al. 2008). An example is the *Q. suber* and *Q. ilex* acorns, traditionally used for feeding the native black Iberian pig (*Sus scrofa mediterraneus*), which produces delicious and renowned hams and pork sausages (Rodríguez-Estevez et al. 2012; García-Gómez et al. 2017; Rodríguez-Estévez et al. 2008). Nowadays, new products made from acorns for human consumption are under development, and some are already on the market. Among the new uses, acorn chocolates, acorn bread, acorn liquors, beer and other beverages, acorn caramels, and acorn honey stand out (García-Gómez et al. 2017). The flour with high starch content is the most common form of acorn consumption by humans since prehistoric times.

Quercus trees are monoecious species, with separate male and female flowers, that produce acorn annually or biannually (Rodríguez-Estevez et al. 2012). The flowering flush happens in spring (April-May), although in some years a small secondary flowering occurs in autumn, the acorn growth occurs in summer, and the ripening and dispersion occur during the autumn (October to November) (Gea-Izquierdo et al. 2006). However, the productivity of acorns is extremely variable, both between and within years, associated to different factors. Those include density of adult trees, masting phenomenon, soil characteristics, individual characteristics of trees (genetic potential, age, canopy surface, etc.), tree mass handling (tilling, pruning, and grazing), meteorological conditions (mainly during flowering), and sanitary status (insect infestation by *Lymantria, Tortrix, Curculio, Cydia, Balaninus,* and *Brenneria*) (Rodríguez-Estevez et al. 2007). The great morphology and composition variability depending on the phylogenetic, as well as ecological factors, may result in considerable changes in the acorn sensory and nutritional attributes (Gea-Izquierdo et al. 2006; Vinha et al. 2016).

The valorization of this nut is important to world sustainability since population growth requires the use of food in a more sustainable way, either environmentally or socially. Since acorn has properties that allow it to be a functional food, its reintroduction in the human food system must be considered. Moreover, an increase in their consumption can also have a positive impact on the economic level (Rosenberg 2008). On the other hand, the land-use practices around the globe have resulted in widespread deforestation and forest degradation, including areas previously dominated by the *Quercus* spp. (Löf et al. 2019). So, the scale of opportunity to implement the *Quercus* reforestation is huge, leading to a consequent increase in acorn production.

Nutritional Composition

Acorns are described as being mainly rich in water (30-40%), carbohydrates (particularly starch), and lipids (particularly unsaturated fatty acids) (Costas 2013; León-Camacho et al. 2004), but are also a source of minerals (like Fe, Cu, Zn, and Mn) and important antioxidant compounds (namely phenolic compounds and sterols) with bioactive properties (Custódio et al. 2015; Rakić et al. 2006a). The gluten-free flours are rich in fiber (3-7%) that can be used to develop gluten-free foodstuffs by employing a commonly disregarded and cheap alternative (Silva et al. 2016; Costas 2013). The nutritional composition of several acorn species is shown in Table 1.

Carbohydrates correspond for 70-90% of dry matter, with starch being the most important carbohydrate in the acorn, constituting on average 50% of the total acorn dry matter. In the study of Correia et al. (2013), isolated acorn starches from *Q. suber* and *Q. rotundifolia* presented high amylose content (53-59%) and resistant starch content (30.8-41.4%). The resistant starch acts as a substrate for bacterial fermentation (prebiotic) in the large intestine and has been reported as one of the many health benefits for humans (Man et al. 2013). In particular, acorn flour has a low and slow glycemic response, compared to potatoes and processed cereals, which can make the release of glucose into the blood to occur evenly throughout the day, avoiding problems of diabetes, obesity and cardiovascular disease (Lin et al. 2018; Miao et al. 2015). Despite not being as nutritionally rich as other nuts, acorns represent a good alternative to other high-starch content products.

The lipid fraction of acorns (5-10%) shows high values of both monounsaturated and polyunsaturated fatty acids (Silva et al. 2016). Oleic acid (about 65% of total fatty acid content) is the main fatty acid, followed by linoleic and palmitic acids. In general, acorn fatty acid profiles are similar to those typically detected in olive oil (Lopes et al. 2005) Comparing with olive oil, the amounts of the remaining fatty acids are significantly higher in the acorn oil, highlighting linolenic acid, part of the family of omega-3, which is an important polyunsaturated fatty acid in the prevention of cardiovascular diseases (Weschenfelder et al. 2020).

Table 1. Nutritional composition of several acorn spp.

Quercus spp.	Processing	Humidity (%)	Ash (%)	Protein (%)	Lipids (%)	Carbohydrates (%)	Fiber (%)	Starch (%)	References
ilex	Fresh	38.4 (FW)	-	-	4.7 (FW)	-	-	-	Akcan et al. 2017
ilex	Lyophilized	30.0 (FW)	1.2	4.0	8.0	87.0	-	-	Vinha et al. 2016
ilex	Dried (45°C)	-	1.3-2.0	3.9-5.9	9.1-15.0	-	-	-	Galván et al. 2012
ilex	Dried (45°C)	5.4	1.8	4.6	13.7	75.2	10.9	57.8	Silva et al. 2016
ilex	Dried (50°C)	10.4	1.9	-	-	-	-	34.2	Amina et al. 2018
ilex	Roasted (200°C)	9.7	1.8	5.0	13.9	78.8	13.0	54.5	Silva et al. 2016
ilex	-	-	-	-	9.0	-	-	-	Charef et al. 2008
ilex	-	-	-	-	6.9	-	-	-	León-Camacho et al. 2004
suber	Fresh	44.3 (FW)	-	-	2.1 (FW)	-	-	-	Akcan et al. 2017
suber	Lyophilized	46.0 (FW)	2.5	9.0	3.0	86.0	-	-	Vinha et al. 2016
suber	Dried (40°C)	42.0	1.7	4.2	5.2	86.2	2.7	49.0	Correia et al. 2009
suber	Dried (60°C)	8.0	1.7	4.1	5.4	86.0	2.8	31.4	Correia et al. 2012
suber	Dried (60°C)	11.7	1.7	4.1	5.4	86.1	2.7	31.4	Correia et al. 2010
suber	Dried (60°C)	-	-	-	6.0	-	-	-	Lopes et al. 2005
suber	Dried (75°C)	-	-	-	4.4-6.6	-	-	-	Ferreira-Dias et al. 2003b
suber	-	-	-	-	9.0	-	-	-	Charef et al. 2008
suber	-	-	-	-	-	-	-	57.9-59.4	Correia et al. 2013
suber	-	-	-	-	8.3	-	-	-	León-Camacho et al. 2004
rotundifolia	Dried (40°C)	37.6	1.9	4.8	11.1	79.2	3.0	48.0	Correia et al. 2009

Quercus spp.	Processing	Humidity (%)	Ash (%)	Protein (%)	Lipids (%)	Carbohydrates (%)	Fiber (%)	Starch (%)	References
rotundifolia	Dried (45°C)	7.5	2.0	4.6	8.4	84.1	17.9	51.8	Silva et al. 2016
rotundifolia	Dried (60°C)	7.7	1.9	4.6	11.1	79.5	2.9	31.6	Correia et al. 2010
rotundifolia	Dried (60°C)	-	-	-	12.1	-	-	-	Lopes et al. 2005
rotundifolia	Roasted (200°C)	22.1	2.0	4.3	10.2	83.2	16.8	52.1	Silva et al. 2016
rotundifolia	Dried (75°C)	-	-	-	7.3-10.9	-	-	-	Ferreira-Dias et al. 2003b
rotundifolia	-	-	-	-	-	-	-	53.7-54.5	Correia et al. 2013
aegilops	Dried (50°C)	-	-	-	7.5	-	-	-	Al-Rousan et al. 2013
aegilops	Dried (60°C)	12.8	0.6	9.5	1.1	74.7	1.31	-	Ajo 2018
aegilops	-	12.0	8.6	5.4	8.3	-	17.5	49.4	Skendi et al. 2018
faginea	Fresh	39.7 (FW)	-	-	3.8 (FW)	-	-	-	Akcan et al. 2017
faginea	Lyophilized	35.0 (FW)	2.0	6	2.0	89.0	-	-	Vinha et al. 2016
faginea	-	-	-	-	4.1	-	-	-	León-Camacho et al. 2004
pyrenaeca	Fresh	41.0 (FW)	-	-	2.5 (FW)	-	-	-	Akcan et al. 2017
pyrenaeca	Dried (75°C)	-	-	-	3.8-4.4	-	-	-	Ferreira-Dias et al. 2003b
coccifera	Fresh	42.3 (FW)	-	-	1.3 (FW)	-	-	-	Akcan et al. 2017
coccifera	Dried (105°C)	-	2.4	4.2	-	-	28.4 (NF) 14.5 (AF)	-	Üniv et al. 2015
mongolica	Fresh	-	1.5	4.4	1.7	90.3	2.1	-	Shimada 2001
serrata	Fresh	-	2.8	4.5	2.5	88.3	1.9	-	Shimada 2001
brantii	Dried (RT)	6.5	1.8	4.1	7.7	-	2.5	-	Parsaei et al. 2018
nigra	Lyophilized	39.0 (FW)	1.2	5.0	2.0	92.0	-	-	Vinha et al. 2016
glandurifera	Dried (40°C)	7.6 (FW)	0.03	10.2	4.2	82.7	3.0	-	Li et al. 2015

Table 2. (Continued)

Quercus spp.	Processing	Humidity (%)	Ash (%)	Protein (%)	Lipids (%)	Carbohydrates (%)	Fiber (%)	Starch (%)	References
calliprinos	Dried (50°C)	-	-	-	3.4	-	-	-	Al-Rousan et al. 2013
infectoria	Dried (50°C)	-	-	-	6.6	-	-	-	Al-Rousan et al. 2013
cerris	Roasted (200°C)	6.7	2.4	6.3	4.0	60.4	26.9	-	Pinto et al. 2019
robur	Dried (60°C)	-	3.3	7.1	-	-	15.9 (NF) 9.5 (AF)	-	Kilic et al. 2010

FW: Fresh weight.

NF: Neutral fiber; AF: Acid fiber; RT: Room temperature.

Proteins represent 2-8% of the acorn dry matter (Özcan 2006). More specifically, Galván et al. (2011) studied the protein yield among *Q. ilex* acorn populations and verified that it varied between 2.92-5.92 mg/g dry weight. On the other hand, their potential nutritive value has great importance as a relatively rich source of some essential amino acids. Particular attention is paid to the lysine needs of adults, since this indispensable amino acid is more likely to be limiting in cereal-based diets, characteristic of the population in large areas of the third world (Pellett et al. 2004).

The mineral content in acorns is also noteworthy since minerals are important for certain essential functions of the body (Vinha et al. 2016; Costas 2013). Rakić et al. (2006) described considerable amounts of Fe, Cu, Zn, and Mn, besides Ca, Mg, P, and K in lower levels, in *Q. robur* acorn samples.

Tocopherols (vitamin E) are the main substances with antioxidant properties naturally present in seed oils. The antioxidant action of vitamin E has a double effect: protection *in vivo*, protecting cellular lipids from oxidation, and *in vitro*, protecting the oil and food from rancidity oxidative (Butterfield et al. 2002; Bramley et al. 2000). Acorn oil also shows high amounts of tocopherol (vitamin E), with a predominance of γ-tocopherol (90% of total tocopherol content), although there is a wide variation between species (León-Camacho et al. 2004). In general, γ-tocopherol reaches levels 4.6-8.7-fold higher than those detected for α-tocopherol. Acorns are also an excellent source of provitamin A since it has been reported that a small number of acorns would guarantee the recommended daily requirements of vitamin A, which might be a great advantage in areas where vitamin A deficiency is a common problem, particularly in some African and Southeast Asian low-income countries (Bainbridge 2001).

Equally relevant are the detected levels of phytosterols, sterols that are found naturally in plants, among which β-sitosterol is the major compound (> 90% of sterols) in acorns. The values reported for some acorn sterols are higher than those obtained in other vegetable oils (León-Camacho et al. 2004). The β-sitosterol chemical structure is very similar to that of cholesterol and studies have also indicated that a diet high in this and other phytosterols can inhibit cholesterol absorption and reduce serum cholesterol levels by competing for intestinal absorption (Lei et al. 2017).

In addition to the tocopherols, phenolic compounds, such as phenolic acids, flavonoids, and tannins, are considered as being the primary bioactive compounds in acorn fruits and that act as strong natural antioxidants, reducing rancidity. Cantos et al. (2003) reported 32 phenolic compounds isolated from *Q. rotundifolia*, *Q. ilex* and *Q. suber* acorns harvested in the south of the Iberian Peninsula and the vast majority correspond to hydrolyzable tannins and their precursors (gallic and ellagic acid) that are compounds containing a central core of glucose or another polyol esterified with gallic acid (Mojzer et al. 2016; Cantos et al. 2003). The presence of these type of compounds provides health benefits, which are mainly correlated to their high antioxidant activity, besides having important functions in decreasing the risk of several disorders such as cardiovascular, neurological and neoplastic diseases (Ullah et al. 2008).

Besides the potential effects of their direct consumption, the inclusion of acorns in animal feeding have been proced to improve the quality of their meat (Lopez-Bote 1998). It can also be considered the indirect effect on human health through the consumption of pork products whose diet incorporates the consumption of acorn.

Bioactive Compounds, Bioactivities and Safety

The high nutritional value of the acorn and the richness in bioactive phytochemicals with biological action led to the development of some studies to increase the use of the acorn and their by-products. An example is the published research by Coelho et al. (2018), where a coffee-like beverage obtained from *Q. ilex* and *Q. suber* acorns was studied. The beverage presented antioxidant capacity and antimutagenic activity related to their phenolic content, and the manufacturing process did not generate genotoxic or cytotoxic compounds. So, this beverage is a promising healthy substitute for coffee. In the same line, Custódio et al. (2015) presented a comparative evaluation of the antioxidant potential and inhibitory activity of leaf and acorn extracts of *Q. suber* on key enzymes relevant for hyperglycemia and neurodegenerative diseases. In another study, Custódio et al. (2013) reported that extracts from *Q. suber* and *Q. ilex* acorns are endowed with bio-compounds (namely phenolic compounds) that act synergistically in relieving symptoms associated with neurological disorders, by exhibiting antioxidant activity, inhibitory activity, and potential to prevent oxidative stress-induced cell damage. Jin et al. (2005) investigated the protective effect of *Q. aliena* acorn extracts against CCl_4-induced hepatotoxicity in rats. These results suggest that the mechanism underlying the protective effects are related to its antioxidant activity and effect on the expression of CYP2E1 (membrane protein expressed in high levels in the liver and involved in the metabolism of xenobiotics in the body). These and other examples of the potential benefits of acorn towards human health are presented in Table 2.

The acorns of all *Quercus* are edible. However, the direct consumption of some acorn species requires prior processing to their use as food (Rosenberg 2008). This processing aims the elimination or reduction of tannins, which are compounds responsible for the bitterness and astringency, through simple treatments like drying, roasting, boiling or leaching (García-Gómez et al. 2017; Costas 2013). Tannins are phenolic compounds that have the ability to combine with proteins and other polymers such as polysaccharides, causing the sensation of astringency, which is no more than the loss of the lubricating effect of saliva (Prinz et al. 2008).

Furthermore, excessive tannin consumption can be associated with digestibility problems and reduction of the nutritional value of foods. An example is tannins minerals chelating capacity throughout the digestive tract, which means they bind to the essential minerals and reduce their absorption (Chung et al. 1998). On the other hand, normally there is no interest in eliminating all the tannins present (only the excess to reduce astringency) in the acorn because they also have health benefits. The hydrolyzable tannins are associated with antimutagenic, anticarcinogenic and antimicrobial properties (Chung et al. 1998). Thus, it is necessary to maintain as much as possible the nutritional and phytochemical profiles of acorns, whereby further studies should be performed to conclude the best processing practices to counteract astringency without compromising the acceptability of the food.

Table 3. Beneficial effects of the acorn or their products

Quercus spp.	Food matrix/Extract	Beneficial effects	Reference
suber and ilex	Coffee-like beverage	• Antioxidant capacity associated as well as anti-mutagenic activities	Coelho et al. 2018
convallata and arizonica	Oak leaves infusions and fermented beverages	• Hypoglycemic and antioxidant effects	Gamboa-Gómez et al. 2017
serrata	Extract	• Anti-inflammatory action	Huang et al. 2016
suber	Extract	• Hypoglycemic and Alzheimer's disease protection	Custódio et al. 2015
suber and ilex	Extract	• Neuroprotective properties	Custódio et al. 2013
aliena	Extract	• Free radical scavenging and hepatoprotective actions	Jin et al. 2005

Extraction Technologies Applied to Acorn

Classical Techniques

These techniques rely on the extraction ability of various solvents and usage of agitation and/or temperature in order to increase the mass transfer rates and thus increasing the solubility of the compounds. The extraction is dependent on the solvent used thus the polarity of the compounds is the most important parameter to be concerned. Additionally, the solvents' toxicity, safety for the environment and human health have to be taken into consideration during solvent selection (Azmir et al. 2013).

Soxhlet extraction is an old but efficient extraction technique being a reference method for other extraction procedures. Soxhlet extraction with hexane or petroleum-ether has been used to obtain oils from *Q. ilex*, *Q. infectoria*, *Q. pyrenaeca*, *Q. rotundifolia* and *Q. suber* kernels (Al-Rousan et al. 2013; Bernardo-Gil et al. 2007; Charef et al. 2008; Ferreira-Dias et al. 2003a; Lopes et al. 2005; Pérez-Serradilla et al. 2007; Rabhi et al. 2015). Phenolic, flavonoids and tannins compounds from *Q. ilex*, *Q. cerric*, *Q. coccifera*, *Q. rotundifolia*, *Q. suber*, *Q. shumardii*, *Q. falcata*, *Q. nigra* and *Q. phellos* kernels and leaves have been successfully extracted with hexane, ethanol, acetone, chloroform, ethyl acetate, and methanol, which had provided higher extraction yields (Rakić et al. 2004; Ofcarcik et al. 1971; Custódio et al. 2015; Custódio et al. 2013; Asghari et al.2011; Hayouni et al. 2007). One of the main disadvantages of this technique is related to the need for large volumes of solvent that are harmful to the environment and human health. It also requires long extraction times and high temperatures, which can lead to degradation of thermal-label compounds. However, the repeated contact of the sample with fresh portions of the solvent increases the compounds' displacement from the matrix and is not necessary additional filtration steps (García-Ayuso et al. 1998).

Maceration extraction is a very simple and low-cost technique to obtain compounds from different matrixes and has been used to recover phenolic, flavonoids and tannins compounds from *Q. aegilops*, *Q. brantii*, *Q. coccifera*, and *Q. suber* (Hayouni et al. 2007; Ebrahimi et al. 2016; Papoti et al. 2018; Zazoua et al. 2018). Since this technique is performed at room temperature, it can be used to extract thermo-sensitive compounds (Zhang et al. 2018). Additionally, maceration can be done with water, but frequently uses methanol, chloroform, acetone, or mixtures to increase the extraction yields that are usually low for some compounds, if they have low solubility at room temperature and often requires long extraction times (Seidel 2006).

Alkaline extraction has been used to extract starch from *Q. ilex*, *Q. rotundifolia*, and *Q. suber* kernels with high purity, because many proteins are soluble in NaOH solutions. The usage of NaOH solutions higher than 0.24% can lead to alteration in the granular organization of starch, which causes alterations in starch structure and properties. Lower concentration than 0.2% NaOH solubilize fibers with low sedimentation ability and denature/flocculate proteins without the occurrence of starch gelatinization (Luciano et al. 2017). In order to isolate the starch, the acorn flour is homogenized with NaOH, centrifuged and the mucilaginous layer is scraped off. The precipitated is suspended in water and dried (Correia et al. 2010; Correia et al. 2012; Correira et al. 2014). Some authors such as Correia et al. (2010) also neutralized the filtrate with HCl, which prevents starch fraction alteration (Daiuto et al. 2005). However, this is a very time-consuming technique.

Enzymatic extraction has been used to extract starch from *Q. rotundifolia* and *Q. suber* kernels. In order to liberate the starch granules that are trapped within the matrix, an acorn flour suspension is usually prepared with water and the protease is then added. After the enzymatic reaction happened at the desired conditions, the suspension is centrifuged, suspended in water, sieved, and finally dried (Correia et al. 2010; Correia et al. 2012; Correira et al. 2014). The enzymatic extractions provide high purity starches from acorn, are more time saving and can still be performed at relatively lower temperatures.

Emerging Techniques

In general, the main inconvenient aspects of the classical extraction technologies are the long extraction times; requirement of high quantities of sample and solvents with high purity, which make the extraction process very costly; the application of high temperatures, which increases costs due to high consumption of energy, and can deteriorate thermolabile compounds; the extraction solvents can have a low selectivity, which leads to low extraction yields and low concentrations of the extracted target compounds (Alexandre et al. 2017). To overcome these limitations, new and promising extraction techniques such as microwave extraction (ME), supercritical carbon dioxide extraction (SCE), pressurized fluid extraction (PFE) and ultrasound extraction (USE) have been applied to extract compounds (Armenta et al. 2019). These technologies are usually greener and environmental-friendlier since they use fewer amounts of organic solvents, can reduce the extraction time, increase the extraction yields and the extracts have higher purity.

Microwave Extraction

This emergent technology is based in the use of microwaves, which are composed of perpendicularly oscillating magnetic ($\vec{B}$) and electric ($\vec{E}$) fields frequency ranged from 0.3 to 300 GHz (Figure 2). Heating is based upon the effects of these fields on polar materials and the energy is converted in heat by dipole rotation and ionic conduction mechanisms (Sagar et al. 2018). The interaction between the microwave and the material can lead to magnetic and dielectric losses, which results in heat.

The dielectric losses have more relevance for non-metal materials and are based on the ionic conduction and dipole rotation principles. The first relates to the migration of charged particles under an electrical field of a microwave generating friction between the moving particles and the surrounding medium, thus causing heat. The dipole rotation states that the dipolar molecules try to align themselves with the alternating electric field produced by microwaves (Figure 3). The oscillation leads to collisions between them and surrounding molecules generating heat (Vinatoru et al. 2017). During ME, the energy is quickly delivered to the solvent and matrix and then absorbed by some substances, especially polar compounds. Consequently, the internal temperature of cells increases rapidly leading to liquid vaporization of the cellular content, which can cause disruption of the cellular walls and/or plasmatic membranes. This cellular disruption may favor the solvent mass transfer phenomena increasing the extraction efficiency (Kala et al. 2016; Zhang et al. 2011). Some advantages of ME over the classical thermal methods are related to higher extraction yields, higher compounds purity, decreased thermal degradation and selective heating. It is a more effective, feasible, and reliable way to extract compounds, and can improve extraction time and energy consumption. It is also considered a green technology since it reduces the consumption of organic solvents, replacing then by cheaper and non-toxic solvents (Alupului et al. 2012; Zhang et al. 2018). Asghari et al. (2011) extracted tannic acid, an antibacterial compound, from *Q. infectoria* galls using ME. The authors concluded that ME provided higher extraction yields, led to faster extraction rates, increased extract purity, and the separation is facilitated comparing to Soxhlet extracts. In another study, Papoti et al. (2018) extracted phenolic compounds from *Q. aegilops* acorns also by ME. These authors observed that ME allowed higher yields of phenolic compounds in comparison to the conventional maceration and reduce drastically the extraction time.

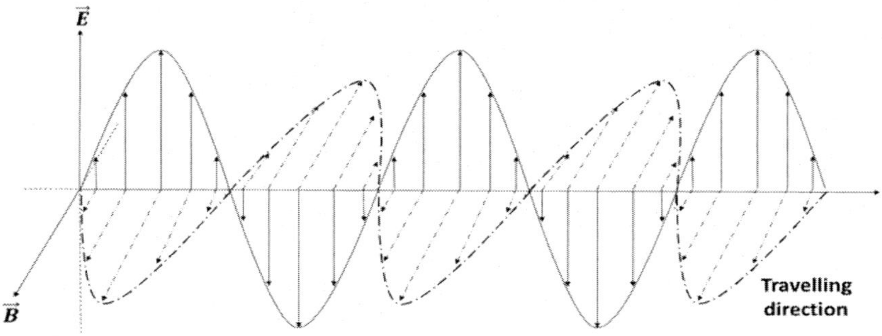

Figure 2. Schematic representation of the propagation of a microwave.

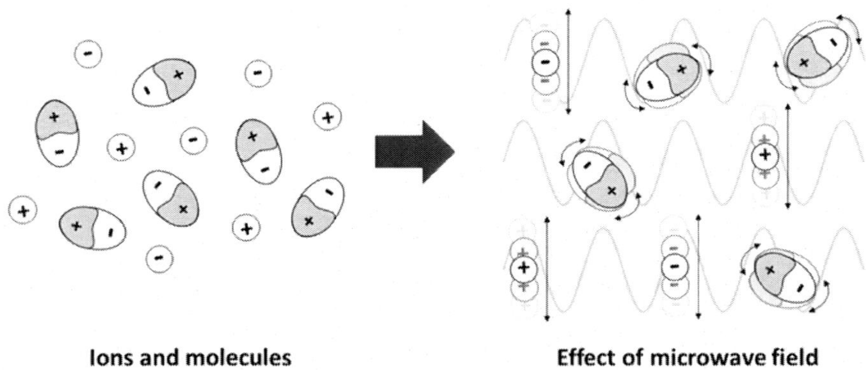

Figure 3. Schematic representation of the microwave effect.

Ultrasounds Extraction

Ultrasound is defined as pressure waves with a frequency of 20 kHz or more, higher than the one human ear can perceive (Alexandre et al. 2011). Their propagation induces a sequence of compressing and rarefaction zones.

Molecules shift temporarily from its original place, shocking and getting apart from each other. If the ultrasound power is enough high, the rarefaction can surpass the forces between the molecules present in the medium and creates cavitation bubbles. These bubbles grow during the rarefaction phases due to the entrance of air from the medium, which is not completely released during compression. After several compression/rarefaction cycles, bubbles hit a critical unstable size, they collapse, and a zone of extreme temperature and pressure conditions is created (Figure 4).

When the bubbles collapse on a surface, the cell deforms into a doughty like shape. In other words, the collapse is asymmetrical and the formed high-speed jets impact the surface, sweeping particles away and damages are induced depending on material friability (Chemat et al. 2017; Vinatoru et al. 2017). Some advantages of USE in relation to the classical extraction techniques are its simplicity, easy operation, reduction of multiple process steps, low cost, effective mixing, can be performed at lower temperatures, reduces extraction times, extraction is more selective, and the extraction yields can be higher. Some of the disadvantages are related with the wave attenuation for a very concentrated dispersed phase, lack of uniformity for dispersed extract materials following the decline in the transmission of ultrasound away from the vicinity of the generator and high energetic consumption when associated at high temperatures (Armenta et al. 2019; Panda et al. 2019; Chemat et al. 2017). USE has been used for herbal and oil extraction, but also to extract proteins, polysaccharides and several bioactive compounds from different sources (Vilkhu et al. 2008; Panda et al. 2019). Regarding the acorns, Kilicarislan et al. (2012) extracted tannin from acorn cups using USE from 40-100 W at 20 kHz for 4-8 h. The extraction yields increased with the increase of power and time, but time did not increase significantly the yields at 100 W. For tannin content, the increment in extraction time led to higher contents at 40 and 60 W. However, tannin content decreased at 80 and 100 W. By comparison, USE at 100 W during 8 h increased the yield by 17.16% in relation to the conventional method at 90°C for 8h. Similarly, USE at 100 W during 6 h increased the amounts of extracted tannin by 6.61% in relation to the conventional method at 90°C for

8 h. Therefore, USE can be a possible alternative to the conventional extraction of tannins from acorn cups.

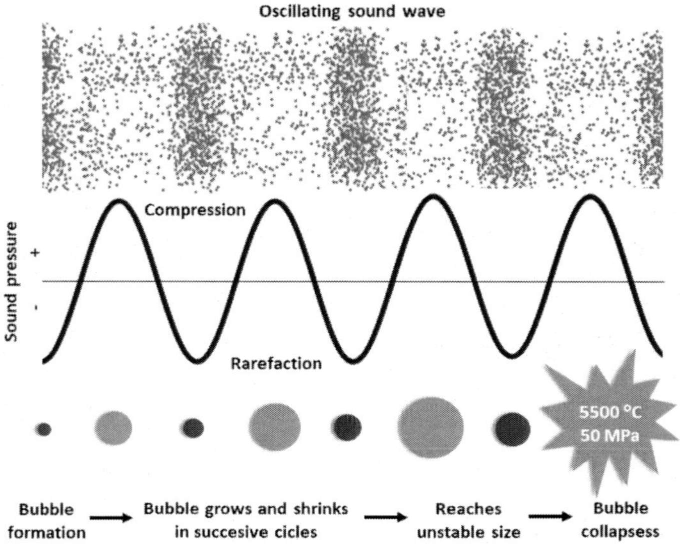

Figure 4. Schematic representation of the ultrasound extraction principle.

Supercritical Carbon Dioxide Extraction

The supercritical state is a special state that can only be reached beyond the critical point of a given material. This point is defined as the pressure and temperature values above which the specific gas and liquid properties have vanished. Thus, the materials possess both, liquid and gas properties, to be attained if a substance is subjected to temperature and pressure beyond its critical point (Azmir et al. 2013). Carbon dioxide is the most used solvent due to its low critical temperature and pressure (31°C and 37 bar), low cost, non-inflammable and is easy to remove. SCE allows performing extractions at lower temperatures, which favor the extraction of thermal label compounds (Figure 5).

This is an environmentally friendly and safe technique since it reduces the consumption of organic solvents, decreases extraction time, and increases yields due to the improvement of the mass transference rates. However, CO_2 is not a good solvent for extraction compounds with high molecular weight and polar compounds due to its non-polar nature. However, this can be overcome with the usage of chemical modifiers, also called co-solvents, but on the other hand, it complicates thermodynamics and increases capital costs (Abbas et al. 2008; Zhang et al. 2018; Armenta et al. 2019; Ahmad et al. 2019).

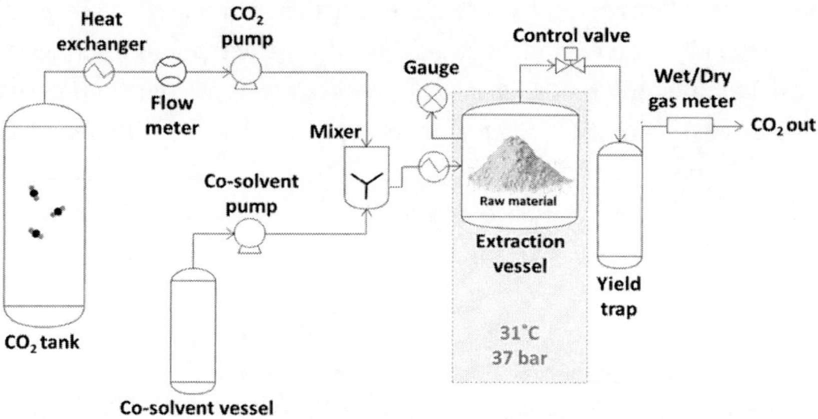

Figure 5. A simplified diagram of the extraction of compounds by supercritical carbon dioxide.

Bernardo-Gil et al. (2007) extracted oil from *Q. rotundifolia* acorns using SCE from 35-60°C and pressures from 12 to 21 MPa. The influence of CO_2 density, flow rate, residence time, feed particle size, and extractor geometry were evaluated. The authors verified that yield increased with the increase of pressure at constant temperature due to increased solubility of oils. The oil recovery increased linearly with the increase of CO_2 density. Additionally, for a given temperature and the same particle size, the initial extraction rate increased with the pressure, but when pressures reached 18 and 21 MPa no differences in oil recovery were observed. For the same CO_2 density and superficial velocity, the initial rates of extraction were dependent on the particle size, i.e., small particle size provided more oil extraction. Additionally, the authors verified that at the beginning of the extraction, the CO_2 density is the most important factor. However, after some extraction time, the oil fraction exposed to CO_2 becomes more important. Concerning the effect of the solvent flow rate and the geometry of the extractor, the authors reported that for the lowest bed high/extractor diameter ratio, the velocity varied, and the initial extraction rate increased with this velocity. For the same extraction time, as the bed high/extractor diameter ratio decreased, the oil recovery decreased. The high/extractor diameter ratio is the preponderant factor when similar velocities are used, decreasing oil recovery when it increases. Regarding the residence time, the loading curves were independent of it until 80% of the oil was extracted, after which it decreased. Xia et al. (2010) extracted the oil content of some subgenera of acorn fruit also using SCE. The authors reported that the oil content from the *Quercus* spp. ranged between 0.7% and 18%. Furthermore, the contents from the *Cyclobalanopsis* (0.7% and 3.8%) was significantly lower than the *Quercus* (1.5% to 18%). No differences were found between the *Quercus* and *Cerris* sectors, but the *Lobatae* had the highest oil content (18%). In a different study, Lopes et al. (2005) extracted oil from *Q. rotundifolia* acorns by SCE (18 MPa at 39.85°C) and Soxhlet (hexane) and studied the composition of the extracted oils. The authors reported that fatty acid (mainly palmitic, oleic, and linoleic fatty acids), triglyceride (where the triglyceride composed of 3 oleic fatty acid molecules was the one in higher amount), sterol (mainly β-sitosterol), and tocopherol (α, γ, and δ isomers) profiles and contents were similar in the oils extracted by both methodologies. However, the SCE oils had undetectable amounts of phospholipids. Oil extracted by SCE had a smaller acidity index in relation to the oil obtained by Soxhlet, indicating that the SCE provides some oil refinement. Moreover, these oils had higher vulnerability to oxidation when compared to its conventional counterpart and the presence of air induced oxidation more rapidly than the light during storage. This indicated that the oil obtained by Soxhlet was more stable than the one collected by SCE.

Pressurized Fluid Extraction

Pressurized fluid extraction (also called pressurized liquid extraction, pressurized solvent extraction, accelerated solvent extraction, or enhanced solvent extraction) uses high temperatures (100-374°C) and pressures up to 20 MPa to maintain the extraction of solvents liquid, above the boiling point.

When water is used as a solvent, this technique is mentioned as subcritical water extraction, superheated water extraction, or pressurized hot water extraction. This method increases the solubility and mass transference phenomena improving the extraction kinetics (Mena-García et al. 2019). The extraction mechanism can be divided into a few stages. Firstly, the compound of interest needs to be desorbed from the active place in the sample matrix under pressure and high temperatures to reach the matrix-fluid border. The position and localization of the compound within the matrix sample is a significant extraction step and so far different positions have been reported (Mustafa et al. 2011). The compound is dispersed in the extraction phase, is diffused through the extraction phase found in the pore, and finally reaches the extraction phase. At the final stage, the extracted compound is collected (Figure 6).

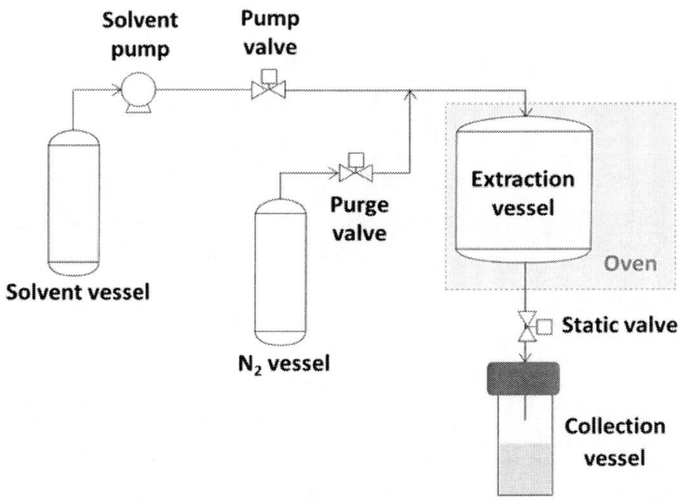

Figure 6. A simplified diagram of the extraction of compounds by pressurized liquid extraction.

According to Teo et al. (2010), the solvent capacity to solubilize a compound is superior by modifying the solvent properties. The solvent viscosity is reduced and has a higher capacity to penetrate the matrix sample. The increased temperatures can affect the non-covalent bonds between the compound of interest and its active place, thus decreasing the energy needed to deserve the compounds. However, compounds can be degraded if they are thermally labile. The use of pressure keeps the solvent liquid above the boiling point and increases solvent solubility and penetration rates. This facilitates the extraction because forces the solvent to reach areas that under atmospheric conditions it would not. Thus, PFE decreases the extraction time and requires less solvent quantity (Onem et al. 2015; Zhang et al. 2018; Azmir et al. 2013). The extraction efficiency also depends on the sample source, the compound of interest, and its location in the matrix sample.

Valonea is a high-value hydrolyzable tannin obtained from acorns and has been used to extract ellagitannins as a tanning agent (Onem et al. 2014). Conventional extraction of valonea uses long extraction times. However, non-tanning compounds are collected decreasing extract purity and higher quantities of the extract are needed. In this sense, Onem et al. (2015) studied its isolation from acorn cups by PFE using a binary of H_2O/CO_2 and ternary system composed of $H_2O/CH_3OH/CO_2$. The authors evaluated the differences in the tannin phenolic compounds using the binary H_2O/CO_2 and ternary $H_2O/CH_3OH/CO_2$ systems under 10 and 200 bars during 3 h for a solvent/feed rate of 100 (w/w). The binary system was chosen since no statistical differences were observed on the tannin phenolic contents when comparing both binary and ternary systems. Considering the critical temperature and pressure conditions of CO_2, the authors evaluated the changes in the tanning phenolic content at 10 to 200 bar and reported that the content increased up to 100 bar and further decreased at 200 bar in relation to the extraction performed at atmospheric conditions with water. Thus, the best extraction pressure was found to be 100 bar. In order to optimize the extraction time, two additional trials were performed during 2 and 3 h under 100 bars at a solvent/feed ratio of 100 g solvent/g feed. As result, the PLE optimum extraction condition of valonea tannin was 34.73% at 100 bars during 3 h and a solvent/feed ration of 100 g solvent/g feed at 80°C with only 2.92% of insoluble solids in the extract. These results showed that PLE can provide extracts with higher tannin phenolic content, reducing extraction times and load waste, which provides a great advantage for the leather industry that uses this compound.

Combined Technologies

Microwave-Assisted Soxhlet Extraction

Soxhlet can be modified to facilitate the accommodation of the sample cartridge compartment in the irradiation zone of a microwave oven. Therefore, the sample is irradiated at the desired power and time intervals conditions, while the sample gets in contact with fresh solvent repeatedly and the target compounds are continually collected (Luque-García et al. 2004). This combined methodology reduces extraction times, improves extraction due to easier cleavage of the bounds of the strongly retained compounds in the matrix by interactions with me MW energy, allows recycling almost all solvent and the solvent distillation is reached by electrical heating, making it independent of the solvent polarity. Additionally, it allows operating with water or and high boiling point extractant. This technique has been used to extract oils and fatty acids from different sources (Priego-Capote et al. 2005; García-Ayuso et al. 2000; Priego-Capote et al. 2007; Priego-López et al. 2003), including oils from *Q. ilex* acorns. Pérez-Serradilla et al. (2007) compared some conventional technologies (namely the Folch, Stirring, ISO, and Soxhlet extraction methods) with FMSE method regarding the acorn oil composition, extraction time, and efficiency of the extraction. Briefly, the FMSE was performed using hexane for a total extraction time of 30 min (15 cycles) at a power of 300 W; the Folch extraction was performed by stirring acorn with trichloromethane–methanol for 45 min; the Stirring extraction was performed by stirring acorn with hexane at 40ºC for 56 h; the ISO was performed by stirring acorn with hexane for 4 h; and the Soxhlet extraction was performed using hexane for a total extraction time of 16 h. The authors reported that the extraction efficiency of the FMSE, Folch, Stirring, ISO, and Soxhlet techniques was 9.22, 9.27, 9.17, 9.26, and 9.32 mg of fatty acid/100 g of acorn, respectively but no differences were found among results ($p>5\%$). Trans fatty acids (*trans,trans*-linoleic, *cis,trans*-linoleic, and elaidic) were only detected on the extracts obtained from the Soxhlet and ISO extraction methods. This may have been due to thermal isomerization. Palmitoleic and *trans, cis*-linoleic fatty acids were not detected in the extracts. Further analysis of the relationship between fatty acid profiles and the extraction methods by principal component analysis revealed that the first component (fatty acid profiles) explained 77.5% of the population variability and enable the separation of the ISO and Soxhlet methods from the Stirring, FMSE, and Folch one. The second component (extraction methods) explained further 12.7% of the variability between ISO and Soxhlet methods. Moreover, the behavior of the Folch, FMSE, and Stirring were similar in relation to the methods where the heptadecanoic, pentadecanoic, palmitoleic, and oleic fatty acid proportions were higher and were statistically equivalent. Thus, the ISO and Soxhlet performance was responsible for 90.2% of variability in the fatty acid profiles, while the differences in the behavior of the Folch, FMSE, and Stirring cause only 9.8% of the variability. These three methods led to higher quality oils, despite the similar extraction efficiency to the ISO and Soxhlet.

Ultrasounds-Enzymatic Extraction

Several authors have been reported that the combination of enzymes with ultrasonic extraction can improve the extraction yields. Luo et al. (2019) reported for the first time the application of this technology to extract tannins from acorns. The authors optimized the extraction conditions, namely the sonication time from 2-3 h, power from 70-130 W, cellulase concentration from 2.40-4.00 g/L, and temperature from 30-40°C in order to archive the optimum yield by means of a Bex-Behnken design. Statistical analysis revealed that the interaction between temperature-cellulase concentration, power-cellulase concentration, and power-temperature was significant. Interactions coefficients involving the extraction time were not significant. Authors reported that for the interaction between temperature-cellulase concentration, at lower cellulase concentrations for a fixed temperature, the yields increased with the increase of the cellulase concentration but then decreased with further higher concentrations. Similar results were observed when the cellulose concentration was fixed, indicating a strong correlation between these two variables. Analysis of the interaction between the power-cellulase concentration showed that when fixing of the cellulase concentration, the extraction yield increases from 70-104.78 W and then decreases with higher power values. Fixing the power level, the extraction yield increases from 2.40-3.32 g/L and then decreases with higher cellulose concentrations. Similar results were obtained for the interaction between power-temperature. The optimum tannin concentration of 63.16% was found to be for an extraction time of 2.51 h at 38ºC with a power of 98 W and using a cellulose concentration of 3.44 g/L.

Ultrasounds-Maceration Extraction

Hadidi et al. (2017) determined for the first time the anti-radical activity of nitric oxide and superoxide anion radicals of hydromethanolic and hot water extracts of *Q. ilex* leaves, acorns, cotyledons, and tegument. The authors verified that the total phenolic compound concentration from leaves and teguments in the hydromethanolic extracts was superior to the hot water extract counterparts, while the total phenolic compounds of the hot water extracts from acorns and cotyledons were superior that the hydromethanolic extract counterparts ($p<0.05$). Concerning the anti-radical activity of the superoxide anion radical, all extracts had significantly lower EC_{50} in relation to the ascorbic acid and no significant differences were bound between the hydromethanolic extracts and its hot water extract counterparts. However, the leaves had the lowest values when compared to the ascorbic acid, while the acorn and cotyledon extracts had higher values ($p<0.05$). Regarding the anti-radical activity of the nitric oxide radical, teguments and leaves also had the lowest EC_{50} values in relation to the ascorbic acid, while no differences were found when comparing the acorn extracts and the ascorbic acid. However, the EC_{50} of the hydromethanolic extract from cotyledons was statistically inferior regarding water extract. The principal component analysis revealed that the first component (plant parts), which explained 60.31% of the variance, separated the leaves extracts from the acorn and tegument ones. Additionally, the leaves extracts were positively correlated with all analysed compounds, except the gallic acid, protocatechuic acids and total phenolic compounds. The second component (extraction solvent) explained 19.69% of the variance and was able to correlate positively the water leaves extracts with protocatechuic acid, epicatechin, ellagic acid, and its derivative and negatively with the total phenolic compounds, gallic acid and remaining flavonoids. Low efficacy was found for both hydromethanolic (22.37% at 830 μg/mL) and how water extracts (26.42% at 500 μg/mL) for acetylcholinesterase inhibition. However, Custódio et al. (2013) obtained an acetylcholinesterase inhibition of 18.5 and 26.5% at 1 and 10 mg/mL, respectively for the hot water extracts. A significant difference was observed in both hydromethanolic (22.90% at 500 μg/mL) and hot water extracts (14.43% at 500 μg/mL) values for the inhibition of butyrylcholinesterase. In the same study, Custódio et al. (2013) reported higher of butyrylcholinesterase inhibition values for the hot water extracts, i.e., 27.8 and 18.7% at 1 and 10 mg/mL, respectively. Custódio et al. (2013) also verified that inhibition values of *Q. ilex* were superior to the *Q. suber* ones. Additionally, these *Q. suber* inhibition values were inferior to those reported by Custódio et al. (2015) for *Quercus suber* leaves.

ACORN AS A SOURCE OF FOOD INGREDIENTS AND FOOD PRODUCTS

Acorn Extracts

Some authors have evaluated the effectiveness of ethanolic extracts of *Q. petraea*, *Q. ithaburensis*, and *Q. ilex* acorns on the physicochemical and antioxidative properties of poultry during refrigerated storage for up to 14 days (Özünlü et al. 2018). The authors dipped the chicken in an acorn extract solution and stored the samples at 2°C. The results revealed that the lightness of samples decreased after 14 days of storage, which could have been due to the dark natural colour of the extracts. Except for days 0 and 1, redness values were inferior on all storage days ($p>0.05$), but the redness values did not vary with the extract concentration. The pH did not vary significantly during storage. The phenolic content of all samples decreased during the storage period but dipped samples had higher content (77.75-82.02 mg gallic acid Eq./100 g meat) than the control (25.75 mg gallic acid Eq./100 g meat), which indicates that acorn can be used as a source of natural phenolic compounds for meat products. Similar results were found for the scavenging antioxidant activity, which also decreases during storage, but the control samples had lower activity values (9.25%) than dipped samples (25.32-29.53%). The authors indicated that the observed decrease could have been caused by the usage of antioxidant compounds and hydrolysis to inhibit oxidation. Samples treated with acorn extracts also had a lower degree of lipid oxidation (0.14-0.18 mg malonaldehyde/kg meat) in relation to the control (0.26 malonaldehyde/kg meat) at the end of the storage. Additionally, samples at the end of the storage had lower carbonyl contents (1.15-1.50 nmol carbonyl/mg protein) in relation to the control (2.40 nmol carbonyl/mg protein), meaning that acorn extracts can be used to reduce lipid and protein oxidation in meats. Acorn extract application did not affect negatively the colour, juiciness, astringency and overall acceptability of the treated meats. Overall, acorn extracts have the potential to be used in the food industry. In a different study, Ferreira et al. (2017) evaluated the effect of the combination of food-grade acetonic extract from *Q. ilex* with low-oxygen modified atmosphere packaging on the lipid and protein oxidative stability of ready-to-eat chicken patties during cooking in an electric oven, cold storage at 4°C, and reheating using a microwave. The results showed that the control patties had a significantly higher degree of lipid oxidation after being cooked and then chilled. However, the reheating did not have a significant contribution to the degree of lipid oxidation. When acorn extract was added to the patties, the degree of lipid oxidation was dramatically decreased in relation to the control patties and no significant differences in the degree of lipid oxidation were found throughout processing for the patties with extract incorporation in relation to the control raw patty treated with the extract. Similar results were observed for the degree of protein oxidation. Only the heating steps (cocking and reheating) increased significantly the carbonyl content in the control patties in relation to the raw control patties. When acorn extract was added to the patties, the carbonyl content decreased significantly for every processing step. Concerning the patties' hardness, only significant differences were found in the reheating step where the patties with acorn extract had lower hardness in relation to the patties without acorn extract. The volatile compound analysis revealed that in relation to the control patties, hexanal and 1-octen-3-ol compounds were found in basal levels and the deca-2,4-dienal compound was not detected during the entire processing. Lastly, the colour and odor assessment showed that the patties incorporated with extract had significantly higher acceptance in relation to the patties without extract. Rakić et al. (2004) observed that the roasting treatment at 200°C for 15 min decreased the tannin content of *Q. cerris* kernels by 3.14%, but the antioxidant activity of the ethanolic extracts was not affected by the thermal treatment. Furthermore, the addition of 0.02% extract to lipids from porcine increased lipid maintenance. Rakić et al. (2006) performed a similar study for *Q. rubor* aqueous extracts and concluded that extracts had antioxidant activity on pig fat in relation to the control, which had a peroxide value of 30.5 mmol/kg on the second day. Sample with extract had a value of 2.47 mmol/kg on the second day, but the BHA effect, a synthetic antioxidant, was more evident (0.81 mmol/kg). Extract concentration also had a great influence, i.e., the peroxide value for the 0.02 and 0.04% concentration on the fifth day was 35.1 and 3.53 mmol/kg, respectively.

Acorn Beverages

Recently, Sanchez-Bel et al. (2015) studied the incorporation of wild edible plants, including *Q. ballota* fruit, in orange juice and kiwifruit cremogenate matrices to improve its nutritional properties and stabilize or improve the organoleptic properties. Authors concluded that in general the addition of wild edible plants did not affect, or can even improve, the consumers' acceptance. The matrixes were less acid, the aroma assessment varied insignificantly, and sweetness increased after the incorporation of *Q. ballota* acorn. The addition of *Q. ballota* fruit did not alter the contents of vitamin C nor its stability during storage for 24 h at 20°C. The polyphenol content did not vary significantly after addition nor after 24 h at 20°C. The matrices' antioxidant capacity did not increase significantly after incorporation nor after 24 h at 20°C.

However, several beverages such as *Raccahout,* that is a Turkish beverage alike hot chocolate; *Eichelkaffee*, an acorn coffee-like beverage; and *Licor de Bolota*, which is a sweet Portuguese liqueur, had been produced (Silva et al. 2016). Coelho et al. (2018) was able to develop a coffee-like beverage using *Q. ilex* and *Q. suber* fresh and roasted cotyledons. The results showed that the coffee-like beverages from fresh cotyledons had higher antioxidant activity in relation to the roasted counterparts. Rakić et al. (2004) reported that the roasting treatment at 200°C for 15 min decreased the tannin content of *Q. cerris* kernels by 3.14%. Similar results were observed for the phenolic compounds content, but in the roasted beverage of *Q. suber* phenolic content was superior to *Q. ilex.* No mutagenicity against *S. typhimurium* TA 98 using increasing concentrations of *Quercus* coffees in the Ames test nor cytotoxicity in hamster cells in the MTT assay was detected. Additionally, the coffee beverages of *Q. suber* and *Q. ilex* (roasted) coffee had antimutagenic properties at concentrations 100 µg/plate (71%) and 50 µg/plate (36%), respectively. These results showed the potential of acorn beverages as a functional food with several benefits to human health, being as well as caffeine and gluten-free, despited be a high source of polyphenols. Pinto et al. (2019) also prepared a coffee-like beverage from *Q. cerris* roasted kernels. However, authors tested the cytotoxicity in Caco-2 and HT29-MTX intestinal cells in the MTT assay against different beverages concentrations (0.1-1000 µg/mL), reporting that the optimal range of non-cytotoxicity was from 0.1 to 10 µg/mL.

Acorn Flours

Dabija et al. (2019) studied the effect of the incorporation of 2% acorn flour (AF) on the rheological, physicochemical, and sensory properties of natural yogurt elaborated from cow milk. Results indicated that the rheological properties of yogurt were improved, and the viscosity increased. Additionally, the syneresis with AF (44%) decreased in relation to the control yogurt (52%), which could be explained by the high carbohydrate content of the AF (55%). The water holding capacity of the yogurt samples also increased from 34 to 43% after incorporation of AF in relation to the control yogurt. Regarding the sensory evaluations, no significant differences in acceptance were observed between the control yogurt and AF yogurt.

Besides being used to elaborate beverages, acorns can also be used to replace regular cornflour in the manufacture of biscuits, cookies and bread. As a matter of fact, acorns are a staple food in various Mediterranean countries when food is scarce (Claudia 2013; Silva et al. 2016). Korus et al. (2017) studied the effect of partial replacement of corn flour by debittered AF at 20, 40, and 60% on the moisture's hydration properties of biscuits, quality, nutritional value, and sensory properties. Results revealed that the swelling volume, water absorption and oil capacity retention increased with the AF percentage in relation to the control biscuits. Protein and fat content did not vary significantly, but the ash and fiber contents increased with the increase of AF percentage in relation to the control. Opposite results were found for the energetic values ($p<0.05$). The total carbohydrate contents decreased in relation to the control biscuits. The total phenolic contents and antioxidant activity increased significantly with the increase of AF percentage in relation to the control biscuits. Parsaei et al. (2018) studied the effect of the usage of AF from *Q. brantii* kernels as a source of natural antioxidant compounds on the stability of fat in biscuits. AF in several percentages was used in combination with corn flour or wheat flour. Results revealed that increasing the percentage of AF, the antioxidant activity of biscuits increased, which also had lower peroxide values due to the high content of phenolic compounds. After 28 days of storage, the peroxide and antioxidant activity values of all biscuits increased and decrease, respectively ($p<0.05$). Sensory analysis revealed that the incorporation of AF did not alter significantly the flavour, texture, after taste, colour, and overall acceptability of the biscuits in relation to those made of corn flour or wheat flour. However, these results are different from those reported by Korus et al. (2017), whose sensory evaluation revealed that AF biscuits had lower punctuation scores for colour, smell, taste, consistency, and appearance in relation to the cornflour control biscuits. Pasqualone et al. (2019) evaluated the possibility of the usage of AF as an ingredient in wheat-based biscuits. Two different AF concentrations from *Q. coccifera* were used to prepare the biscuits. The total phenolic compounds, flavonoids, and antioxidant contents of biscuits increased with the AF concentration used in relation to the control. Sensory analysis revealed that biscuits elaborated with AF had superior browning indexes (lower lightness values) in relation to the control biscuits due to the fraction of the phenolic compound that underwent oxidation. Regarding the colour, Korus et al. (2017) verified that the lightness and yellowness values decreased significantly with the increase of the AF content, however, the redness values decreased in relation to the corn flour control biscuits. Pasqualone et al. (2019) verified that the friability of the AF biscuits was statistically superior in relation to the control biscuits and the fracture force of the AF biscuits decreased significantly as the AF concentration used increased in relation to the control biscuits. However, Korus et al. (2017) reported that biscuits hardness increased with the increase of AF replacement in relation to the corn flour control biscuits but was only significant at 60% substitution. Additionally, Pasqualone et al. (2019) reported that graininess and astringency increased with the use of AF concentration, but no bitterness was detected. AF biscuits were also perceived as less sweet than the control ones, but they had a stronger odor sensation of brain-like and caramels notes as the AF concentration increased. Additionally, these biscuits had superior furfural content, which is a caramel-like odorant deriving by Maillard's reaction.

Ajo (2018) evaluated the replacement of wheat flour by AF at 5, 10 and 15% from *Q. aegilops* acorns in order to produce Arabic flatbread and compare the physiochemical characteristics between bread. Arabic flatbread produced with AF had an increased nutritional value in relation to the control flatbread, i.e., higher protein and fiber contents as well as higher humidity and ash levels. Fat and carbohydrate contents did not differ significantly in relation to the control bread. Skendi et al. (2018) evaluated the replacement of corn and rice flours by different concentrations of AF from *Q. aegilops* acorns. The authors verified that the total phenolic compounds content in bread increased with the percentage of AF replaced. Korus et al. (2015) also applied and evaluated the nutritional benefits of the incorporation of debittered AF on gluten-free bread and to check its effects on rheological properties of the dough and bread quality. From a nutritional standpoint, acorn supplemented bread had increased protein, minerals, and dietary fiber contents and reduced total carbohydrate content in relation to the control bread. Korus et al. (2015) verified that partial replacement of starch with AF increased both elastic and viscous moduli and decreased the phase shift tangent values of the bread doughs when compared to the control one. These results indicated that doughs were more elastic and less susceptible to applied stress. Bread volume also increased, and crumb physical properties were improved when starch was replaced with a lower level of AF, but higher amounts decreased bread volume. The incorporation of AF reduced bread lightness and increased red- and yellowness, making bread more sensory attractive as verified by Skendi et al. (2018) due to AF its natural dark colour. For Ajo (2018), AF bread was less white and become darker as the AF percentage increased in relation to the control of wheat flour bread. Bread red- and yellowness did not differ in relation to the control bread, but values decreased after 24 h storage. Regarding the texture properties, Skendi et al. (2018) verified that AF bread increased with the AF percentage, but the increase in hardness at 70% humidity was not so noticeable as 65 or 75%. Bread with 25% AF and 65% water content had the highest hardness values. At 70% water content, the springiness of AF bread was similar in relation to the control when the increase of AF percentage. An increase of springiness was observed at 75% water and 25% AF. The chewiness of AF bread increased with the percentage of AF. However, the specific volume of bread significantly decreased with increasing the concentration of acorn meal at the same water content. Ajo (2018) verified that crumb texture (softness) varied insignificantly between bread after 1-hour storage, but softness decreased with the increase of AF percentage. Additionally, the staling rate decreased with the AF percentage, thus indicating that bread substituted with AF hardness increased with aging. The specific volume of bread did not differ significantly. Skendi et al. (2018) reported that increasing the AF percentage can increase the dilution of gelatinized starch, decreasing the availability for starch to recrystallize. This leads to lower crystal formation and the reduction of the retrogradation enthalpy values, decreasing bread retrogradation. Korus et al. (2015) reported that the replacement with AF also decreased the amylopectin recrystallization enthalpy values and consequently bread loaves had slower staling. Regarding sensory assessment, the replaced bread had greater consumer acceptance in relation to the control bread. However, the sensory analysis performed by Ajo (2018) revealed that the sensory attribution scores did not differ significantly for the AF bread except for the reduced crust colour, blister, and crumb colour in relation to the control bread. No significant overall difference in quality compared was found between the AF bread and the control one.

CONCLUSION

This chapter starts by presenting the potential of the *Quercus* acorns as functional foodstuffs since it present beneficial properties including to be gluten-free. In times of scarcity, the *Quercus* acorns served as fundamental nuts for human consumption, supported the local economy, and had an important role in folk medicine. However, the acorns are currently far from being as widely used. Nevertheless, with more consumer information and investment in more studies about the potential benefits of acorn and its by-products, it will be possible to make the most of this valuable resource. In this chapter, a special focus is given to the existing research on conventional and emerging extraction methodologies, which allow obtaining compounds of greater interest. The classical extraction techniques have been used to extract a large variety of compounds from compound variety several acorn species. Maceration has been used for the extraction of phenolic compounds and Soxhlet for both oils and phenolic compounds isolation, while the alkaline and enzymatic technologies are used to extract the starch. According to the literature, the emergent technologies namely microwave, ultrasounds, and pressure liquid were used to extract phenolic compounds, while supercritical extraction was used to extract oils from acorns. The use of combined technologies has shown promising evidences to surpass the disadvantages of the classical extraction technologies reducing the extraction times and improving extraction yields, but also providing new and more diverse knowledge. However, the literature available is scarce and very limited technologies have been applied to acorns yet. Thus, more research must be performed, and new emergent technologies applied to acorns must be tested in order to comprehend its impact on the compound's extraction from acorns. This will allow uncovering possible new applications for acorns. Moreover, promising results can be found in the literature regarding the reincorporation of acorn in human nutrition in order to take advantage of acorns' nutritional value, but also from its phenolic compound composition and their health benefits. In this sense, acorn extracts have already shown evidence to be a safer natural alternative to the synthetic food antioxidants used by the food industry, contributing positively to the physicochemical, antioxidant and sensory characteristics of chicken meat. These extracts have also been shown to contribute positively to the processing of ready-to-eat foods. In addition to the extracts, acorns can also be used in the preparation of various functional drinks to maintain health and assist in disease prevention. Moreover, the incorporation of acorn flour into yogurt, biscuits and bread has a positive contribution not only from a nutritional point of view but also in improving the properties of food without negatively affecting consumers' perception of these novel foods as well as their sensory characteristics.

REFERENCES

Abbas, K. A., A. Mohamed, A. S. Abdulamir, and H. A. Abas. 2008. 'A Review on Supercritical Fluid Extraction as New Analytical Method'. *American Journal of Biochemistry and Biotechnology* 4 (4): 345–353.

Ahmad, T., F. A. Masoodi, S. A. Rather, S. M. Wani, and A. Gull. 2019. 'Supercritical Fluid Extraction: A Review'. *Journal of Biological and Chemical Chronicles* 5 (1): 114–122.

Ajo, R. Y. 2018. 'Effect of Acorn Flour Substitution on Arabic Bread Properties'. *Pakistan Journal of Agricultural Sciences* 55 (4): 913–919.

Akcan, T., R. Gökçe, M. Asensio, M. Estévez, and D. Morcuende. 2017. 'Acorn (Quercus Spp.) as a Novel Source of Oleic Acid and Tocopherols for Livestock and Humans: Discrimination of Selected Species from

Mediterranean Forest'. *Journal of Food Science and Technology* 54 (10): 3050–3057.

Al-Rousan, W. M., R. Y. Ajo, K. M. Al-Ismail, A. Attlee, R. R. Shaker, and T. M. Osaili. 2013. 'Characterization of Acorn Fruit Oils Extracted from Selected Mediterranean Quercus Species'. *Oils and Fats* 64 (5): 554–560.

Aldrich, P. R., and J. Cavender-Bares. 2011. 'Chapter 6 - Quercus'. In *Wild Crop Relatives: Genomic and Breeding Resources: Temperate Fruits*, 89–129.

Alexandre, E. M. C., M. G. Castro, S. A. Moreira, and J. A. Saraiva. 2017. 'Comparison of Emerging Technologies to Extract High-Added Value Compounds from Fruit Residues: Pressure- and Electro-Based Technologies'. *Food Engineering Reviews* 9 (September 2017): 190–212.

Alexandre, E. M. C., D. M. Santos-Pedro, T. R. S. Brandão, and C. L. M. Silva. 2011. 'Study on Thermosonication and Ultraviolet Radiation Processes as an Alternative to Blanching for Some Fruits and Vegetables'. *Food and Bioprocess Technology* 4 (6): 1012–1019.

Allard, G., N. Berrahmouni, C. H. Besacier, D. Boglio, M. Briens, A. Brizay, A. Camia, et al. 2013. *State of Mediterranean Forests 2013*.

Alupului, A., I. Călinescu, and V. Lavric. 2012. 'Microwave Extraction of Active Principles from Medicinal Plants'. *UPB Scientific Bulletin, Series B: Chemistry and Materials Science* 74 (2): 129–142.

Amina, M., F. Djamel, and H. Djamel. 2018. 'Influence of Fermentation and Germination Treatments on Physicochemical and Functional Properties of Acorn Flour'. *Bulgarian Journal of Agricultural Science* 24 (4): 719–726.

Armenta, S., S. Garrigues, F. A. Esteve-Turrillas, and M. de la Guardia. 2019. 'Green Extraction Techniques in Green Analytical Chemistry'. *Trends in Analytical Chemistry* 116: 248–253.

Asghari, J., B. Ondruschka, and M. Mazaheritehrani. 2011. 'Extraction of Bioactive Chemical Compounds from the Medicinal Asian Plants by Microwave Irradiation'. *Journal of Medicinal Plants Research* 5 (4): 495–506.

Axelrod, Daniel I. 1983. 'Biogeography of Oaks in the Arcto-Tertiary Province'. *Annals of the Missouri Botanical Garden* 70 (4): 629–657.

Azmir, J., I. S.M. Zaidul, M. M. Rahman, K. M. Sharif, A. Mohamed, F. Sahena, M. H. A. Jahurul, K. Ghafoor, N. A. N. Norulaini, and A. K. M. Omar. 2013. 'Techniques for Extraction of Bioactive Compounds from Plant Materials: A Review'. *Journal of Food Engineering* 117 (4): 426–436.

Bainbridge, D. 2001. *Acorns as Food: History, Use, Recipes, and Bibliography. Sierra Nature Prints.*

Bakiş, Y., and M. T. Babaç. 2014. 'Morphological Variability of Acorns and Its Taxonomic Significance in Quercus L. from Turkey'. *Bangladesh Journal of Botany* 43 (3): 293–299.

Bernardo-Gil, M., I. M. G. Lopes, M. Casquilho, M. A. Ribeiro, M. M. Esquível, and J. Empis. 2007. 'Supercritical Carbon Dioxide Extraction of Acorn Oil'. *The Journal of Supercritical Fluids* 40 (3): 344–348.

Bramley, P. M., I. Elmadfa, A. Kafatos, F. J. Kelly, Y. Manios, H. E. Roxborough, W. Schuch, P. J. A. Sheehy, and K-h Wagner. 2000. 'Review - Vitamin E'. *Journal of the Science of Food and Agriculture* 80 (December 1999): 913–938.

Butterfield, D. A., A. Castegna, J. Drake, and G. Scapagnini. 2002. 'Vitamin E and Neurodegenerative Disorders Associated with Oxidative Stress'. *Nutritional Neuroscience* 5 (4): 229–239.

Cantos, E., J. C. Espín, C. López-Bote, L. D. L. Hoz, J. A. Ordóñez, and F. A. Tomás-Barberán. 2003. 'Phenolic Compounds and Fatty Acids from Acorns (Quercus Spp.), the Main Dietary Constituent of Free-Ranged Iberian Pigs'. *Journal of Agricultural and Food Chemistry* 51: 6248–6255.

Charef, M., M. Yousfi, M. Saidi, and P. Stocker. 2008. 'Determination of the Fatty Acid Composition of Acorn (Quercus), Pistacia Lentiscus Seeds Growing in Algeria'. *Journal of the American Oil Chemists' Society* 85 (10): 921–924.

Chemat, F., N. Rombaut, A. G. Sicaire, A. Meullemiestre, A. S. Fabiano-Tixier, and M. Abert-Vian. 2017. 'Ultrasound Assisted Extraction of Food and Natural Products. Mechanisms, Techniques, Combinations,

Protocols and Applications. A Review'. *Ultrasonics Sonochemistry* 34: 540–560.

Chung, K. T., T. Y. Wong, C. I. Wei, Y. W. Huang, and Y. Lin. 1998. 'Tannins and Human Health: A Review'. *Critical Reviews in Food Science and Nutrition* 38 (6): 421–464.

Claudia, P. 2013. 'Acorn Bread: A Traditional Food of the Past in Sardinia (Italy)'. *Journal of Cultural Heritage* 14S (3): S71–S74.

Coelho, M., S. Silva, L. M. Rodríguez-Alcalá, A. Oliveira, E. M. Costa, A. Borges, C. Martins, A. S. Rodrigues, and M. M. E. Pintado. 2018. 'Quercus Based Coffee-like Beverage: Effect of Roasting Process and Functional Characterization'. *Journal of Food Measurement and Characterization* 12 (1): 471–479.

Correia, P. R., and M. L. Beirão-da-Costa. 2010. 'Chestnut and Acorn Starch Properties Affected by Isolation Methods'. *Starch* 62 (8): 421–428.

Correia, P. R., and M. L. Beirão-Da-Costa. 2012. 'Starch Isolation from Chestnut and Acorn Flours through Alkaline and Enzymatic Methods'. *Food and Bioproducts Processing* 90 (2): 309–316.

Correia, P. R., M. Cristiana, and M. L. Beirão-da-Costa. 2013. 'The Effect of Starch Isolation Method on Physical and Functional Properties of Portuguese Nut Starches. II. Q. Rotundifolia Lam. and Q. Suber Lam. Acorns Starches'. *Food Hydrocolloids* 30 (1): 448–455.

Correia, P. R., A. E. Leitão, and M. L. Beirão-da-Costa. 2009. 'Effect of Drying Temperatures on Chemical and Morphological Properties of Acorn Flours'. *International Journal of Food Science & Technology* 44 (9): 1729–1736.

Correira, P. M. R., S. Gonçalves, H. Gil, and J. Coelho. 2014. 'The Wettability and Surface Free Energy of Acorn Starch Gels Isolated by Alkaline and Enzymatic Methods'. In *FoodBalta*, 280–283.

Costas, C. 2013. *Acorn Cooking Manual for the Post Oil Era.*

Custódio, L., J. Patarra, F. Albericio, N. R. Neng, J. M. F. Nogueira, and A. Romano. 2013. 'Extracts from Quercus Sp. Acorns Exhibit in Vitro Neuroprotective Features through Inhibition of Cholinesterase and Protection of the Human Dopaminergic Cell Line SH-SY5Y from

Hydrogen Peroxide-Induced Cytotoxicity'. *Industrial Crops and Products* 45: 114–120.

Custódio, L., J. Patarra, F. Alberício, R. Neng, J. M. F. Nogueira, and A. Romano. 2015. 'Phenolic Composition, Antioxidant Potential and in Vitro Inhibitory Activity of Leaves and Acorns of Quercus Suber on Key Enzymes Relevant for Hyperglycemia and Alzheimer ' s Disease'. *Industrial Crops & Products* 64: 45–51.

Dabija, A., G. G. Codină, S. Ropciuc, and S. G. Stroe. 2019. 'Studies Regarding the Production of a Novel Yogurt Using Some Local Plant Raw Materials'. *Journal of Food Processing and Preservation* 43 (6): e13826.

Daiuto, É., M. Cereda, S. Sarmento, and O. Vilpoux. 2005. 'Effects of Extraction Methods on Yam (Dioscorea Alata) Starch Characteristics'. *Starch* 57 (3–4): 153–160.

Ebrahimi, A., I. Mehregan, T. N. Sattari, M. Assadi, and K Larijani. 2016. 'Population Variability in Quercus Brantii Lindl. Based on the Acorn Morphometry and Composition of Phenolic Compounds'. *Applied Ecology and Environmental Research* 14 (3): 215–231.

Ferreira-Dias, S., D. G. Valente, and J. M. F. Abreu. 2003a. 'Comparison between Ethanol and Hexane for Oil Extraction from Quercus Suber L. Fruits'. *Grasas y Aceites* 54 (4): 378–383.

———. 2003b. 'Pattern Recognition of Acorns from Different Quercus Species Based on Oil Content and Fatty Acid Profile'. *Oils and Fats* 54 (4): 384–391.

Ferreira, V. C. S., D. Morcuende, S. H. Hérnandez-López, M. S. Madruga, F. A. P. Silva, and M. Estévez. 2017. 'Antioxidant Extracts from Acorns (Quercus Ilex L.) Effectively Protect Ready-to-Eat (RTE) Chicken Patties Irrespective of Packaging Atmosphere'. *Journal of Food Science* 82 (3): 622–631.

Galván, J. V., J. J. J. Novo, A. G. Cabrera, D. Ariza, J. García-Olmo, and R. M. N. Cerrillo. 2012. 'Population Variability Based on the Morphometry and Chemical Composition of the Acorn in Holm Oak (Quercus Ilex Subsp. Ballota [Desf.] Samp.)'. *European Journal of Forest Research* 131 (4): 893–904.

Galván, J. V., L. Valledor, R. Mª. N. Cerrillo, E. G. Pelegrín, and J. V. Jorrín-Novo. 2011. 'Studies of Variability in Holm Oak (Quercus Ilex Subsp. Ballota [Desf.] Samp.) through Acorn Protein Profile Analysis'. *Journal of Proteomics* 74 (8): 1244–1255.

Gamboa-Gómez, C. I., L. E. Simental-Mendía, R. F. González-Laredo, E. J. Alcantar-Orozco, V. H. Monserrat-Juarez, J. C. Ramírez-España, J. A. Gallegos-Infante, M. R. Moreno-Jiménez, and N. E. Rocha-Guzmán. 2017. 'In Vitro and in Vivo Assessment of Anti-Hyperglycemic and Antioxidant Effects of Oak Leaves (Quercus Convallata and Quercus Arizonica) Infusions and Fermented Beverages'. *Food Research International* 102 (September): 690–699.

García-Ayuso, L. E., M. Sánchez, A. F. de Alba, and M. D. L. de Castro. 1998. 'Focused Microwave-Assisted Soxhlet: An Advantageous Tool for Sample Extraction'. *Analytical Chemistry* 70 (11): 2426–2431.

García-Ayuso, L. E., J. Velasco, M. C. Dobarganes, and M. D. L. de Castro. 2000. 'Determination of the Oil Content of Seeds by Focused Microwave-Assisted Soxhlet Extraction'. *Chromatographia* 52: 103–108.

García-Gómez, E., R. Pérez-Badia, J. Pereira, and R. K. Puri. 2017. 'The Consumption of Acorns (from Quercus Spp.) in the Central West of the Iberian Peninsula in the 20th Century'. *Economic Botany* 71 (3): 256–268.

Gea-Izquierdo, G., I. Cañellas, and G. Montero. 2006. 'Acorn Production in Spanish Holm Oak Woodlands'. *Agricultural research. Forest systems and resources* 15 (3): 339–354.

Hayouni, E. A., M. Abedrabba, M. Bouix, and M. Hamdi. 2007. 'The Effects of Solvents and Extraction Method on the Phenolic Contents and Biological Activities in Vitro of Tunisian Quercus Coccifera L. and Juniperus Phoenicea L. Fruit Extracts'. *Food Chemistry* 105 (3): 1126–1134.

Huang, J., Y. Wang, C. Li, X. Wang, and X. He. 2016. 'Anti-Inflammatory Oleanolic Triterpenes from Chinese Acorns'. *Molecules* 21 (5): 669.

Jin, Y., Y. Jin, S. Heo, Y. Jin, S. Heo, J. Lee, H. Rhee, et al. 2005. 'Free Radical Scavenging and Hepatoprotective Actions of Quercus Aliena

Acorn Extract against CCl4-Induced Liver'. *Free Radical Research* 39 (12): 1351–1358.

Johnson, P. S., and S. R. Shifley. 2001. 'Oak-Dominated Ecosystems'. In *The Ecology and Silviculture of Oaks*, 1–53.

Kala, H. K., R. Mehta, K. K. Sen, R. Tandey, and V. Mandal. 2016. 'Critical Analysis of Research Trends and Issues in Microwave Assisted Extraction of Phenolics: Have We Really Done Enough'. *Trends in Analytical Chemistry* 85: 140–152.

Kilic, U., M. Boga, and I. Guven. 2010. 'Chemical Composition and Nutritive Value of Oak (Quercus Robur) Nut and Leaves'. *Journal of Applied Animal Research* 38 (1): 101–104.

Kilicarislan, C., and H. Ozgunay. 2012. 'Ultrasound Extraction of Valonea Tannin and Its Effects on Extraction Yield'. *Journal of the American Leather Chemists Association* 107 (November 2012): 394–403.

Korus, A., D. Gumul, M. Krystyjan, L. Juszczak, and J. Korus. 2017. 'Evaluation of the Quality, Nutritional Value and Antioxidant Activity of Gluten-Free Biscuits Made from Corn-Acorn Flour or Corn-Hemp Flour Composites'. *European Food Research and Technology* 243 (8): 1429–1438.

Korus, J., M. Witczak, R. Ziobro, and L. Juszczak. 2015. 'The Influence of Acorn Flour on Rheological Properties of Gluten-Free Dough and Physical Characteristics of the Bread'. *European Food Research and Technology* 240 (6): 1135–1143.

Lei, L., H. Zhu, C. Zhang, X. Wang, K. Y. Ma, L. Wang, Y. Zhao, and Z. Y. Chen. 2017. 'Dietary β-Sitosterol Is More Potent in Reducing Plasma Cholesterol than Sesamin in Hypercholesterolemia Hamsters'. *European Journal of Lipid Science and Technology* 119 (7): 1600349.

León-Camacho, Manuel, Isabel Viera-Alcaide, and Isabel M Vicario. 2004. 'Acorn (Quercus Spp.) Fruit Lipids: Saponifiable and Unsaponifiable Fractions: A Detailed Study'. *Journal of the American Oil Chemists' Society* 81 (5): 447–453.

Li, S., Y. Zhou, M. Liu, Y. Zhang, and S. Cao. 2015. 'Nutrient Composition and Starch Characteristics of Quercus Glandulifera Bl. Seeds from China'. *Food Chemistry* 185 (October): 371–376.

Lin, Y., Y. Lu, Z. Song, and D. Huang. 2018. 'Characterizations of the Endogenous Starch Hydrolase Inhibitors in Acorns of Quercus Fabri Hance'. *Food Chemistry* 258 (February): 111–117.

Löf, M., J. Castro, M. Engman, A. B. Leverkus, P. Madsen, J. A. Reque, A. Villalobos, and E. S. Gardiner. 2019. 'Tamm Review: Direct Seeding to Restore Oak (Quercus Spp.) Forests and Woodlands'. *Forest Ecology and Management* 448 (April): 474–489.

Lopes, I. M. G., and M. G. Bernardo-Gil. 2005. 'Characterisation of Acorn Oils Extracted by Hexane and by Supercritical Carbon Dioxide'. *European Journal of Lipid Science and Technology* 107 (1): 12–19.

Lopez-Bote, C. J. 1998. 'Sustained Utilization of the Iberian Pig Breed'. *Meat Science* 49: 17–27.

Luciano, C. G., C. M. L. Franco, G. A. Valencia, P. J. A. Sobral, and I. C. F. Moraes. 2017. 'Evaluation of Extraction Method on the Structure and Physicochemical Properties of Starch from Seeds of Two Jackfruit Varieties'. *Starch* 69 (11–12): 1700078.

Luo, X., R. Bai, D. Zhen, Z. Yang, D. Huang, H. Mao, X. Li, et al. 2019. 'Response Surface Optimization of the Enzyme-Based Ultrasound-Assisted Extraction of Acorn Tannins and Their Corrosion Inhibition Properties'. *Industrial Crops and Products* 129 (December 2018): 405–413.

Luque-García, J.L., and M.D. L. de Castro. 2004. 'Focused Microwave-Assisted Soxhlet Extraction: Devices and Applications'. *Talanta* 64 (3): 571–577.

Man, J., Y. Yang, J. Huang, C. Zhang, F. Zhang, Y. Wang, M. Gu, Q. Liu, and C. Wei. 2013. 'Morphology and Structural Properties of High-Amylose Rice Starch Residues Hydrolysed by Amyloglucosidase'. *Food Chemistry* 138 (4): 2089–2098.

Mena-García, A., A. I. Ruiz-Matute, A. C. Soria, and M. L. Sanz. 2019. 'Green Techniques for Extraction of Bioactive Carbohydrates'. *Trends in Analytical Chemistry* 119: 115612.

Miao, M., B. Jiang, S. W. Cui, T. Zhang, and Z. Jin. 2015. 'Slowly Digestible Starch - A Review'. *Critical Reviews in Food Science and Nutrition* 55 (12): 1642–1657.

Mojzer, E. B., M. K. Hrnčič, M. Škerget, Ž. Knez, and U. Bren. 2016. 'Polyphenols: Extraction Methods, Antioxidative Action, Bioavailability and Anticarcinogenic Effects'. *Molecules* 21 (7): 901–939.

Mustafa, A., and C. Turner. 2011. 'Pressurized Liquid Extraction as a Green Approach in Food and Herbal Plants Extraction: A Review'. *Analytica Chimica Acta* 703 (1): 8–18.

Ofcarcik, R. P., and E. E. Burns. 1971. 'Chemical and Physical Properties of Selected Acorns'. *Physical and Chemical Methods* 36 (4): 576–578.

Onem, E., G. Gulumser, S. Akay, and O. Yesil-Celiktas. 2014. 'Optimization of Tannin Isolation from Acorn and Application in Leather Processing'. *Industrial Crops and Products* 53 (February): 16–22.

Onem, E., G. Gulumser, M. Renner, A. L. Oelbermann, and O. Yesil-Celiktas. 2015. 'Pressurized Fluid Extraction (PFE) of Valonea Tannin with Binary H2O–CO2 and Ternary H2O–CH3OH–CO2 Systems and Phase Equilibrium Studies'. *Separation and Purification Technology* 146 (May): 101–107.

Özcan, T. 2006. 'Total Protein and Amino Acid Compositions in the Acorns of Turkish Quercus L. Taxa'. *Genetic Resources and Crop Evolution* 53 (2): 419–429.

Özünlü, O., H. Ergezer, and R. Gökçe. 2018. 'Improving Physicochemical, Antioxidative and Sensory Quality of Raw Chicken Meat by Using Acorn Extracts'. *Food Science and Technology* 98 (July): 477–484.

Panda, D., and S. Manickam. 2019. 'Cavitation Technology - The Future of Greener Extraction Method: A Review on the Extraction of Natural Products and Process Intensification Mechanism and Perspectives'. *Applied Sciences* 9 (4): 766.

Papoti, V. T., N. Kizaki, A. Skaltsi, P. D. Karayannakidis, and M. Papageorgiou. 2018. 'The Phytochemical Rich Potential of Acorn (Quercus Aegilops) Products and by Products'. *Food Science and Biotechnology* 27 (3): 819–828.

Parsaei, M., M. Goli, and H. Abbasi. 2018. 'Oak Flour as a Replacement of Wheat and Corn Flour to Improve Biscuit Antioxidant Activity'. *Food Science & Nutrition* 6 (2): 253–258.

Pasqualone, A., F. Z. Makhlouf, M. Barkat, G. Difonzo, C. Summo, G. Squeo, and F. Caponio. 2019. 'Effect of Acorn Flour on the Physico-Chemical and Sensory Properties of Biscuits'. *Heliyon* 5 (8): e02242.

Pellett, P. L., and S. Ghosh. 2004. 'Lysine Fortification: Past, Present, and Future'. *Food and Nutrition Bulletin* 25 (2): 107–113.

Pérez-Serradilla, J. A., M. C. Ortiz, L. Sarabia, and M. D. L. de Castro. 2007. 'Focused Microwave-Assisted Soxhlet Extraction of Acorn Oil for Determination of the Fatty Acid Profile by GC–MS. Comparison with Conventional and Standard Methods'. *Analytical and Bioanalytical Chemistry* 388 (2): 451–462.

Pinto, D., S. D. Franco, A. M. Silva, S. Cupara, M. Koskovac, K. Kojicic, S. Soares, et al. 2019. 'Chemical Characterization and Bioactive Properties of a Coffee-like Beverage Prepared from Quercus Cerris Kernels'. *Food & Function* 10 (4): 2050–2060.

Priego-Capote, F., and M. D. L. de Castro. 2005. 'Focused Microwave-Assisted Soxhlet Extraction: A Convincing Alternative for Total Fat Isolation from Bakery Products'. *Talanta* 65 (1): 81–86.

Priego-Capote, F., J. Ruiz-Jiménez, and M. D. L. de Castro. 2007. 'Identification and Quantification of Trans Fatty Acids in Bakery Products by Gas Chromatography-Mass Spectrometry after Focused Microwave Soxhlet Extraction'. *Food Chemistry* 100 (2): 859–867.

Priego-López, E., J. Velasco, M. C. Dobarganes, G. Ramis-Ramos, and M. D. L. de Castro. 2003. 'Focused Microwave-Assisted Soxhlet Extraction: An Expeditive Approach for the Isolation of Lipids from Sausage Products'. *Food Chemistry* 83 (1): 143–149.

Prinz, J. F., and P. W. Lucas. 2008. 'Saliva Tannin Interactions'. *Journal of Oral Rehabilitation* 27 (11): 991–994.

Rabhi, F., M. Narváez-Rivas, S. Boukhchina, and M. León-Camacho. 2015. 'Authentication of Quercus Species According to Their N-Alkanes Profile by Off-Line Combination of High-Performance Liquid

Chromatography and Gas Chromatography'. *Food Analytical Methods* 8 (7): 1710–1717.

Rakić, S., R. Maletić, M. Perunović, and G. Svrzić. 2004. 'Influence of Thermal Treatment on Tannin Content and Antioxidation Effect of Oak Acorn Quercus Cerris Extract'. *Journal of Agricultural Sciences* 49 (1): 97–107.

Rakić, S., D. Povrenović, V. Tešević, M. Simić, and R. Maletić. 2006. 'Oak Acorn, Polyphenols and Antioxidant Activity in Functional Food'. *Journal of Food Engineering* 74 (3): 416–423.

Rodríguez-Estevez, V., A. Garcia, A. Perea, C. Mata, and A. G. Gómez. 2007. 'Acorn Production at the Dehesa: Influential Factors'. *Archives of Zootechnics* 56 (R): 25–43.

Rodríguez-Estévez, V., A. García Martínez, C. M. Moreno, J. M. Perea Muñoz, and A. G. Gómez Castro. 2008. 'Measures and Nutritional Characteristics of Quercus Acorns From the Dehesa'. *Archives of Zootechnics* 57 (R): 1–12.

Rodríguez-Estevez, V., M. Sanchez-Rodriguez, C. Arce, A. R. García, J. M. Perea, and A. G. Gómez-Castro. 2012. 'Consumption of Acorns by Finishing Iberian Pigs and Their Function in the Conservation of the Dehesa Agroecosystem'. In *Agroforestry for Biodiversity and Ecosystem Services - Science and Practice*, 1–22.

Rosenberg, D. 2008. 'The Possible Use of Acorns in Past Economies of the Southern Levant: A Staple Food or a Negligible Food Source?' *Levant* 40 (2): 167–175.

Sagar, N. A., S. Pareek, S. Sharma, E. M. Yahia, and M. G. Lobo. 2018. 'Fruit and Vegetable Waste: Bioactive Compounds, Their Extraction, and Possible Utilization'. *Comprehensive Reviews in Food Science and Food Safety* 17 (3): 512–531.

Sanchez-Bel, P., A. Romojaro, I. Egea, and M. T. Pretel. 2015. 'Wild Edible Plants as Potential Antioxidant or Nutritional Supplements for Beverages Minimally Processed'. *Food Science and Technology* 62 (1): 830–837.

Seidel, V. 2006. 'Initial and Bulk Extraction'. In *Natural Products Isolation*, edited by S. D. Sarker, Z. Latif, and A. I. Gray, 2nd ed., 27–46.

Shimada, T. 2001. 'Nutrient Compositions of Acorns and Horse Chestnuts in Relation to Seed-Hoarding'. *Ecological Research* 16 (4): 803–808.

Silva, S., E. M. Costa, A. Borges, A. P. Carvalho, M. J. Monteiro, and M. M. E. Pintado. 2016. 'Nutritional Characterization of Acorn Flour (a Traditional Component of the Mediterranean Gastronomical Folklore)'. *Journal of Food Measurement and Characterization* 10 (3): 584–588.

Skendi, A., P. Mouselemidou, M. Papageorgiou, and E. Papastergiadis. 2018. 'Effect of Acorn Meal-Water Combinations on Technological Properties and Fine Structure of Gluten-Free Bread'. *Food Chemistry* 253 (January): 119–126.

Teo, C. C., S. N. Tan, J. W. H. Yong, C. S. Hew, and E. S. Ong. 2010. 'Pressurized Hot Water Extraction (PHWE)'. *Journal of Chromatography A* 1217 (16): 2484–2494.

Ullah, M. F., and W. Khan. 2008. 'Food as Medicine: Potential Therapeutic Tendencies of Plant Derived Polyphenolic Compounds'. *Asian Pacific Journal of Cancer Prevention* 9: 187–196.

Üniv, H., V. Fak, A. Kamalak, K. G. Hassan, S. M. Ameen, H. M. Zebari, A. H. Hasan, and F. Aslan. 2015. 'Determination of Chemical Composition, Potential Nutritive Value and Methane Emission of Oak Tree (Quercus Coccifera) Leaves and Nuts'. *Harran University Journal of the Faculty of Veterinary Medicine* 4 (1): 1–5.

Valencia-A., S. 2004. 'Diversity of the Genus Quercus (Fagaceae) in Mexico'. *Bulletin of the Botanical Society of Mexico* 75: 33–53.

Vilkhu, K., R. Mawson, L. Simons, and D. Bates. 2008. 'Applications and Opportunities for Ultrasound Assisted Extraction in the Food Industry - A Review'. *Innovative Food Science and Emerging Technologies* 9 (2): 161–169.

Vinatoru, M., T. J. Mason, and I. Calinescu. 2017. 'Ultrasonically Assisted Extraction (UAE) and Microwave Assisted Extraction (MAE) of Functional Compounds from Plant Materials'. *Trends in Analytical Chemistry* 97: 159–178.

Vinha, A. F., C. M. Barreira, A. S. G. Costa, and M. B. P. P. Oliveira. 2016. 'A New Age for Quercus Spp. Fruits: Review on Nutritional and Phytochemical Composition and Related Biological Activities of

Acorns'. *Comprehensive Reviews in Food Science and Food Safety* 15: 947–981.

Vinha, A. F., A. S. G. Costa, J. C. M. Barreira, R. Pacheco, and M. B. P. P. Oliveira. 2016. 'Chemical and Antioxidant Profiles of Acorn Tissues from Quercus Spp.: Potential as New Industrial Raw Materials'. *Industrial Crops and Products* 94 (December): 143–51.

Weschenfelder, C., A. S. De Quadros, J. L. Dos Santos, S. B. Garofallo, and A. Marcadenti. 2020. 'Adipokines and Adipose Tissue-Related Metabolites, Nuts and Cardiovascular Disease'. *Metabolites* 10 (32): 1–26.

Xia, K., C. E. Seal, W-Y Chen, Z-K Zhou, and H.W. Pritchard. 2010. 'Fruit Oil Contents of the Genus Quercus (Fagaceae): A Comparative Study on Acorns of Subgenus Quercus and the Asian Subgenus Cyclobalanopsis'. *Seed Science and Technology* 38 (1): 136–145.

Zazoua, A., S. Bouraoui, R. Halim, N. Khedimallah, and N. Jaffrezic-Renault. 2018. 'Polyphenolic Natural Products for the Electrochemical Determination of Cadmium'. *Analytical Letters* 51 (3): 359–370.

Zhang, H. F., X. H. Yang, and Y. Wang. 2011. 'Microwave Assisted Extraction of Secondary Metabolites from Plants: Current Status and Future Directions'. *Trends in Food Science and Technology* 22 (12): 672–688.

Zhang, Q. W., L. G. Lin, and W. C. Ye. 2018. 'Techniques for Extraction and Isolation of Natural Products: A Comprehensive Review'. *Chinese Medicine* 13 (20): 1–26.

In: Quercus: Classification, Ecology and Uses ISBN: 978-1-53618-026-8
Editor: Bent J. Steffensen

Chapter 3

PHENOLOGY AND AEROBIOLOGY OF THE *QUERCUS* GENUS: REPRODUCTIVE IMPLICATIONS

Herminia García-Mozo*, PhD, Rocío López-Orozco and Carmen Galán, PhD
Department of Botany, Ecology and Plant Physiology,
University of Cordoba, Córdoba, Spain

ABSTRACT

Quercus genus includes around 600 species of trees and shrubs that can be found in a wide variety of habitats, including temperate deciduous forests, temperate and sub-tropical evergreen forests, sub-tropical and tropical savannah and subtropical woodland in North Hemisphere. *Quercus* comprises arboreal dominant species of many different ecosystems. Their ecological but also economic importance is unquestionable, mainly regarding to timber industry in temperate forests, and livestock feed and cork management in Mediterranean areas. All these species are monoecious with unisexual wind pollinated flowers. Each

* Corresponding Author's Email: bv2gamoh@uco.es.

spring male flowers placed in catkins, produce large amounts of pollen grains to ensure fertilization. This anemophilous characteristic allows to study airborne pollen as a bioindicator of the reproductive phenology of this taxon and as a valuable source of information for clinicians and population allergic to this pollen type. In this chapter we review main literature related to *Quercus* aerobiology and phenology, and their contribution to a deeper understanding of the ecological behavior and the complex reproductive performance of these species. In this sense, we analysis some researches relating pollination season characteristics and final acorn production in determined species of this taxon, also pointing out *Quercus* aerobiological data as an important aspect to be considered in reproduction and ecological oaks studies.

Keywords: *Quercus*, oaks, aerobiology, phenology

1. Introduction

Quercus genus includes around 600 species of trees and shrubs that can be found in a wide variety of habitats, including temperate deciduous forests, temperate and sub-tropical evergreen forests, sub-tropical and tropical savannah and subtropical woodland in North Hemisphere. *Quercus* comprises arboreal dominant species of many different ecosystems where provide a high diversity of ecosystem services to the environment and society: supply services (wood, firewood, cork, livestock feed); regulation services (carbon sequestration and climate change mitigation, soil formation and protection, improvement of air and water quality) and even cultural ones (recreational, landscape, cultural identity). For all these reasons, *Quercus* species and the organisms that interact with them have become clear protagonists of endless research and actions related to the ecology and management of forests and savannahs. In the present monograph we try to collect a broad and updated vision of the state of the scientific art that surrounds *Quercus* species, offering some brushstrokes on different aspects related to regeneration, physiology, genetic diversity and distribution of this genus So emblematic of plants. In addition, the main trends in the current research related to *Quercus* are highlighted and the applicability of new approaches and tools that are potentially useful for the conservation and management of ecosystems dominated by these species is highlighted. Therefore, this monograph collects a wide collection of articles where *Quercus* species have been investigated at different hierarchical levels, from molecular biology to community ecology. A total of 13449 publications related to *Quercus* species appear in indexed journals (Figure 1), showing exponential growth over time. In last years, *Quercus* related article have suffered an annual growth rate of 4.7%.

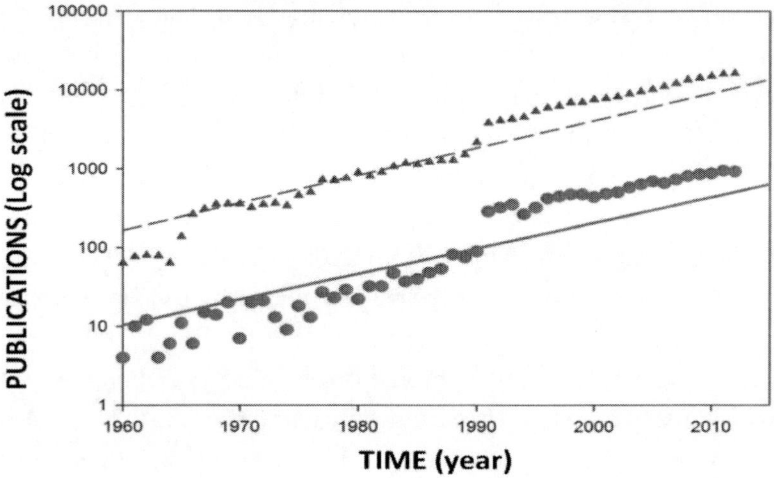

Source: Marañón et al. 2014.

Figure 1. Temporal trends in the number of publications including the topics "*Quercus*" (circles) and "Forest" (triangles) for journals indexed by the Web of Science (Thompson Reuters). Values are presented in logarithmic scale and regression lines are drawn.

2. Taxonomy and Distribution

The genus *Quercus* is one of the most important clades of woody angiosperms in the northern hemisphere belong to Fagales. Cronquist (1988) defined the Fagales as amentiferous members of the Hamamelidae. He circumscribed the taxon to include four families: Balanopaceae, Fagaceae, Nothofagaceae and Betulaceae; before his death he accepted a fifth family, the Ticodendraceae. Nee and Takhtajan (1998) place all the above in his

Faganae, adding the family Corylaceae and dividing the superorder into two units, the Fagales and the Corylales. Thorne (1992) basically agrees with Takhtajan but retains Corylaceae in the Betulaceae and uses the ordinal name, Betulales. The families currently included in Fagales are as follows: Betulaceae, Casuarinaceae, Fagaceae, Juglandaceae, Myricaceae, Nothofagaceae, Rhoipteleaceae, Ticodendraceae (Judd et al. 2007). *Quercus* genus is a member of the family Fagaceae, into the order Fagales. Oaks appeared on Earth over 56 million years ago and have radiated and expanded subsequently across the Northern Hemisphere (Hipp et al. 2010; Manos and Stanford 2001). Today they extend from the equator (Colombia and Indonesia) to the boreal regions up to 60° latitude N in Europe, and from sea level to 4000 meters in the Yunnan province in China (Camus 1936; 1938; 1952; Menitsky 2005; de Beaulieu and Lamant 2010). Oaks' habitats vary from dense bottomland forests to open grasslands and savannas, and from alkaline to acidic soils (Abrams 1992; Nixon and Muller 1997). The genus has diversified into numerous species in America and Asia, with highest diversity at 15–30° N in Mexico and east Asia (Valencia 2004; Menitsky 2005; Nixon 2006). Europe exhibits lower species richness (up to 30 species), but the genus is nearly as widespread there as it is in North America and east Asia, as a limited number of European species have expanded across the continent (Camus 1936). The genus is in any case among the most species-rich tree genera in the Northern Hemisphere together with *Salix*, and the most important in terms of species and biomass in the Americas (Cavender-Bares et al. 2016; Cavender-Bares 2019).

These studies also suggest that divergent selection has shaped genetic variation across species ranges. A clinal variation tracking geographic temperature-driven trends is often observed for growth and phenological traits (Soularue and Kremer 2014). However, the direction of the cline may differ between species or continents. Budburst, for example, tends to occur later in spring for European oak populations from higher elevation or latitude than for more southerly populations (Ducousso, Guyon, and Kremer 1996; Vitasse et al. 2009), whereas the opposite pattern is observed for northern red oak (McGee 1970; 1974; Kriebel et al. 1976; Daubree and Kremer 1993). This raises the question of whether the abiotic or biotic factors driving divergent selection for flushing date differ between the two continents or between different species. More recent studies have focused on physiological traits thought to contribute to adaptation to climate in *Quercus petraea* (Matt.), Liebl. (Torres-Ruiz et al. 2019) and *Quercus suber* L. (Ramírez-Valiente et al. 2014). Interestingly, in these two cases, population differentiation was less pronounced for leaf traits and traits relating to water metabolism, suggesting that divergent selection has disproportionately targeted more integrated traits, such as overall growth. The genetic architecture of the traits contributing to climate adaptation is therefore highly polygenic (Kremer and Hipp 2019).

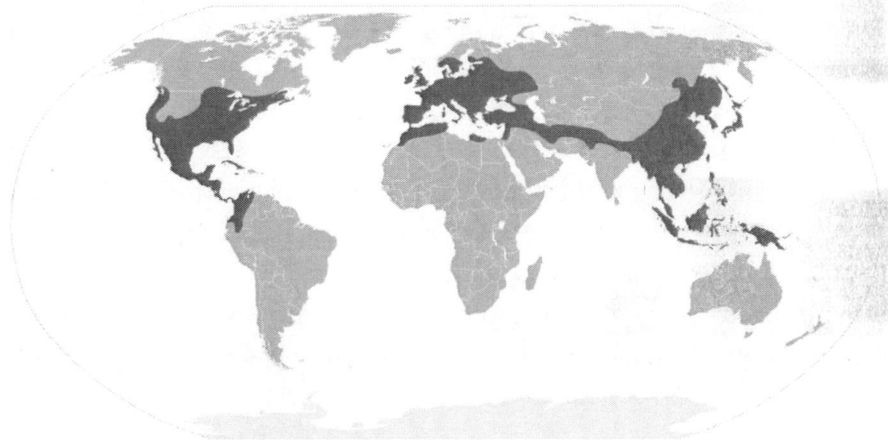

Figure 2. Fagaceae family distribution.

The Neotropics, particularly southern Mexico, harbors the greatest diversity of oak species in the New World. These oaks are often among the tallest trees in the forest areas in which they occur, ranging from low elevation to high montane forests. The clearest relationships of Central American oaks are with lower-elevation Mexican oaks. Several species are widespread from Mexico to Costa Rica and Panama. In addition, at least one distinctive clade of red oaks with annual maturation is common in Central America and Colombia and includes members as *Quercus seemannii* Liebm., *Quercus rapurahuensis* Pittier and *Quercus gulielmi-treleasei*

C.H.Mull., *Quercus costaricensis* Liebm., *Quercus eugeniifolia*, Liebm. and *Quercus humboldtii* Bonpl. (Nixon 2006).

3. Ecology

In terms of species diversity, ecological dominance, and economic value it includes around 600 species of trees and shrubs that can be found in a wide variety of habitats. Because of the dominance of oak in many ecosystems, it is the subject of a vast number of ecological studies that focus on interactions between oaks and fungi, plants and animals. The species of genus *Quercus* play an important role in maintaining the structure and functioning of various forest communities therefore have a significant ecological role including temperate deciduous forests and temperate and sub- tropical evergreen forests in North Hemisphere (Nixon 2006). In the current literature, oaks are dealt with either as a solitary genus with two sub genera (*Quercus* and *Cyclobalanopsis*), or them as two separate genera. However, in the Western Hemisphere, the genus *Quercus* has been divided into three distinct groups; the white oaks, the red oaks and the golden oaks. In the Northern Hemisphere this genus is extending from cool mild to tropical scopes in America, Asia and Europe. North America contains the biggest number of oak species.

In the Mediterranean are *Quercus* comprises arboreal dominant species of many different ecosystems. Perennial and marcescent *Quercus* appear in different vegetation areas including chaparral, maqui, oak woodland, and evergreen oak forest (Nixon 1993; 2002; Nixon and Muller 1997; Kappelle et al. 1995). Their ecological but also economic importance in the Mediterranean areas are unquestionable, mainly regarding to timber industry in temperate forests and livestock feed and cork management in Mediterranean areas. In the Mediterranean area, *Quercus ilex* subsp. *ballota* [Desf.] Samp (holm oak) and *Quercus suber* L. (cork oak) are the two species with more economic importance, since their acorns are a major food source for certain animal species vital to the region's economy: pigs, bulls and deer. Holm oak is the dominant tree species in natural forest ecosystems over large areas of the Western Mediterranean Basin (Benito Garzón, Sánchez de Dios, and Sainz Ollero 2008). In Spain, it is widely used for forest systems and forest practice conservation, covering a surface of 2489000 ha approximately, which present estimated values of over 120 million Euro in fruit acorn production per year (Gea-Izquierdo, Cañellas, and Montero 2006). Holm oak acorns are a major component in the feeding systems of many Mediterranean wild and livestock species and, at the same time, they are the basic feed ingredient for domestically-bred pigs to obtain high quality meat (Galván et al. 2012). The economic value of *Q. suber*, present in mountainous and wet areas, is enhanced by cork production. Also in the Mediterranean area appears *Quercus coccifera* L., which is typical of degraded soils.

Regarding the *Quercus* genus in the Neotropic areas, several species are widespread from Mexico to Costa Rica and Panama. The country where more *Quercus* species grow in the tropical area is Mexico with 161 species, 109 autochthonous from the country. The species with the widest distribution in this country are *Quercus candicans* Née, *Quercus castanea* Née, *Quercus crassifolia* Bonpl., *Quercus laeta* Liebm., *Quercus microphylla*, Née, *Quercus obtusata* Bonpl.and *Quercus rugosa* Née (Valencia 2004). Neotropic oaks are part of different vegetation associations: subtropical evergreen forest, subtropical and tropical savannah, subtropical woodland, oak-pine forest, oak-pine-juniper woodlands, cloud forest, tropical premontane and montane forest and summer rain chaparral.

Oaks also enter, and are important, along the margins of various other vegetation types, such as coniferous forests, prairies, tropical grasslands, desert and semi-desert scrublands, dry (deciduous) tropical forest, and in some evergreen tropical forests (Barbour and Billings 2000).

In addition to ecological and aesthetic value, one crucial role of oaks is that they help in maintaining watershed integrity. The extensive root system of oaks reduces soil erosion, stabilizes slopes and permit groundwater storage. The capacity of oaks and different trees to decrease air contamination and trap airborne particulates is very much archived. The basic qualities of the wood make oak a standout amongst the most adaptable hardwoods, esteemed by numerous businesses. On account of their immense interest in oak seed creation, oaks assume a basic part in supporting natural life and keeping up local biodiversity. Numerous transient types of feathered creatures and bats perch or home in them, notwithstanding utilizing the nourishment asset (Tantray, Wani, and Hussain 2017).

Therefore, seed production in oaks has been an issue largely studied by different studies. Most of them coincide in describing sexual ecology of temperate and tropical oaks dominated by a high variable and mast-seeding behavior. Mast-seeding implies a synchronous production of large seed crops within a *Quercus* population every two or more years. Authors have analyzed *Quercus* seed production in different ecosystems in base to different hypotheses trying to explain the evolutionary ecology of mast-seeding in oak species, both at temperate and Mediterranean areas (Sork 1984; Koenig et al. 2003; Liebhold et al. 2004; García-Mozo et al. 2007). The three non-mutually exclusive hypotheses are: 1) mast-seeding as a consequence of mast-flowering years with high pollination efficiency in this anemophilous genus; 2) mast-seeding evolved as an anti-predator adaptation by satiating seed predators during mast and allow survival of some seeds; 3) populations with larger seeds present more masting events due to large seeds require more time to accumulate reserves to mature. Results obtained both in Mediterranean and North-American *Quercus* studies support the "wind pollination" hypothesis. These researches pointed out pollination as one of the main factors explaining acorn production fluctuations, together with meteorological variables. Integration of aerobiological, field phenological and meteorological represented an important step forward in forest fruit production research (García-Mozo et al. 2007; García-Mozo, Domínguez Vilches, and Galán 2012).

4. Reproduction

Female flowers appear in the axils of leaves produced in the same year. They are produced on a short stalk and become visible a few days after the emergence of the male catkins (Sharp and Sprague 1967). Inflorescence primordia are difficult to distinguish from lateral bud primordial before late summer, hence the exact time of the initiation of pistillate inflorescences is difficult to determine. As hermaphrodite flowers are known to occur occasionally, (Bonnet-Masimbert 1978) has hypothesized that their initiation may occur in late spring, when the staminate inflorescences develop. Female flowers develop in late winter or early spring. After fertilization, the acorns mature within about 3 months, then fall. Not only do species differ in dormancy, but within a species differences can be due to geography or altitude.

The most definitive studies on acorn dormancy are those of (Peterson 1983) on water oak. He concluded that delayed germination (dormancy) was the result of at least three factors: (1) Mechanical strength of the pericarp; (2) chemical inhibition by the pericarp, which was alleviated by stratification for approximately 4 weeks; and (3) slow increase in capacity to imbibe the water required for pericarp rupture.

Depending fundamentally on wind fertilization, huge amounts of pollens are produced in the male blooms (25-100 for every catkin) each spring. The female blooms tucked unnoticeably in the hubs of pivotal twigs develop somewhat later, staying away from self-fertilization. Either single or bunches of a few oak seeds start forming. Most trees start oak seed creation following 20 years.

Seed production depends on available airborne pollen (Campbell and Halama 1993; Minero et al. 1998; Galán et al. 2000; García-Mozo et al. 2002). During pollen grain development within the anther, microsporogenesis may be affected by both genetic (Kaul 2012) and environmental stress (Neilson and Wullstein 1980; García-Mozo et al. 2001). Viability mainly depends on relative atmospheric humidity at shedding and during pollen transport (Frankel, 1977; Pacini 1990; Bellani and Paoletti 1992). Pollen germination also decreases in cases where the mother plant has been damaged by atmospheric pollution, since this affects the pollen membrane (Bellani and Paoletti 1992).

Individual reproduction success variability can be attributed to a variety of physical factors of pollen dispersion and to oaks flowering traits (Streiff et al. 1999). This could be a reproductive strategy for enhancing crosses between species and individuals (Vázquez 1998); intra-species variation may thus be related to environmental diversity (Latorre and Bianchi 1998; Grime 1979; Askeyev et al. 2005). Reproductive success and seed formation at a global woodland level is largely determined by the difference in the start of flowering (Vázquez 1998) and heterogeneity of pollen dispersal (Streiff et al. 1999). Díaz-Fernández, Climent, and Gil (2004) reported a relationship between phenology and acorn maturation as influenced by environmental conditions, which determine the annual or biennial pattern. The cause of this biennial rhythm and its synchronicity over large areas is not clear. However, some authors suggest that it is physiological factors that are mostly involved: the high pollen production usually results in high fruit production, which takes a lot energy, inhibiting the development of new flowers and pollen for the next year. The high fruit production also inhibits other processes, for example the vegetative growth of the tree (Latałowa, Miętus, and Uruska 2002). Moreover, in recent years some researchers have demonstrated a relationship between the amount of airborne pollen and seed production (Allison 1990; Campbell and Halama 1993; Minero et al. 1998; Galán et al. 2000; García-Mozo et al. 2002). In fact, *Quercus* pollen production and dispersion has been used as a tool in crop forecast studies in the *Quercus* genus (García-Mozo et al. 2007; García-Mozo, Domínguez Vilches, and Galán 2012). These studies determined that *Quercus* pollen emission during flowering season was the main factor determining the final acorn harvest, but also some meteorological variables played an important role in explaining acorn crop variations, especially humidity and temperature during the months of April and September. The reliability of the proposed model was externally validated using data not included in its construction; validation yielded acceptable results, with a minimum error of estimation. Results appear to be very useful for planning cropping and pig feeding strategies in the Mediterranean "dehesas." Further research could extend the use of airborne pollen counts in forest studies relating to anemophilous species, in order to optimize agricultural policies.

Effective pollination in anemophilous species has been suggested to depend on pollen shedding, which is related to diurnal patterns of temperature, relative humidity, and probably micro-environmental electrostatic conditions (Jones and Little 1983), as well as pollen production (Allison 1990). Pollen production reportedly depends on rainfall and temperature over the period prior to flowering (Fornaciari et al. 1997; Galán et al. 2001). These results suggest that, in view of the wind intensity, humidity, rainfall, and temperature recorded in 2001, the number of pollen grains released into the air was greater than in 2000 since these parameters also influence pollen dispersal (Galán et al. 1998). The amount of pollen released by individual members of each species in the Mediterranean area can be summarized approximately in the following proportions: 1:3:3:6 (*Q. coccifera: Q. ilex ssp. ballota: Q. faginea: Q. suber*).

Effective pollination depends not only on pollen production but also on pollen viability.

Compatible and receptive stigmas are fundamental requisites for pollination. Stigma receptivity lasts from a few hours up to one month. Brewbaker (1967) observed, as a general trend, that trinucleate pollen loses its viability much faster than binucleate pollen. Pollen of the *Quercus* genus has been reported as binucleate (Stairs 1964). These findings would account for the longevity observed here in *Quercus* species pollen. After 48 hours, *Quercus suber* pollen grains were found to germinate rapidly, usually within a few hours after pollination *Q. suber* displays a high percentage of viability over the first 48 hours *Quercus coccifera* has the lowest viability index but a higher level of pollen germination.

Although the potential capability of pollen grains to germinate is high, *Q. suber* displays a high percentage of viability over the first 48 hours, but it is the species with the lowest real range of germination. *Quercus coccifera* has the lowest viability index but a higher level of pollen germination (Gómez-Casero et al. 2004).

5. Proteomics

Q. ilex subsp. *ballota*, like other oak species, has some features that favor a high intra-provenance genetic diversity: long life span, allogamy, wind pollination, monoecy, and a continuous geographical distribution

(Michaud, Lumaret, and Romane 1992). Spanish Holm oak provenances have showed a prolonged isolation, in spite of being native and distributed throughout diverse forest areas with complex geographic patterns of genetic variation (Lumaret et al. 2002; Petit et al. 2002). Some provenances have survived under the extreme dry climate conditions that prevail in southern Spain. However, and despite these differences, the classification of individual trees according to a given provenance based on phenotypes is quite challenging. In this context, proteomic studies have already proved their value to assess provenance variability in Andalusia Holm oaks. Proteomic approaches have been carried out to study in this species the following elements: i) the leaf proteome (Jorge et al. 2005); ii) the leaf proteome at several plant developmental stages with different provenances and under drought conditions (Echevarría-Zomeño et al. 2009; Jorge et al. 2006); and iii) the variability through acorn protein profile analysis (Galván et al. 2011). Our results generally showed a remarkable difference between the different Andalusia provenances. According to these findings, we started a study focusing on the Holm oak pollen proteome, expected to be less variable than the leaf or acorn proteomes.

Several pollen proteomic studies have been generally carried out in herbaceous plants (Dai et al. 2006; Holmes-Davis et al. 2005; Sheoran et al. 2006; 2007) and to a lesser extent in forest trees and woody crops, focusing on allergen proteins (Egger et al. 2008; Wallner et al. 2009; Napoli et al. 2008). An analysis of *Quercus ilex* subsp. *ballota* pollen proteome would help to gain insight into the molecular nature of allergens, as well as into other proteins involved in the functional specialization of oak pollen.

A comprehensive analysis of holm oak pollen proteome has been performed by Galván et al. (2012) using complementary gel-based (1-DE and 2-DE with MALDI-TOF-TOF) and gel-free (nLC-LTQ Orbitrap MS) approaches. The available protein, DNA and EST sequences accessible in public databases for protein identification. The potentiality of these approaches in assessing provenance variability has also been evaluated. Multivariate statistical analyses were performed to compare the 1- and 2-DE protein profiles of four provenances, and to establish phylogenetical distances among them. Holm oak pollen proteome has been partially described, with 355 protein species identified from peptide spectra techniques, allowing a functional protein characterization, as well as with the identification of provenance-specific protein markers.

In *Quercus*, seventeen differentially expressed proteins have been identified. These proteins are involved in a variety of cellular processes, most of which had been neither previously associated with embryo development nor identified in the genus *Quercus*. Some of these proteins are involved in stress and pollen development and others play a role in the metabolism of tannins and phenylpropanoids, which represent two of the major pathways for the synthesis of cork chemical components. Furthermore, the augmented expression levels found for specific proteins are probably related to the homozygous state of a doubled-haploid sample. In the case of cork-oak proteins involved in synthesis of cork components can be detected at such early stages of development, showing the potential of the method to be useful in searching for biomarkers related to cork quality.

6. PHENOLOGY

Regarding the reproductive characteristics, *Quercus* are overwhelmingly monoecious species with unisexual wind pollinated flowers, male and female blossoms borne on two different types of inflorescences, very occasionally they bear hermaphroditic flowers. Each spring male flowers placed in catkins, produce large amounts of pollen grains to ensure fertilization. The anemophilous pollination of *Quercus* genus allow us to analysis airborne pollen as a bioindicator of the reproductive phenology of this taxon. Classical phenology is the art of subdividing the development of biological taxa into identifiable sections along a time axis (Lieth 1997). These sections are known as phenophases and are used to monitor the development of the life cycle of an organism. The occurrence of a phenophase is determined by the biotic characteristics of the plant species, and by local climate characteristics (Orsham 1989). At the same time, climate variations influence flowering periods in many species (Zerboni et al. 1991; Branzi and Zanotti 1992). Nowadays, floral phenology includes the study of periodic events as influenced by the environment, especially temperature changes driven by weather and climate (Schwartz 2003). Floral phenological changes in anemophilous communities appear to fit a sigmoid pattern that may be approximated by a logistic equation (Latorre 1997). Phenology has emerged as an important focus of ecological research (Schwartz 1999).Budburst could be brought forward in time, rendering buds more vulnerable to late spring frosts; alternatively, budburst could be delayed, thereby possibly not making full use of the available growing season (Myking and Heide 1995; Kramer 1996; Sparks, Jeffree, and Jeffree 2000). Moreover, definition of the phenological phases of male and female flowering in dioecius species is essential in order to highlight the role of reproductive biology in breeding and gene-conservation programs (Varela 1996).

Quercus species, are dichogamous, and more particularly protandrous, which enhances successful crossing between individuals of the same population and even between different species (Vázquez 1998). The study of Mediterranean vegetation based on growth forms and phenological phases can make a great contribution to existing classifications (ecological, physiognomic, phytosociological, etc. (Latorre and Cabezudo 2006) and understanding phenological vegetative and reproductive behaviors and adaptations. The temporal distribution of phases discerned in the development of reproductive structures represents floral phenological development. Each of these stages is clearly differentiated and defined, both morphologically and physiologically, based on inflorescence development from undifferentiated bud to catkin senescence or fruit formation (Varela 1996; Vázquez 1998). The reproductive biology in this group includes two different types of seed cycles: the "annual biotype," that requires only one season to complete its reproductive cycle, and the "biennial biotype" which requires two full years. From an ecological standpoint, forest dynamics and structure depend on the reproductive phenology of woody species. In the "dehesas" – traditional, semi-natural, man-made systems found throughout the Iberian Peninsula and especially in southwestern Spain – predominant evergreen *Quercus* are associated with cereal crops and grasslands that play a major role in avoiding soil erosion, increasing soil fertility and providing for genetic variability (Campos Palacín 1991; Joffre 1992).

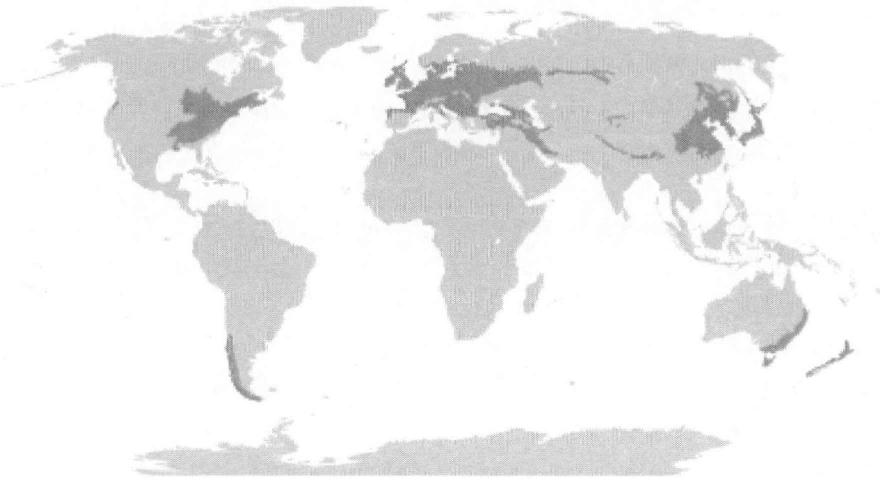

Figure 3. Mediterranean forest distribution.

Field work usually consisted in weekly visits to different zones at the start of the flowering season, and twice-weekly visits during maximum flowering and the fruit-formation period in order to take note of all the phenological changes. Male flowers are grouped in catkins which develop in the axils of either the inner bud scales or the first leaves, in the lower part of the branches produced in the same year. Staminate inflorescences are initiated in late spring, flowers develop in early summer and meiosis occurs in the following spring, giving rise to binucleate pollen grains immediately prior to the emergence of catkins. For a given tree, if weather conditions are suitable, catkin growth is achieved 1-2 weeks after bud opening, and pollination is completed in 2-4 days. In deciduous oaks, leaf expansion ceases during the release of pollen, which allows free movement of pollen. Catkin length varied considerably catkins in *Q. ilex* subsp. *ballota* were up to 10-12 cm long, compared to 15-17 cm in *Q. suber* and 4-7 cm in *Q. coccifera*.. The catkin elongation phase usually overlapped with the female flower differentiation stage, while full flowering and receptivity in the female overlapped with the start of flowering in the male, coinciding with the start of pollen release. Moreover, not all trees in each population developed at the same speed. The flowering period started later in zones at a higher altitude phenological development was also determined by relative position; trees facing south flowered before those facing north. Female flowers and fruits were observed both on new branches and on previous-year branches, since fruit requires two years to ripen. Inter-species and inter-annual differences in phenological range were observed for every phenophase.

There was generally some overlap between full receptivity in a single female and the start of male flowering. However, not all members of a given population flowered simultaneously. Although the tendency varied among species and years, the general phenological tendency seemed to fit a sigmoid curve in all cases, rising with time. The flowering period in *Q. ilex* subsp. *ballota* and *Q. coccifera* was earlier than that of *Q. suber*. Flowering is more strongly influenced by temperature than by rainfall (Spano et al. 1999; Garau, de Delfino, and Berrondo 2000; Reeves and Coupland 2000).

Although double flowering is not a general event, in some occasion perennial oaks as *Q. ilex* subsp. *ballota* and *Q. suber* displayed a second flowering period at the start of autumn. This phenomenon has been described for Mediterranean *Quercus* communities after a so dry spring. First autumn rainfalls seems to favor a second flowering period to compensate low pollination in adverse springs (Vázquez 1998). This second flowering was probably due to mild temperatures and rainfall after a stress period, harsh summer conditions that may act breaking the dormancy.

Weather conditions prior to flowering and vegetative growth are decisive factors in plant development and it has been observed in *Quercus* pollen season (García-Mozo et al. 2002; 2006). Prolonged dry periods may weaken trees, causing loss of leaves and reducing vegetative growth and thus the possibility of vegetative and floral bud formation (Nieto Caldera et al. 1993; Latorre et al. 1996) a competitive role exists between reproductive and vegetative phenophases to acquire plant resources (Latorre and Cabezudo 2002). Most studies in temperate-zone trees species assume that temperature regulates phenological development induced by a chilling period, depending on the species and geographical areas, that induces budburst followed by a period with forcing temperatures (also variable) that releases dormancy and thus triggers the onset of growth (Aron 1983; Hari and Häkkinen 1991; Frenguelli et al. 1992; Kramer 1996; Wielgolaski 1999; Chuine 2000; Fulton, Hall, and Catley 2001; Galán et al. 2001). The catkin elongation phenophase is highly sensitive to sharp change in humidity (Sharp and Chisman 1961). Adverse temperature conditions recorded during that phase in *Quercus gambelii* Nutt. (Neilson and Wullstein 1980) and *Q. suber* (Galán et al. 2001) are reported to have a negative effect on the reproductive cycle of the tree. Trees located in sunny areas displayed a shorter total phenological range. Moreover, shorter ranges were recorded in years with warmer springs; temperature and solar radiation have been shown to influence the flowering process in these *Quercus* species (Vázquez 1998). In general, female phases displayed a shorter range. Moreover, slow development in the initial phases was followed by fast changes during the flowering period, with a subsequent slowing down in the final phases (Gómez-Casero, Galán, and Domínguez-Vilches 2007).

Techniques of remote sensing (such as MODerate resolution Imaging Spectroradiometer (MODIS) and Advanced Very High-Resolution Radiometer (AVHRR), among others) are being used to develop phenological studies (Liu et al. 2017; Walker, de Beurs, and Henebry 2015; Zhang 2015). The satellite sensors capture the annual variability and the phenological dynamics of vegetation by land surface phenology (LSP) (de Beurs and Henebry 2004). The phenology of the vegetation and the pollen grain emissions to the air have been widely studied, showing a close relationship (Romero-Morte et al. 2018; Tormo-Molina et al. 2015; Velasco-Jiménez et al. 2015). Satellite-derived sources, such as the Normalized Difference Vegetation Index (NDVI), have been used to capture the phenology of different types of vegetation (oak and grass) in the Mediterranean environment (Liu et al. 2017).

7. AEROBIOLOGY

Anemophily is the pollination method used by the *Quercus* species. It is based on an extremely indiscriminate and random dispersal mechanism; for that reason, anemophilous plants have developed an increase in pollen production in order to compensate for limited efficiency. Wind pollination is much less precise than biotic pollination, and many pollen grains are required to ensure pollination (Proctor, Yeo, and Lack 1996). Successful pollination depends in part on pollen production (Allison 1990). The capacity to produce pollen is under genetic and physiological control. Pollen production per plant is determined by the number of pollen grains per anther, number of anthers per flower, number of flowers, and plant size. Nevertheless climate influences pollen production (Stanley and Linskens 2012; Fornaciari et al. 1997; Galán et al. 2001). The knowledge of potential pollen production per plant and of production dates is a good measure of the health of the Mediterranean forest. As Tormo-Molina et al. (1996) have reported for anemophilous tree species, there is a constant value for pollen production between species. In the case of *Quercus*, differences have been observed between species within the same year; the evergreen shrub *Q. coccifera* produced the highest amount of pollen per stamen and a higher number of groups of inflorescences per m^2 than the other species. These results would suggest that this is a method of compensating for the smaller size of pollen grains, lower individual height and smaller crown diameter or size in this species. Total pollen production per tree was higher for *Q. suber* than for the other species studied. *Q. ilex ssp. ballota* was over one hundred times lower than that reported for the same species by Tormo-Molina et al. (1996), probably because the trees studied in this area were smaller. the number of pollen grains per stamen was similar since the number appears to be under genetic and physiological control (Reddi and Reddi 1986; Stanley and Linskens 2012).

Pollen viability and germinability depends on environmental factors, such as photoperiod, irradiance, day or night temperature, water stress, or any interaction of these (Cecich 1997). Moreover, it can be negatively affected by the impact of atmospheric pollution on the mother plants, so that at higher pH, pollen germination is inhibited (Bellani and Paoletti 1992). The final pollen production indicates the relative potential ability of each species to charge the ambient air with their pollen (Reddi and Reddi 1986). Aerobiological studies provide an estimate of the partial contribution of each species to the total amount of pollen grains in the air. Pollen grain shape is the same (stenopalynous) for all four *Quercus* species, although grains differ slightly in size (Colombo, Lorenzoni, and Grigoletto 1983; Solomon 1983). they are all classified as *Quercus* type, since they are impossible to differentiate under light microscopy (Grant 1990).

From an aerobiological standpoint, data on total pollen production per plant, supported by knowledge of the coverage of the various species in a given area, could enable us to determine the partial contribution of each species to the total amount of pollen in the atmosphere (Tormo-Molina et al. 1996; Hidalgo, Galán, and Domínguez 2003; Prieto-Baena et al. 2003). Moreover, *Quercus* pollen presents a certain allergenicity, so aerobiological pollen data are of a great importance for allergic population. In spring, tree pollen grains of the Fagales order which includes *Quercus* genus, are a main allergen source in the northern hemisphere, with cross-reactivity occurring between pollen within the same order (Mothes and Valenta 2004; Matthiesen, Ipsen, and Løwenstein 1991). Several studies consider *Quercus* pollen type responsible for allergy in areas with abundant *Quercus* vegetation (Butland et al. 1997; Luscri et al. 1996; Negrini and Arobba 1992; Ross, Corden, and Fleming 1996; Subiza, Subiza, and Jerez 1986; Subiza et al. 1995).

The relationship between airborne Fagales pollen behavior and climate also is of great interest for aerobiology and forestry sciences. The relationships among the airborne pollen spectrum, pollination phenology, plant distribution and land cover modifications provoke alterations in the aerobiological content. Aeropalinological studies providing data on exposure of people to allergens in the air are of substantial importance in pollen allergy diagnosis (Rapiejko et al. 2004). Pollen is the main source of aeroallergens from trees in springtime (D'Amato 2001). In northern, central and eastern Europe, allergic airway diseases from early springtime are caused by pollen from Betulaceae. The beginning of Fagales pollination in

Europe follows the same pattern all the years: February for *Alnus* and *Corylus*, March for *Betula, Quercus* and *Carpinus,* April for *Juglans.* The start of pollen release by *Quercus* coincided with the end of the *Betula.* (Oei, Spieksma, and Bruynzeel 1986; Matthiesen, Ipsen, and Løwenstein 1991; Spieksma et al. 2003; Radišič et al. 2003; Piotrowska 2004; Peternel et al. 2007; Ianovici 2009). Airpollen do not present the same allergenic significance for humans. *Betula*, followed by *Alnus* and *Corylus*, has the greatest allergenic potency in this group of allergenic trees.

Other pollen types recorded in this study and cited as allergenic, do not account for high concentrations and are moderately allergenic significance (*Alnus, Corylus*) and slightly allergenic pollen (*Carpinus, Quercus, Juglans*) (Radišič et al. 2003).

During the last years, the behavior of the Fagaceae pollen seasons and their fluctuation in the annual totals over several years has been studied (Corden and Millington 1999; Jato et al. 2006; García-Mozo et al. 2002; Rizzi-Longo, Pizzulin-Sauli, and Ganis 2005; de la Cruz, Reyes, and Sánchez 2008). Rodríguez-Rajo, Frenguelli, and Jato (2003) reported a threshold value for *Quercus* of 50 pollen grains/m^3 in Spain. The APS of Fagales trees was relatively long Correlation analysis with weather parameters demonstrated that the relative humidity and wind are the main factors influencing the daily pollen concentrations in the atmosphere. Because Fagales pollen emission is a springtime phenomenon, pollen season dynamics are sensitive to changes in phenology (García-Mozo et al. 2006; Jablonska and Rapiejko 2010).

Linear trends in elevated annual Fagales pollen counts may be influenced by impacts of urbanization (the supplementary planting of ornamental species, soil nutrification, habitat fragmentation, disturbance, and the urban heat island effect) (Yli-Panula et al. 2009). The size of the residential areas around town is an important influence on Fagales pollen levels.

The release and dispersion of pollen depend on the microclimate, which explains the differences in the timing of flowering and occurrence of pollen in the air within the same species (Kasprzyk et al. 2004; Palacios, Tormo-Molina, and Rodríguez 2007; Puc 2012). The response to temperature is clear in the *Quercus* genus, and it controls to a large extent the flowering time, especially flowering start (Garcia-Mozo et al. 2000; 2006).

The daily average relative humidity showed negative correlations with flowering dates. The negative influence of relative humidity on airborne pollen dispersion is a global finding in aeropalynological studies (Skjøth et al. 2009; Estrella et al. 2006). Although aerobiological data are often used in phenological research as an indicator of flowering, airborne pollen concentrations are influenced by a number of factors that could affect pollen curves. This paper reports on a study of various aspects of reproductive biology in *Q. ilex* subsp. *ballota*, together with environmental factors influencing pollen release and transport, with a view of achieving reliable interpretation of *Quercus* pollen curves in Ourense (NW Spain). Aerobiological data were recorded from 2002 to 2004 at two sites in the province of Ourense. From 1st February to the end of the flowering period, phenological observations were carried out on 19 trees from the *Q. ilex* subsp. *ballota* population found in the Ourense area. Pollen production was calculated for the same trees. The chilling and heating requirements for triggering development were also calculated. The mean flowering period lasted 11-15 days. Reduced pollen output per catkin and, especially, a reduced number of catkins per tree in 2003 and 2004, prompted a marked decline in overall pollen production. Major differences observed in *Q. ilex* subsp. *ballota* pollen curves were attributed to the considerable influence both of weather conditions during pollination and pollen production. In years with high pollen production and weather conditions favouring pollen release, Q. *ilex* subsp. *ballota* contributed almost 10% to the total *Quercus* pollen curve. Around 20% of the pollen trapped was captured before or after flowering periods (Jato, Rodríguez-Rajo, and Aira 2007).

8. Climate Change

Effects of climate change on these forests depend on how changes in water availability and temperature will take place. Global average surface temperature has increased 0.6ºC since the late 19th century. Climate change has been evaluated through plant phenology as an indicator of the response of ecosystems to the increase in temperature (García-Mozo et al. 2008; Ma et al. 2013; Parmesan and Yohe 2003). In view of the effect of climate change (IPCC, 2017), forecasting is increasingly important, and it is necessary to use statistics (González-Naharro et al. 2019). The Special Report on Emissions Scenarios (SRES) produced a series of scenarios of CO_2 emissions representing outcomes of distinct narratives of economic development (Nakicenovic et al. 2000). Scenarios are plausible future states of the climate system. They are derived from predictions of climate change from Global Climate Models (GCMs) based on CO_2 emissions scenarios. GCMs are mathematical descriptions of the physical elements and processes in the atmosphere, oceans and land surface (winds and ocean currents, clouds, rainfall, soils). However, in areas where bio-geography has a significant effect on wheather, scenarios based on global models will fail to capture local details for impact assessment. The best method for adding these details to global models is to use a Regional Climate Model (RCM), essentially a higher-resolution version of a GCM covering a limited area (Meteorological Office 2001). Simulation RCM can be replicated by using forecasting phenological models, and allows assessment of the potential variation in vegetation dynamics.

Spring temperatures in the Mediterranean countries have risen by 2ºC over the last two decades (Jones et al. 1999; Moreno et al. 2005). The increase has been shown to prompt earlier leaf bud bursting and flowering in temperate Mediterranean tree species such as oak (Galán et al. 2005; García-Mozo et al. 2002).

Drought is one of the main factors causing the deterioration of *Quercus* forests in many regions of the world (Johnson, Shifley, and Rogers 2009; Shifley et al. 2006) and in particular in the Mediterranean region. Pérez-Ramos et al. (2013) conclusions suggested in this paper is that the increase in aridity predicted by climate change models will further limit the natural regeneration capacity of these species, alter their patterns of relative abundance in the community and enhance the facilitating role of scrub as a key element for the successful establishment of seedlings and seedlings. From Valladares et al. (2014), The results of this work suggest that fragmentation processes could mitigate some negative effects of climate change, specifically the decrease in oak productivity associated with increasing water stress. However, since climate change operates directly on other processes such as the phenology of trees and seed dispersers, it could lead to decouples between the vital rhythms of both groups of organisms and negatively affect mutualism between them.

Particular attention was paid to the effect of the detected temperature increase in later years on the *Quercus* pollen season, and to the potential consequences of the global climate change. In recent years, the rise in winter and spring temperatures as a consequence of global climate change is causing earlier leafing and flowering in many species, in both Europe and the USA (Bradley et al. 1999; Emberlin et al. 1997; Menzel and Fabian 1999). Changes in the pollination period features, both timing and intensity, of plants producing allergenic pollen would have important consequences on allergy sufferers (Beggs 2004; Emberlin 1994). Knowledge and forecasting of the pollination dynamics provides useful information for managing and preventing allergic symptoms. The effect of recent climate warming on vegetation phenology and aerobiology is an important issue which is being taken into account in climate-change studies (Giannini and Magnani, 1994; Schwartz 1999). Because *Quercus* pollen emission is a springtime phenomenon, pollen season dynamics are sensitive to changes in phenology. *Quercus* pollen season dynamics and their relationship with meteorological factors were studied at 14 sites of the Iberian Peninsula (Spain) (García-Mozo et al. 2006).

Regarding changes on vegetation due to climate change, Díaz (2014) offers an updated overview of the spatial configuration of the woodland and landscape in the adhered systems and its effects on the biological diversity and the natural regeneration process of the dominant *Quercus* species. One of the main conclusions reached is that the thinning favors the transitions of the predispersive phases of the regenerative cycle, but collapses the process of seed dispersal, presumably due to the negative effects of the risk of predation on efficacy (both quantitative and qualitative) of acorn dispersers. With regard to the management of this type of anthropized ecosystems, it proposes some measures aimed at the conservation of pastures that allow contributing to their economic sustainability while guaranteeing the satisfactory regeneration of the woodland Perea, Miguel, and Gil (2014), which reviews the role that plant-animal interactions play in the demography of Mediterranean *Quercus* species. there are multi-trophic interactions (eg acorn-predator-disperser) and facilitation phenomena (thicket seedling) that reveal the importance of multiple relationships in the regeneration of the *Quercus* forests and that should be considered in the management and restoration plans of said forest masses Leverkus, Castro, and Benayas (2014), where they evaluate the colonization by holm oaks of a burning repopulation pine forest. The presence of holm oaks favours the recovery of the vegetation after the fire both for its regrowth capacity and for the post-fire recruitment due to the scattering of acorns by jays.

ACKNOWLEDGMENTS

The authors thank gratefully the Project CLIMAQUER (Reference 1260464) granted by the University of Córdoba and supported by the Council of Economy and Knowledge of the Andalusia Region Government trough the European Regional Development Funds (ERDF).

REFERENCES

Abrams, Marc D. 1992. "Fire and the Development of Oak Forests." *BioScience* 42 (5): 346–53.

Allison, Taber D. 1990. "Pollen Production and Plant Density Affect Pollination and Seed Production in Taxus Canadensis." *Ecology* 71 (2): 516–22. https://doi.org/10.2307/1940305.

Aron, Robert. 1983. "Availability of Chilling Temperatures in California." *Agricultural Meteorology* 28 (4): 351–63.

Askeyev, O. V., D. Tischin, T. H. Sparks, and I. V. Askeyev. 2005. "The Effect of Climate on the Phenology, Acorn Crop and Radial Increment of Pedunculate Oak (*Quercus robur*) in the Middle Volga Region, Tatarstan, Russia." *International Journal of Biometeorology* 49 (4): 262–66.

Barbour, Michael G., and William Dwight Billings. 2000. *North American Terrestrial Vegetation*. Cambridge University Press.

Beaulieu, Antoine Le Hardÿ de, and Thierry Lamant. 2010. *Guide Illustré Des Chênes*. Edilens.

Beggs, P. J. 2004. "Impacts of Climate Change on Aeroallergens: Past and Future." *Clinical & Experimental Allergy* 34 (10): 1507–13. https://doi.org/10.1111/j.1365-2222.2004.02061.x.

Bellani, Lorenza M., and E. Paoletti. 1992. "New Type of Damage to *Quercus Ilex*: Pollen Germination and Hydration Ability." *European Journal of Forest Pathology* 22 (5): 284–90.

Benito Garzón, Marta, Rut Sánchez de Dios, and Helios Sainz Ollero. 2008. "Effects of Climate Change on the Distribution of Iberian Tree Species." *Applied Vegetation Science* 11 (2): 169–78.

Beurs, Kirsten M de, and Geoffrey M Henebry. 2004. "Land Surface Phenology, Climatic Variation, and Institutional Change: Analyzing Agricultural Land Cover Change in Kazakhstan." *Remote Sensing of Environment* 89(4): 497–509. https://doi.org/10.1016/j.rse.2003.11.006.

Bonnet-Masimbert, M. 1978. "Biologie Florale Des Chênes Pédonculés et Sessiles (*Quercus pedunculata* Ehrh et *Q Sessiliflora* Sal)." [Floral Biology of Pedunculate and Sessile Oaks ("*Quercus pedunculata Ehrh and Q Sessiliflora Sal)*."]. In *Symp, Régénération et Le Traitement Des Forêts Feuillues de Qualité En Zone Tempérée. CNRF [*In *Symp, Regeneration and Treatment of Quality Hardwood Forests in Temperate Zone. CNRF*]., 11–15.

Bradley, Nina L., A. Carl Leopold, John Ross, and Wellington Huffaker. 1999. "Phenological Changes Reflect Climate Change in Wisconsin." *Proceedings of the National Academy of Sciences* 96 (17): 9701–4. https://doi.org/10.1073/pnas.96.17.9701.

Branzi, Giovanna Puppi, and Anna Letizia Zanotti. 1992. "Estimate and Mapping of the Activity of Airborne Pollen Sources." *Aerobiologia* 8 (1): 69–74.

Brewbaker, James L. 1967. "The Distribution and Phylogenetic Significance of Binucleate and Trinucleate Pollen Grains in the Angiosperms." *American Journal of Botany* 54 (9): 1069–83. https://doi.org/10.1002/j.1537-2197.1967.tb10735.x.

Butland, Barbara K., David P. Strachan, Sarah Lewis, John Bynner, Neville Butler, and John Britton. 1997. "Investigation into the Increase in Hay Fever and Eczema at Age 16 Observed between the 1958 and 1970 British Birth Cohorts." *BMJ* 315 (7110): 717–21.

Campbell, Diane R., and Kenneth J. Halama. 1993. "Resource and Pollen Limitations to Lifetime Seed Production in a Natural Plant Population." *Ecology* 74 (4): 1043–51.

Campos Palacín, P. 1991. "*Economía y Conservación de Sistemas Agrosilvopastorales Mediterráneos de Dehesas y Montados. Valores Económicos y Ambientales de Las Dehesas Espeñolas*." ["*Economy and Conservation of Mediterranean Agrosilvopastoral Systems of Dehesas and Mounted Mediterranean. Economic and Environmental Values of Las Dehesas Espeñolas*."] Seminario CEE/CSIC/SIA/A. Ayuntamiento de Jerez-Badajoz. Jerez de la ….

Camus, A. 1936. "*Monographie Du Genre Quercus*." *Tome I. Genre Quercus. Sous-Genre Cyclobalanopsis et Sous-Genre Euquercus (Section Cerris et Mesobalanu).* Paris: Editions Paul Lechevalier. [*Quercus Genus Monograph." Volume I. Genus Quercus. Cyclobalanopsis Subgenus and Euquercus Subgenus (Cerris and Mesobalanu Section)*. Paris: Editions Paul Lechevalier].

Camus, A. 1938. "Les Chênes: Monographie Du Genre Quercus. Tome II: Genre *Quercus*. Sous-Genre Euquercus (Sections Lepidobalanus et Macrobalanus)." *Encyclopédie Economique de Sylviculture VII.* Paul Lechevalier, Editeur, Paris, France. ["The Oaks: Quercus Genus Monograph. Volume II: Genus Quercus. Euquercus Subgenus (Lepidobalanus and Macrobalanus Sections)." *Economic Encyclopedia of Forestry VII*. Paul Lechevalier, Publisher, Paris, France.].

Camus, A. 1952. *Les Chênes: Monographie Du Genre Quercus. Tome III (2ième Partie). Genre Quercus. Sous-Genre Euquercus (Sections Protobalanus et Erythrobalanus). Monographie Du Genre Lithocarpus.*

Editions Paul Lechevalier, Paris. [*The Oaks: Quercus Genus Monograph. Volume III (Part 2). Genus Quercus. Subgenus Euquercus* (*Protobalanus and Erythrobalanus Sections*). *Monograph of the Genus Lithocarpus*. Editions Paul Lechevalier, Paris].

Cavender-Bares, Jeannine, David D. Ackerly, Sarah E. Hobbie, and Philip A. Townsend. 2016. "Evolutionary Legacy Effects on Ecosystems: Biogeographic Origins, Plant Traits, and Implications for Management in the Era of Global Change." *Annual Review of Ecology, Evolution, and Systematics* 47: 433–62.

Cavender-Bares, Jeannine. 2019. "Diversification, Adaptation, and Community Assembly of the American Oaks (*Quercus*), a Model Clade for Integrating Ecology and Evolution." *New Phytologist* 221 (2): 669–92.

Cecich, Robert A. 1997. "Pollen Tube Growth in *Quercus*." *Forest Science* 43 (1): 140–46.

Chuine, Isabelle. 2000. "A United Model for Budburst of Trees." *J Theor Biol* 2007: 337–47.

Colombo, Paola Mariani, Francesca Chiesura Lorenzoni, and Francesco Grigoletto. 1983. "Pollen Grain Morphology Supports the Taxonomical Discrimination of Mediterranean Oaks (*Quercus*, Fagaceae)." *Plant Systematics and Evolution* 141 (3–4): 273–84.

Corden, Julie, and Wendy Millington. 1999. "A Study of *Quercus* Pollen in the Derby Area, UK." *Aerobiologia* 15 (1): 29–37.

Cronquist, Arthur. 1988. "*1988 The Evolution and Classification of Flowering Plants*." 2d Ed. New York Botanical Garden, New York."

Cruz, David Rodríguez de la, Estefanía Sánchez Reyes, and José Sánchez Sánchez. 2008. "Aerobiological Study of Fagaceae Pollen in the Middle-West of Spain." *Aerobiologia* 24 (2): 67.

Dai, Shaojun, Lei Li, Taotao Chen, Kang Chong, Yongbiao Xue, and Tai Wang. 2006. "Proteomic Analyses of Oryza Sativa Mature Pollen Reveal Novel Proteins Associated with Pollen Germination and Tube Growth." *Proteomics* 6 (8): 2504–29.

D'Amato, Gennaro. 2001. "Airborne Paucimicronic Allergen-carrying Particles and Seasonal Respiratory Allergy." *Allergy* 56 (12): 1109–11.

Daubree, J. B., and A. Kremer. 1993. "Genetic and Phenological Differentiation between Introduced and Natural Populations of *Quercus Rubra* L." In *Annales Des Sciences Forestières*, 50:271s–80. EDP Sciences.

Díaz, Mario. 2014. "Distribución del arbolado y persistencia a largo plazo de las dehesas: patrones y procesos." ["Woodland distribution and long-term persistence of grasslands: patterns and processes."] *Revista Ecosistemas* 23 (2): 5-12–12. https://doi.org/10.7818/re.2014.23-2.00.

Díaz-Fernández, Pedro M., José Climent, and Luis Gil. 2004. "Biennial Acorn Maturation and Its Relationship with Flowering Phenology in Iberian Populations of *Quercus Suber*." *Trees* 18 (6): 615–21.

Ducousso, A., J. P. Guyon, and A. Kremer. 1996. "Latitudinal and Altitudinal Variation of Bud Burst in Western Populations of Sessile Oak (*Quercus Petraea* (Matt) Liebl)." In *Annales Des Sciences Forestieres*, 53:775–82. EDP Sciences.

Echevarría-Zomeño, Sira, David Ariza, Inmaculada Jorge, Christof Lenz, Antonio Del Campo, Jesús V. Jorrín, and Rafael M. Navarro. 2009. "Changes in the Protein Profile of *Quercus Ilex* Leaves in Response to Drought Stress and Recovery." *Journal of Plant Physiology* 166 (3): 233–45.

Egger, C., M. Focke, A. J. Bircher, K. Scherer, N. Mothes-Luksch, F. Horak, and R. Valenta. 2008. "The Allergen Profile of Beech and Oak Pollen." *Clinical & Experimental Allergy* 38 (10): 1688–96.

Emberlin, J. 1994. "The Effects of Patterns in Climate and Pollen Abundance on Allergy." *Allergy* 49 (s18): 15–20. https://doi.org/10.1111/j.1398-9995.1994.tb04233.x.

Emberlin, Jean, John Mullins, Julie Corden, Wendy Millington, Marion Brooke, Mike Savage, and Sandra Jones. 1997. "The Trend to Earlier Birch Pollen Seasons in the U.K.: A Biotic Response to Changes in Weather Conditions?" *Grana* 36 (1): 29–33. https://doi.org/10.1080/00173139709362586.

Estrella, Nicole, Annette Menzel, Ursula Krämer, and Heidrun Behrendt. 2006. "Integration of Flowering Dates in Phenology and Pollen Counts in Aerobiology: Analysis of Their Spatial and Temporal Coherence in

Germany (1992–1999)." *International Journal of Biometeorology* 51 (1): 49–59.

Fornaciari, M., L. Pieroni, P. Ciuchi, and B. Romano. 1997. "A Statistical Model for Correlating Airborne Pollen Grains (*Olea Europaea* L.) with Some Meteorological Parameters." *Agricoltura Mediterranea* 127: 134–37.

Frankel, R. 1977. "Pollination Mechanisms, Reproduction and Plant Breeding." *Monographs on Theoretical and Applied Genetics* 2.

Frenguelli, Giuseppe, Emma Bricchi, Bruno Romano, Gianfranco Mincigrucci, Francesco Ferranti, and Evasio Antognozzi. 1992. "The Role of Air Temperature in Determining Dormancy Release and Flowering of *Corylus Avellana* L." *Aerobiologia* 8 (3): 415–18.

Fulton, T. A., A. J. Hall, and J. L. Catley. 2001. "Chilling Requirements of Paeonia Cultivars." *Scientia Horticulturae* 89 (3): 237–48.

Galán, C., M. Jose Fuillerat, Paul Comtois, and Eugenio Dominguez-Vilches. 1998. "Bioclimatic Factors Affecting Daily Cupressaceae Flowering in Southwest Spain." *International Journal of Biometeorology* 41 (3): 95–100. https://doi.org/10.1007/s004840050059.

Galán, C., H. Garcıa-Mozo, M. T. Gómez-Casero, and E. Dominguez-Vilches. 2000. "The Use of the Aerobiological and Phenological Data as a Tool to Predict *Quercus* Fruit Production." In *Proceedings of the 2nd European Symposium on Aerobiology, Viena, Austria*, 5–9.

Galán, C., H. García-Mozo, P. Cariñanos, P. Alcázar, and E. Domínguez-Vilches. 2001. "The Role of Temperature in the Onset of the *Olea Europaea* L. Pollen Season in Southwestern Spain." *International Journal of Biometeorology* 45 (1): 8–12.

Galán, C., H. García-Mozo, L. Vázquez, L. Ruiz, C. Díaz De La Guardia, and M. M. Trigo. 2005. "Heat Requirement for the Onset of the *Olea Europaea* L. Pollen Season in Several Sites in Andalusia and the Effect of the Expected Future Climate Change." *International Journal of Biometeorology* 49 (3): 184–88.

Galván, José Valero, Luis Valledor, Rafael Mª Navarro Cerrillo, Eustaquio Gil Pelegrín, and Jesus V. Jorrín-Novo. 2011. "Studies of Variability in

Holm Oak (*Quercus Ilex* Subsp. Ballota [Desf.] Samp.) through Acorn Protein Profile Analysis." *Journal of Proteomics* 74 (8): 1244–55.

Galván, José Valero, Jesús J. Jorrín Novo, Augusto Gómez Cabrera, David Ariza, Juan García-Olmo, and Rafael Mª Navarro Cerrillo. 2012. "Population Variability Based on the Morphometry and Chemical Composition of the Acorn in Holm Oak (*Quercus Ilex* Subsp. Ballota [Desf.] Samp.)." *European Journal of Forest Research* 131 (4): 893–904.

Garau, A., S. Filippini de Delfino, and G. Berrondo. 2000. "Influencia de Factores Climáticos En Las Fechas de Inicio de Floración y Brotación de Clones de Álamo en el Delta Del Paraná, Argentina." ["Influence of Climatic Factors on the Starting Dates of Flowering and Budding of Poplar Clones in the Delta Del Paraná, Argentina."] *Forest Systems* 9 (1): 169–76.

García-Mozo, H., C. Galán, M. T. Gómez-Casero, and E. Domínguez. 2000. "A Comparative Study of Different Temperature Accumulation Methods for Predicting the Start of the *Quercus* Pollen Season in Córdoba (South West Spain)." *Grana* 39 (4): 194–99.

García-Mozo, Herminia, J. Hidalgo Pablo, Carmen Galán, María Teresa Gómez-Casero, and Eugenio Domínguez. 2001. "Catkin Frost Damage in Mediterranean Cork-Oak (*Quercus Suber* L.)." *Israel Journal of Plant Sciences* 49 (1): 42–47.

García-Mozo, H., C. Galán, M. J. Aira, J. Belmonte, C. Díaz de la Guardia, D. Fernández, A. M. Gutierrez, F. J. Rodriguez, M. M. Trigo, and E. Dominguez-Vilches. 2002. "Modelling Start of Oak Pollen Season in Different Climatic Zones in Spain." *Agricultural and Forest Meteorology* 110 (4): 247–57. https://doi.org/10.1016/S0168-1923(02)00003-5.

García-Mozo, Herminia, Carmen Galán, Victoria Jato, Jordina Belmonte, Consuelo Díaz De La Guardia, Delia Fernández, Montserrat Gutiérrez, María Jesús Aira, Joan M. Roure, and Luis Ruiz. 2006. "*Quercus* Pollen Season Dynamics in the Iberian Peninsula: Response to Meteorological Parameters and Possible Consequences of Climate Change." *Annals of Agricultural and Environmental Medicine* 13 (2): 209.

García-Mozo, H., M. T. Gómez-Casero, E. Domínguez, and C. Galán. 2007. "Influence of Pollen Emission and Weather-Related Factors on Variations in Holm-Oak (*Quercus Ilex* Subsp. Ballota) Acorn Production." *Environmental and Experimental Botany* 61 (1): 35–40. https://doi.org/10.1016/j.envexpbot.2007.02.009.

García-Mozo, H., I. Chuine, M. J. Aira, J. Belmonte, D. Bermejo, C. Díaz de la Guardia, B. Elvira, et al. 2008. "Regional Phenological Models for Forecasting the Start and Peak of the *Quercus* Pollen Season in Spain." *Agricultural and Forest Meteorology* 148 (3): 372–80. https://doi.org/10.1016/j.agrformet.2007.09.013.

García-Mozo, H., E. Domínguez Vilches, and Carmen Galán. 2012. "A Model to Account for Variations in Holm-Oak (*Quercus Ilex* Subsp. Ballota) Acorn Production in Southern Spain." *Annals of Agricultural and Environmental Medicine 19 (3), 403-408 (2012)*. http://helvia.uco.es/xmlui/handle/10396/11243.

Gea-Izquierdo, G., I. Cañellas, and G. Montero. 2006. "Acorn Production in Spanish Holm Oak Woodlands." *Forest Systems* 15 (3): 339–54.

Giannini, Raffaello, and Federico Magnani. 1994. "*Impact of Global Change on Pollination Processes and on the Genetic Diversity of Forest tree populations*," 8.

Gómez-Casero, Mª Teresa, Pablo J Hidalgo, Herminia García-Mozo, Eugenio Domínguez, and Carmen Galán. 2004. "Pollen Biology in Four Mediterranean *Quercus* Species." *Grana* 43 (1): 22–30. https://doi.org/10.1080/00173130410018957.

Gómez-Casero, Mª Teresa, Carmen Galán, and Eugenio Domínguez-Vilches. 2007. "Flowering Phenology of Mediterranean *Quercus* Species in Different Locations (Córdoba, SW Iberian Peninsula)." *Acta Botanica Malacitana* 32: 127–46.

González-Naharro, Rocío, Elia Quirós, Santiago Fernández-Rodríguez, Inmaculada Silva-Palacios, José María Maya-Manzano, Rafael Tormo-Molina, Raúl Pecero-Casimiro, Alejandro Monroy-Colin, and Ángela Gonzalo-Garijo. 2019. "Relationship of NDVI and Oak (*Quercus*) Pollen Including a Predictive Model in the SW Mediterranean Region." *Science of the Total Environment* 676: 407–19.

Grant, E. 1990. *Sampling and Identifying Allergenic Pollens and Molds*. Blewstone Press. San Antonio, Texas.

Grime, John Philip. 1979. "Plant Strategies and Vegetation Processes." *Plant Strategies and Vegetation Processes.*

Hari, Pertti, and Risto Häkkinen. 1991. "The Utilization of Old Phenological Time Series of Budburst to Compare Models Describing Annual Cycles of Plants." *Tree Physiology* 8 (3): 281–87.

Hidalgo, Pablo J., Carmen Galán, and Eugenio Domínguez. 2003. "Male Phenology of Three Species of *Cupressus*: Correlation with Airborne Pollen." *Trees* 17 (4): 336–44.

Hipp, Andrew L., Paul S. Manos, Marlene Hahn, Michael Avishai, Cathérine Bodénès, Jeannine Cavender-Bares, Andrew A. Crowl, et al. 2019. "Genomic Landscape of the Global Oak Phylogeny." *New Phytologist* n/a (n/a). https://doi.org/10.1111/nph.16162.

Holmes-Davis, Rachel, Charlene K. Tanaka, William H. Vensel, William J. Hurkman, and Sheila McCormick. 2005. "Proteome Mapping of Mature Pollen of Arabidopsis Thaliana." *Proteomics* 5 (18): 4864–84.

Ianovici, N. 2009. "Analysis of Pollen Counts of Betulaceae in Timisoara, 2001-2004." *Journal of Horticulture, Forestry and Biotechnology* 13: 94–99.

Jablonska, K., and Piotr Rapiejko. 2010. "Using the Results of a Nationwide Phenological Network to Examine the Impact of Changes in Phenology of Plant Species on the Concentration of Plant Pollen in the Air." *Acta Agrobotanica* 63 (2).

Jato, V., F. J. Rodríguez-Rajo, and M. J. Aira. 2007. "Use of *Quercus Ilex* Subsp. Ballota Phenological and Pollen-Production Data for Interpreting *Quercus* Pollen Curves." *Aerobiologia* 23 (2): 91.

Jato, Victoria, F. J. Rodríguez-Rajo, P. Alcázar, P. De Nuntiis, C. Galán, and P. Mandrioli. 2006. "May the Definition of Pollen Season Influence Aerobiological Results?" *Aerobiologia* 22 (1): 13.

Joffre, R. 1992. "The Dehesa: Does This Complex Ecological System Have a Future?" In *Responses of Forest Ecosystems to Environmental Changes*, edited by A. Teller, P. Mathy, and J. N. R. Jeffers, 381–88.

Dordrecht: Springer Netherlands. https://doi.org/10.1007/978-94-011-2866-7_35.

Johnson, P. S., S. R. Shifley, and R. Rogers. 2009. "Oak-Dominated Ecosystems." *The Ecology and Silviculture of Oaks*, no. Ed. 2: 8–57.

Jones, C. Eugene, and R. John Little. 1983. *Handbook of Experimental Pollination Biology*. Scientific and Academic Editions New York.

Jones, Phil D., Mark New, David E. Parker, Seelye Martin, and Ignatius G. Rigor. 1999. "Surface Air Temperature and Its Changes over the Past 150 Years." *Reviews of Geophysics* 37 (2): 173–99.

Jorge, Inmaculada, Rafael M. Navarro, Christof Lenz, David Ariza, Carlos Porras, and Jesús Jorrín. 2005. "The Holm Oak Leaf Proteome: Analytical and Biological Variability in the Protein Expression Level Assessed by 2-DE and Protein Identification Tandem Mass Spectrometry de Novo Sequencing and Sequence Similarity Searching." *Proteomics* 5 (1): 222–34.

Jorge, Inmaculada, Rafael M. Navarro, Christof Lenz, David Ariza, and Jesús Jorrín. 2006. "Variation in the Holm Oak Leaf Proteome at Different Plant Developmental Stages, between Provenances and in Response to Drought Stress." *Proteomics* 6 (S1): S207–14.

Judd, Walter S., Christopher S. Campbell, Elizabeth A. Kellogg, Peter F. Stevens, and M. J. Donoghue. 2007. *Taxonomy [In:] Plant Systematics–A Phylogenetic Approach*. Sinauer Associates, Sunderland.

Kappelle, Maarten, Jan-Gerrit Van Uffelen, and Antoine M. Cleef. 1995. "Altitudinal Zonation of Montane *Quercus* Forests along Two Transects in Chirripó National Park, Costa Rica." *Vegetatio* 119 (2): 119–53.

Kasprzyk, I., A. Uruska, K. Szczepanek, M. Latałowa, J. Gaweł, K. Harmata, D. Myszkowska, A. Stach, and D. Stępalska. 2004. "Regional Differentiation in the Dynamics of the Pollen Seasons of *Alnus*, *Corylus* and *Fraxinus* in Poland (Preliminary Results)." *Aerobiologia* 20 (2): 141–51.

Kaul, Mohan L. H. 2012. *Male Sterility in Higher Plants*. Vol. 10. Springer Science & Business Media.

Koenig, Walter D., Dave Kelly, Victoria L. Sork, Richard P. Duncan, Joseph S. Elkinton, Mikko S. Peltonen, and Robert D. Westfall. 2003.

"Dissecting Components of Population-level Variation in Seed Production and the Evolution of Masting Behavior." *Oikos* 102 (3): 581–91.

Kramer, K. 1996. "Phenology and Growth of European Trees in Relation to Climate Change." Phd, S.l.: Kramer. https://library.wur.nl/WebQuery/wurpubs/33578.

Kremer, Antoine, and Andrew L. Hipp. 2019. "Oaks: An Evolutionary Success Story." *New Phytol.*

Kriebel, H. B., W. T. Bagley, F. J. Deneke, R. W. Funsch, P. Roth, J. J. Jokela, C. Merritt, J. W. Wright, and R. D. Williams. 1976. "Geographic Variation in *Quercus Rubra* in North Central United States Plantations." *Silvae Genetica* 25 (3/4): 118–22.

Latałowa, Małgorzata, Mirosław Miętus, and Agnieszka Uruska. 2002. "Seasonal Variations in the Atmospheric Betula Pollen Count in Gdańsk (Southern Baltic Coast) in Relation to Meteorological Parameters." *Aerobiologia* 18 (1): 33–43.

Latorre, A. V. Pérez, and B. Cabezudo. 2006. "Phenomorphology and Eco-Morphological Characters of Rhododendron Lauroid Forests in the Western Mediterranean (Iberian Peninsula, Spain)." *Plant Ecology* 187 (2): 227–47. https://doi.org/10.1007/s11258-005-6574-0.

Latorre, A. V. Pérez, B. Cabezudo, J. M. Nieto, and T. Navarro. 1996. "Caracterización Fenológica y Ecomorfológica de Alcornocales Andaluces (Málaga, España)." ["Phenological and Ecomorphological Characterization of Andalusian Cork Oak (Málaga, Spain)."] In *Anales Jard. Bot*, 54:554–60.

Latorre, Fabiana. 1997. "Comparison between Phenological and Aerobiological Patterns of Some Arboreal Species of Mar Del Plata (Argentina)." *Aerobiologia* 13 (1): 49.

Latorre, A. V. Pérez, and Baltasar Cabezudo. 2002. "Use of Monocharacteristic Growth Forms and Phenological Phases to Describe and Differentiate Plant Communities in Mediterranean-Type Ecosystems." *Plant Ecology* 161 (2): 231–49.

Latorre, Fabiana, and María Martha Bianchi. 1998. "Relationships between Flowering Development of *Ulmus Pumila* and Fraxinus Excelsior and Their Airborne Pollen." *Grana* 37 (4): 233–38.

Leverkus, Alexandro B., Jorge Castro, and José María Rey Benayas. 2014. "Regeneración post-incendio de la encina en pinares de repoblación mediterráneos." ["Post-fire regeneration of the oak in Mediterranean repopulation pine forests."] *Revista Ecosistemas* 23 (2): 48-54–54. https://doi.org/10.7818/re.2014.23-2.00.

Liebhold, Andrew, Victoria Sork, Mikko Peltonen, Walter Koenig, Ottar N. Bjørnstad, Robert Westfall, Joseph Elkinton, and Johannes MH Knops. 2004. "Within-population Spatial Synchrony in Mast Seeding of North American Oaks." *Oikos* 104 (1): 156–64.

Lieth, Helmut. 1997. "Aims and Methods in Phenological Monitoring." *Phenology in Seasonal Climates I.* Backhuys Publishers, Leiden.

Liu, Yan, Michael J. Hill, Xiaoyang Zhang, Zhuosen Wang, Andrew D. Richardson, Koen Hufkens, Gianluca Filippa, et al. 2017. "Using Data from Landsat, MODIS, VIIRS and PhenoCams to Monitor the Phenology of California Oak/Grass Savanna and Open Grassland across Spatial Scales." *Agricultural and Forest Meteorology* 237–238 (May): 311–25. https://doi.org/10.1016/j.agrformet.2017.02.026.

Lumaret, R., C. Mir, H. Michaud, and V. Raynal. 2002. "Phylogeographical Variation of Chloroplast DNA in Holm Oak (*Quercus Ilex* L.)." *Molecular Ecology* 11 (11): 2327–36.

Luscri, M., E. Li Bianchi, A. Fanti, C. Scarnati, M. Cecconi, and M. Marinelli. 1996. "Tree Pollen Allergy in Rome: P 388." *Allergy: European Journal of Allergy & Clinical Immunology, Supplement* 51.

Ma, Xuanlong, Alfredo Huete, Qiang Yu, Natalia Restrepo Coupe, Kevin Davies, Mark Broich, Piyachat Ratana, et al. 2013. "Spatial Patterns and Temporal Dynamics in Savanna Vegetation Phenology across the North Australian Tropical Transect." *Remote Sensing of Environment* 139 (December): 97–115. https://doi.org/10.1016/j.rse.2013.07.030.

Manos, Paul S., and Alice M. Stanford. 2001. "The Historical Biogeography of Fagaceae: Tracking the Tertiary History of Temperate and

Subtropical Forests of the Northern Hemisphere." *International Journal of Plant Sciences* 162 (S6): S77–93. https://doi.org/10.1086/323280.

Marañón, T., C. M. Padilla Díaz, I. M. Pérez Ramos, and R. Villar. 2014. "Research trends on ecology and management of *Quercus* species." *Ecosistemas* 23 (2): 124–29.

Matthiesen, F., H. Ipsen, and H. Løwenstein. 1991. "Pollen Allergens." *Allergenic Pollen and Pollinosis in Europe*, 36–44.

McGee, Charles E. 1970. "Bud-Break on Red Oak Seedlings Varies with Elevation of Seed Source." *Tree Planters' Notes* 21 (2): 18–19.

McGee, C. E. 1974. "Elevation of Seed Sources and Planting Sites Affects Phenology and Development of Red Oak Seedlings." *Forest Science* 20 (2): 160–64.

Menitsky, Yu L. 2005. *Oaks of Asia*. CRC Press.

Menzel, Annette, and Peter Fabian. 1999. "Growing Season Extended in Europe." *Nature* 397 (6721): 659–659. https://doi.org/10.1038/17709.

Meteorological Office. 2001. "*The Hadley Centre Regional Climate Modelling System. PRECIS: Providing Regional Climates for Impacts Studies*." GOV.UK. 2001. https://www.gov.uk/dfid-research-outputs/the-hadley-centre-regional-climate-modelling-system-precis-providing-regional-climates-for-impacts-studies.

Michaud, H., R. Lumaret, and F. Romane. 1992. "Variation in the Genetic Structure and Reproductive Biology of Holm Oak Populations." In *Quercus Ilex L. Ecosystems: Function, Dynamics and Management*, 107–13. Springer.

Minero, Francisco Jose Gonzalez, Pilar Candau, Julia Morales, and Carmen Tomas. 1998. "Forecasting Olive Crop Production Based on Ten Consecutive Years of Monitoring Airborne Pollen in Andalusia (Southern Spain)." *Agriculture, Ecosystems & Environment* 69 (3): 201–15.

Moreno, José Manuel, Miguel Álvarez Cobelas, Gerardo Benito, Jordi Catalán, M. Ramos, Diego de la Rosa, Fernando Valladares Ros, and Caridad Zazo. 2005. "*Principales Conclusiones de La Evaluación Preliminar de Los Impactos En España Por Efecto Del Cambio*

Climático." ["*Main Conclusions of the Preliminary Evaluation of the Impacts in Spain by Effect of Climate Change*."]

Mothes, Nadine, and Rudolf Valenta. 2004. "Biology of Tree Pollen Allergens." *Current Allergy and Asthma Reports* 4 (5): 384–90.

Myking, Tor, and O. M. Heide. 1995. "Dormancy Release and Chilling Requirement of Buds of Latitudinal Ecotypes of Betula Pendula and B. Pubescens." *Tree Physiology* 15 (11): 697–704.

Nakicenovic, N., J. Alcamo, G. Davis, B. de Vries, J. Fenhann, and S. Gaffin. 2000. *IPCC Special Report on Emissions Scenarios Nakicenovic N, Swart R, Editors*. Cambridge, United Kingdom: Cambridge University Press.

Napoli, Anna, Donatella Aiello, Leonardo Di Donna, Petros Moschidis, and Giovanni Sindona. 2008. "Vegetable Proteomics: The Detection of Ole e 1 Isoallergens by Peptide Matching of MALDI MS/MS Spectra of Underivatized and Dansylated Glycopeptides." *Journal of Proteome Research* 7 (7): 2723–32.

Nee, Michael, and A. Takhtajan. 1998. "Diversity and Classification of Flowering Plants." *Brittonia* 50 (2): 191. https://doi.org/10.2307/2807851.

Negrini, A. C., and D. Arobba. 1992. *An Illustrated Guide to Pollen Analysis*. Hodder and Stoughton eds., London.

Neilson, R. P., and L. H. Wullstein. 1980. "Catkin Freezing and Acorn Production in Gambel Oak in Utah, 1978." *American Journal of Botany* 67 (3): 426–28.

Nieto Caldera, José María, Andrés V. Pérez Latorre, Teresa Navarro, and Baltasar Cabezudo. 1993. "*Estudios Fenomorfológicos En La Vegetación Del Sur de España. II: Alcornocales Mesomediterráneos*. (Montes de Málaga, Málaga)." ["*Phenomorphological studies in the vegetation of southern Spain. II: Mesomediterranean cork oak forests*. (Montes de Málaga, Malaga)."]

Nixon, K. C. 1993. *The Genus Quercus in Mexico.: TP Ramamoorthy, R. Bye, A. Lot, y J. Fa, Eds. Biological Diversity of Mexico: Origins and Distribution*. Oxford University Press. Nueva York, EUA. p.

Nixon, K. C., and C. H. Muller. 1997. *Quercus Linnaeus Sect. Quercus. Flora of North America North of Mexico 3: 471-506.*

Nixon, K. C. 2002. "The Oak (Quercus) Biodiversity of California and Adjacent Regions." In: Standiford, Richard B., et al. *Tech. Editor. Proceedings of the Fifth Symposium on Oak Woodlands: Oaks in California's Challenging Landscape. Gen. Tech. Rep.* PSW-GTR-184, Albany, CA: Pacific Southwest Research Station, Forest Service, US Department of Agriculture: 3-20. Vol. 184.

Nixon, K. C. 2006. "Global and Neotropical Distribution and Diversity of Oak (Genus *Quercus*) and Oak Forests." In *Ecology and Conservation of Neotropical Montane Oak Forests*, 3–13. Springer.

Oei, H. D., F. Th. M. Spieksma, and P. L. B. Bruynzeel. 1986. "Birch Pollen Asthma in the Netherlands." *Allergy* 41 (6): 435–41.

Orsham, G. 1989. *Plant Pheno-Morphological Studies in Mediterranean Type Ecosystems*. Vol. 12. Springer Science & Business Media.

Pacini, E. 1990. "Harmomegathic Characters of Pteridophyta Spores and Spermatophyta Pollen." In *Morphology, Development, and Systematic Relevance of Pollen and Spores*, 53–69. Springer.

Palacios, Inmaculada Silva, Rafael Tormo Molina, and Adolfo F. Muñoz Rodríguez. 2007. "The Importance of Interactions between Meteorological Conditions When Interpreting Their Effect on the Dispersal of Pollen from Homogeneously Distributed Sources." *Aerobiologia* 23 (1): 17.

Parmesan, Camille, and Gary Yohe. 2003. "A Globally Coherent Fingerprint of Climate Change Impacts across Natural Systems." *Nature* 421 (6918): 37–42.

Perea, Ramón, Alfonso San Miguel, and Luis Gil. 2014. "Interacciones planta-animal en la regeneración de *Quercus pyrenaica*: ecología y gestión." ["Plant-animal interactions in the regeneration of Quercus pyrenaica: ecology and management."] *Revista Ecosistemas* 23 (2): 18-26–26. https://doi.org/10.7818/re.2014.23-2.00.

Pérez-Ramos, Ignacio M., Jesús Rodríguez-Calcerrada, Jean M. Ourcival, and Serge Rambal. 2013. "*Quercus Ilex* Recruitment in a Drier World: A Multi-Stage Demographic Approach." *Perspectives in Plant Ecology,*

Evolution and Systematics 15 (2): 106–17. https://doi.org/10.1016/j.ppees.2012.12.005.

Peternel, Renata, Sanja Music Milanovic, Ivana Hrga, Tatjana Mileta, and Josip Culig. 2007. "Incidence of Betulaceae Pollen and Pollinosis in Zagreb, Croatia, 2002-2005." *Annals of Agricultural and Environmental Medicine* 14 (1).

Peterson, J. K. 1983. "Mechanisms Involved in Delayed Germination of Quercus Nigra L. Seeds." *Annals of Botany* 52 (1): 81–92.

Petit, Rémy J., Simon Brewer, Sándor Bordács, Kornel Burg, Rachid Cheddadi, Els Coart, Joan Cottrell, Ulrike M. Csaikl, Barbara van Dam, and John D. Deans. 2002. "Identification of Refugia and Post-Glacial Colonisation Routes of European White Oaks Based on Chloroplast DNA and Fossil Pollen Evidence." *Forest Ecology and Management* 156 (1–3): 49–74.

Piotrowska, Krystyna. 2004. "Comparison of *Alnus*, *Corylus* and *Betula* Pollen Counts in Lublin (Poland) and Skien (Norway)." *Ann Agric Environ Med* 11 (2): 205–8.

Prieto-Baena, José C., Pablo J. Hidalgo, Eugenio Domínguez, and Carmen Galán. 2003. "Pollen Production in the Poaceae Family." *Grana* 42 (3): 153–59.

Proctor, M., P. Yeo, and A. Lack. 1996. "The Natural History of Pollination." *The Natural History of Pollination*. https://www.cabdirect.org/cabdirect/abstract/19960200695.

Puc, Małgorzata. 2012. "Artificial Neural Network Model of the Relationship between Betula Pollen and Meteorological Factors in Szczecin (Poland)." *International Journal of Biometeorology* 56 (2): 395–401.

Radišič, P., Branko Sikoparija, Miklos Juhász, and Nicoleta Ianovici. 2003. "*Betula* Pollen Season in the Danube-Kris-Mures-Tisa Euroregion (2000-2002)." *Annals of Faculty Engineering Hunedoara–Journal of Engineering*, 197–200.

Ramírez-Valiente, José Alberto, Fernando Valladares, David Sánchez-Gómez, Antonio Delgado, and Ismael Aranda. 2014. "Population

Variation and Natural Selection on Leaf Traits in Cork Oak throughout Its Distribution Range." *Acta Oecologica* 58: 49–56.

Rapiejko, P., A. Lipiec, A. Wojdas, and D. Jurkiewicz. 2004. "Threshold Pollen Concentration Necessary to Evoke Allergic Symptoms." *International Review of Allergology and Clinical Immunology* 10 (3): 91–94.

Reddi, C. Subba, and N. S. Reddi. 1986. "Pollen Production in Some Anemophilous Angiosperms." *Grana* 25 (1): 55–61.

Reeves, Paul H., and George Coupland. 2000. "Response of Plant Development to Environment: Control of Flowering by Daylength and Temperature." *Current Opinion in Plant Biology* 3 (1): 37–42. https://doi.org/10.1016/S1369-5266(99)00041-2.

Rizzi-Longo, Loredana, Marialuisa Pizzulin-Sauli, and Paola Ganis. 2005. "Aerobiology of Fagaceae Pollen in Trieste (NE Italy)." *Aerobiologia* 21 (3–4): 217.

Rodríguez-Rajo, F. Javier, Giuseppe Frenguelli, and Victoria Jato. 2003. "The Influence of Air Temperature on the Starting Date of *Quercus* Pollination in the South of Europe." *Grana* 42 (3): 145–52.

Romero-Morte, Jorge, Jesús Rojo, Rosario Rivero, Federico Fernández-González, and Rosa Pérez-Badia. 2018. "Standardised Index for Measuring Atmospheric Grass-Pollen Emission." *Science of The Total Environment* 612 (January): 180–91. https://doi.org/10.1016/j.scitotenv.2017.08.139.

Ross, A. M., Julie M. Corden, and D. M. Fleming. 1996. "The Role of Oak Pollen in Hay Fever Consultations in General Practice and the Factors Influencing Patients' Decisions to Consult." *Br J Gen Pract* 46 (409): 451–55.

Schwartz, M. D. 1999. "Advancing to Full Bloom: Planning Phenological Research for the 21st Century." *International Journal of Biometeorology* 42(3):113–18. https://doi.org/10.1007/s004840050093.

Schwartz, M. D. 2003. *Phenology: An Integrative Environmental Science*. Springer.

Sharp, Ward M., and Henry H. Chisman. 1961. "Flowering and Fruiting in the White Oaks. I. Staminate Flowering through Pollen Dispersal." *Ecology* 42 (2): 365–72.

Sharp, Ward M., and Vance G. Sprague. 1967. "Flowering and Fruiting in the White Oaks. Pistillate Flowering, Acorn Development, Weather, and Yields." *Ecology* 48 (2): 243–51.

Sheoran, Inder S., Kerry A. Sproule, Douglas J. H. Olson, Andrew RS Ross, and Vipen K. Sawhney. 2006. "Proteome Profile and Functional Classification of Proteins in Arabidopsis Thaliana (Landsberg Erecta) Mature Pollen." *Sexual Plant Reproduction* 19 (4): 185–96.

Sheoran, Inder S., Andrew R. S. Ross, Douglas J. H. Olson, and Vipen K. Sawhney. 2007. "Proteomic Analysis of Tomato (*Lycopersicon Esculentum*) Pollen." *Journal of Experimental Botany* 58 (13): 3525–35.

Shifley, Stephen R., Frank R. Thompson III, William D. Dijak, Michael A. Larson, and Joshua J. Millspaugh. 2006. "Simulated Effects of Forest Management Alternatives on Landscape Structure and Habitat Suitability in the Midwestern United States." *Forest Ecology and Management* 229 (1–3): 361–77.

Skjøth, Carsten Ambelas, Matt Smith, Jørgen Brandt, and Jean Emberlin. 2009. "Are the Birch Trees in Southern England a Source of *Betula* Pollen for North London?" *International Journal of Biometeorology* 53 (1): 75–86.

Solomon, Allen M. 1983. "Pollen Morphology and Plant Taxonomy of Red Oaks in Eastern North America." *American Journal of Botany* 70 (4): 495–507.

Sork, Victoria L. 1984. "Examination of Seed Dispersal and Survival in Red Oak, *Quercus Rubra* (Fagaceae), Using Metal-Tagged Acorns." *Ecology* 65 (3): 1020–22.

Soularue, Jean-Paul, and Antoine Kremer. 2014. "Evolutionary Responses of Tree Phenology to the Combined Effects of Assortative Mating, Gene Flow and Divergent Selection." *Heredity* 113 (6): 485–94.

Spano, Donatella, Carla Cesaraccio, Pierpaolo Duce, and Richard L. Snyder. 1999. “Phenological Stages of Natural Species and Their Use as Climate Indicators.” *International Journal of Biometeorology* 42 (3): 124–33.

Sparks, T. H., E. P. Jeffree, and C. E. Jeffree. 2000. “An Examination of the Relationship between Flowering Times and Temperature at the National Scale Using Long-Term Phenological Records from the UK.” *International Journal of Biometeorology* 44 (2): 82–87. https://doi.org/10.1007/s004840000049.

Spieksma, F. Th M., J. M. Corden, M. Detandt, W. M. Millington, H. Nikkels, N. Nolard, C. H. H. Schoenmakers, R. Wachter, L. A. De Weger, and R. Willems. 2003. “Quantitative Trends in Annual Totals of Five Common Airborne Pollen Types (*Betula, Quercus*, Poaceae, *Urtica*, and *Artemisia*), at Five Pollen-Monitoring Stations in Western Europe.” *Aerobiologia* 19 (3–4): 171–84.

Stairs, G. R. 1964. “Microsporogenesis and Embryogenesis in *Quercus*.” *Botanical Gazette* 125 (2): 115–21. https://doi.org/10.1086/336255.

Stanley, Robert G., and Hans F. Linskens. 2012. *Pollen: Biology Biochemistry Management*. Springer Science & Business Media.

Streiff, R., A. Ducousso, Ch. Lexer, H. Steinkellner, J. Gloessl, and A. Kremer. 1999. “Pollen Dispersal Inferred from Paternity Analysis in a Mixed Oak Stand of *Quercus Robur* L. and *Q. Petraea* (Matt.) Liebl.” *Molecular Ecology* 8 (5): 831–41.

Subiza, E., J. Subiza, and M. Palinología Jerez. 1986. “En: Tratado de Alergología e Inmunología Clínica.” [“In: Treatise on Allergology and Clinical Immunology.”] *Vol IV. Madrid: Luzán*, 211–54.

Subiza, Javier, Miguel Jerez, Juan Antonio Jiménez, María José Narganes, Martha Cabrera, Susana Varela, and Eliseo Subiza. 1995. “Allergenic Pollen and Pollinosis in Madrid.” *Journal of Allergy and Clinical Immunology* 96 (1): 15–23.

Tantray, Younas Rasheed, Mohammad Saleem Wani, and Ashiq Hussain. 2017. “*Genus Quercus: An Overview*.”

Thorne, Robert F. 1992. “Classification and Geography of the Flowering Plants.” *The Botanical Review* 58 (3): 225–327. https://doi.org/10.1007/BF02858611.

Tormo-Molina, Rafael, Adolfo Muñoz Rodríguez, Inmaculada Silva Palacios, and Francisco Gallardo López. 1996. "Pollen Production in Anemophilous Trees." *Grana* 35 (1): 38–46.

Tormo-Molina, Rafael, José-María Maya-Manzano, Inmaculada Silva-Palacios, Santiago Fernández-Rodríguez, and Ángela Gonzalo-Garijo. 2015. "Flower Production and Phenology in Dactylis Glomerata." *Aerobiologia* 31 (4): 469–79. https://doi.org/10.1007/s10453-015-9381-7.

Torres-Ruiz, José M., Antoine Kremer, Madeline R. Carins Murphy, Tim Brodribb, Laurent J. Lamarque, Laura Truffaut, Fabrice Bonne, Alexis Ducousso, and Sylvain Delzon. 2019. "Genetic Differentiation in Functional Traits among European Sessile Oak Populations." *Tree Physiology* 39 (10): 1736–49.

Valencia, Susana. 2004. "Diversidad Del Género *Quercus* (Fagaceae) En México." *Botanical Sciences*, no. 75: 33–53.

Valladares, Fernando, Silvia Matesanz, François Guilhaumon, Miguel B. Araújo, Luis Balaguer, Marta Benito-Garzón, Will Cornwell, et al. 2014. "The Effects of Phenotypic Plasticity and Local Adaptation on Forecasts of Species Range Shifts under Climate Change." *Ecology Letters* 17 (11): 1351–64. https://doi.org/10.1111/ele.12348.

Varela, M. C. (Estacao Florestal Nacional-Quinta do Marques. 1996. "Phenological Phases of Quercus Suber L. Flowering." *International Journal of Forest Genetics*. http://agris.fao.org/agris-search/search.do?recordID=SK1997000612.

Vázquez, Francisco María. 1998. *Semillas del género Quercus L.: (biología, ecología y manejo): (de especial interés para la Península Ibérica).* [*Seeds of the Quercus L. genus: (biology, ecology and management): (of special interest to the Iberian Peninsula).*] https://dialnet.unirioja.es/servlet/libro?codigo=184691.

Velasco-Jiménez, María José, Manuel Arenas, Purificación Alcázar, Carmen Galán, and Eugenio Domínguez-Vilches. 2015. "Aerobiological and Phenological Study of *Pistacia* in Córdoba City (Spain)." *Science of The Total Environment* 505 (February): 1036–42. https://doi.org/10.1016/j.scitotenv.2014.10.017.

Vitasse, Yann, Sylvain Delzon, Caroline C. Bresson, Richard Michalet, and Antoine Kremer. 2009. "Altitudinal Differentiation in Growth and Phenology among Populations of Temperate-Zone Tree Species Growing in a Common Garden." *Canadian Journal of Forest Research* 39 (7): 1259–69.

Walker, J. J., K. M. de Beurs, and G. M. Henebry. 2015. "Land Surface Phenology along Urban to Rural Gradients in the U.S. Great Plains." *Remote Sensing of Environment* 165 (August): 42–52. https://doi.org/10.1016/j.rse.2015.04.019.

Wallner, Michael, Anja Erler, Michael Hauser, Eva Klinglmayr, Gabriele Gadermaier, Lothar Vogel, Adriano Mari, Barbara Bohle, Peter Briza, and Fatima Ferreira. 2009. "Immunologic Characterization of Isoforms of Car b 1 and Que a 1, the Major Hornbeam and Oak Pollen Allergens." *Allergy* 64 (3): 452–60.

Wielgolaski, F. E. 1999. "Starting Dates and Basic Temperatures in Phenological Observations of Plants." *International Journal of Biometeorology* 42 (3): 158–68.

Yli-Panula, Eija, Desta Bey Fekedulegn, Brett James Green, and Hanna Ranta. 2009. "Analysis of Airborne Betula Pollen in Finland; a 31-Year Perspective." *International Journal of Environmental Research and Public Health* 6 (6): 1706–23.

Zerboni, R., P. V. Arrigoni, M. Manfredi, M. Rizzotto, L. Paoletti, and C. Ricceri. 1991. "Geobotanical and Phenological Monitoring of Allergenic Pollen Grains in the Florence Area." *Grana* 30 (2): 357–63.

Zhang, Xiaoyang. 2015. "Reconstruction of a Complete Global Time Series of Daily Vegetation Index Trajectory from Long-Term AVHRR Data." *Remote Sensing of Environment* 156 (January): 457–72. https://doi.org/10.1016/j.rse.2014.10.012.

BIOGRAPHICAL SKETCH

Herminia García-Mozo

Affiliation: Department of Botany, Ecology and Plant Physiology, University of Cordoba, Córdoba, Spain. Email: bv2gamoh@uco.es

Education: Doctor in Biological Sciences. University of Cordoba, Spain. Post-doctoral contracts at the CNRS in Montpellier France (2005-, 2006) and at the University of Poznan (Poland) (2003) and being an Assistant Professor at the University Pablo de Olavide of Seville (2004/2005).

Business Address: Department of Botany, Ecology and Plant Physiology, University of Cordoba, 14071, Córdoba, Spain.

Research and Professional Experience: Doctor in Biological Sciences. University of Cordoba, Spain. Post-doctoral contracts at the CNRS in Montpellier France (2005-, 2006) and at the University of Poznan (Poland) (2003) and being an Assistant Professor at the University Pablo de Olavide of Seville (2004/2005). Researcher at the University of Córdoba (2006-2015). Full Professor University of Cordoba from 2015.

Professional Appointments: Currently belongs to the research group RNM-130 of the UCO, has recognized 3 research six years, 65 articles in JCR, 13 in national journals, 6 books and 16 book chapters, Standardized Impact Factor: 1, 24, Index H: 28. Participation in 37 publicly funded research projects, IP in 5 of them, and in 16 research contracts funded by private, national and foreign companies, and public administrations. 95 communications in international congresses, 7 invited papers

Honors: Associate editor of the international scientific journals Aerobiologia, International Journal of Aerobiology (2.02 Impact Index) and Advances in Meteorology (1,577 Impact Index). Member of the Editorial

Board of 7 international scientific journals. Member of National and International Associations of Aerobiology, Phenology and Botany. Organization of International Scientific Congresses.

Publications from the Last 3 Years:

1) Herminia García-Mozo, Rocío López-Orozco, Carolina Canalejo, Jose Oteros “Indoor biological particles inside a train atmosphere.” *Aerobiologia* 2020 In Press.
2) San Miguel, E. G. Hernández-Ceballos, M. A., De Felice, L., García-Mozo, H., Bolívar, J. P.
Evidences of different meteorological patterns governing ^{7}Be and and ^{210}Pb surface levels in the southern Iberian Peninsula. *Journal of Environmental Radioactivity* 198(2019):1-10. doi: https://doi.org/10.1016/j.jenvrad.2018.12.007.
3) García-Mozo H., Hernández-Ceballos M. A., Trigo M. M., Galán C. Wind dynamics' influence on south Spain airborne olive-pollen during African intrusions. *Science of the Total Environment* 609 (2017) 1340–1348. http://dx.doi.org/10.1016/j.scitotenv.2017.08.005.
4) Herminia García-Mozo Poaceae pollen as the leading aeroallergen worldwide: A review. *Allergy*, *72*(12), 1849-1858. 2017 http://dx.doi.org/10.1111/all.13210.
5) Hidalgo, M. D., Herminia García-Mozo, Jose Oteros, & Carmen Galán, Phenological Behaviour of Early Spring Flowering Trees in Spain in Response to Recent Climate Changes. *Theoretical and Applied Climatology*. 2017. 132(1), 263-273 DOI: 10.1007/s00704-017-2089-6.
6) Jose Oteros, Rosa Mª Valencia, Sara del Río, Ana Mª Vega, Herminia García-Mozo, Carmen Galán, Pablo Gutiérrez, Paolo Mandrioli and Delia Fernández-González. Concentric Ring Method for generating pollen maps. Quercus as a case study. *Science of the Total Environment Volume 576*, 15 January 2017, Pages 637–645. https://doi.org/10.1016/j.scitotenv.2016.10.121.

In: Quercus: Classification, Ecology and Uses ISBN: 978-1-53618-026-8
Editor: Bent J. Steffensen

Chapter 4

COMPLEX STUDIES ON THE *QUERCUS* L. (FAGACEAE) GENUS SPECIES DURING THEIR INTRODUCTION IN THE BOTANICAL GARDEN OF THE SOUTHERN FEDERAL UNIVERSITY, ROSTOV-ON-DON, RUSSIA

Vasiliy A. Chokheli**, Boris L. Kozlovsky, Pavel A. Dmitriev, Denis P. Kupriushkin, Michael V. Kuropyatnikov, Vishnu D. Rajput, Tatyana V. Varduny, Maria A. Ignatova, Ekaterina P. Tarik, Igor V. Kornienko and Vladimir S. Lysenko

Botanical Garden of Academy of Biology and Biotechnology,
Southern Federal University, Rostov-on-Don, Russia

* Corresponding Author's Email: vachokheli@sfedu.ru.

ABSTRACT

Deciduous species of *Quercus* L. genus were studied during their introduction in the Botanical garden of Southern Federal University, Russia. Ecological parameters of 17 species and varieties of *Quercus* L., including winter hardiness, drought resistance, phenology, genetic data, resistance to diseases and pests were evaluated in a complex research. In addition, the hyperspectral survey data were tested as a tool for remote identification of oak species. *Q. macrocarpa*, *Q. hartwissiana*, *Q. iberica*, *Q. robur var. tardiflora* Czern. were shown to be the most promising for the regional (North-Caucasus, Russia) cultivation. The results have suggested that hyperspectral vegetation indices REP_Li and CL may be successfully applied for the remote identification of *Q. robur* species using unmanned aerial vehicles. It has also been shown that *Q. robur var tardiflora* samples of different origin have a high degree of genetic similarity when studied using the ISSR method.

Keywords: *Quercus*, plant introduction, plant phenology, winter hardiness, drought resistance, resistance to diseases and pests, hyperspectral survey, vegetation indices, genetic studies, ISSR method

1. INTRODUCTION

Landscaping of inhabited localities, forestry and agriculture in the steppe zone of South Russia need stable tissue samples obtained from long-living species of deciduous trees and suitable for microclonal propagation. They are necessary for the creation of green spaces in cities and villages, for the cultivation of forests which are less flammable compared with the coniferous forests and necessary for cultivation of forests having ameliorative significance.

Such a tissue samples may be found among the plants of *Quercus* L. genus naturally grown in the Rostov region (Zozulin, 1992) and also among the plants of widely cultivated pedunculate oak (*Q. robur* L.) (Kozlovsky, 2009). *Q. robur* L. grows well in floodplain and bayrack forests. However, its size, life longevity and growth rate do not meet the requirements of landscaping in the conditions of flat hill and steppe areas of the region.

Thus, life span of the petiolate oak grown in urban plantings of Rostov-on-Don (Russia) is only 70-80 years. In this consideration, the phenological forms of *Q. robur* are of great interest. Their forms differing in terms of the timing of the onset of phenological phases were described by Chernyaev (1858). In particular, *Q. robur var. praecox* Czern. was found to be the early form and *Q. robur var. tardiflora* Czern – the later form. Significant differences in ecology, growth, and development in these forms were revealed in a number of works (Veresin, 1958; Efimov, 1967; Milenin, 1997; Molchanov, 2011, 2012; Rubtsov, et al., 2007; Silchenko, 2012).

Q. robur var. tardiflora is a promising form for landscaping that is widely used to form tree plantings in European countries (Coutinho et al., 2014). Long-term studies conducted in the Botanical garden of the Southern Federal University (Rostov-on-Don, Russia) showed that *Q. robur var. tardiflora* Czern. is a more promising variety than *Q. robur var. praecox* Czern. for landscaping settlements and afforestation in the region. In this relation, studies on the genotypes of oak phenological forms were suggested as an important task (Pirko et al., 2018, Chokheli et al., 2016). It is obvious that a clear genetic identification of *Q. robur var. tardiflora* will allow to identify other closely related varieties.

Quercus L. plants form broad-leaved forests and they are widely used to create artificial plantings. Wood of many forms of *Quercus* plants is of great economic importance. There are about 10 million hectares of *Quercus* L. trees that have been preserved in the temperate climate zone of Europe, including 3.7 million hectares in Russia. At the same time, *Quercus* trees belong to one of the types of rapidly disappearing forest ecosystems (Tsaralunga, 2005). Therefore, the natural and artificial forests of *Quercus* L. require to be monitored from the point of view of their environmental conditions. Large areas of their plantings allow monitoring using the standard techniques. In general, this is a rather expensive and time-consuming process, and therefore there are great perspectives for using the corresponding remote sensing data. Recent increase in the development of unmanned aerial vehicles (UAVs) and appearance of small-sized remote sensing cameras suitable as platforms for remote sensing occurs due to their fast development (Zarcotejada et al., 2013, Sankaran et al., 2015). At the same time, use of UAVs gives several advantages compared to satellites. In particular, they are: a more rapid data acquisition; a higher spatial and spectral (in the case of using the hyperspectral equipment) resolution.

In this work we presented a complex study on deciduous species of the genus *Quercus* L. and revealed the most promising species for the regional cultivation it terms of stability, life longevity and economic qualities. In addition, we showed, that hyperspectral vegetation indices REP_Li и Cl may be applied for the identification of *Quercus* species and varieties using unmanned aerial vehicles. Parallel genetic studies showed the possibility of the identification of *Q. robur. var. tardiflora* plants of different origin using five ISSR-primers.

2. Materials and Methods

2.1. Research Region and Climate Conditions

The study of samples of the genus *Quercus* L. was conducted in the Botanical garden of the Southern Federal University (SFedU), Russia, Rostov-on-don, located 47^{0}14'02"N 39^{0}38'43"E.

Figure 1. Geographical location of the Rostov region.

Rostov region is located in the area of moderately continental and arid climate. Average annual precipitation is 548 mm, most of which falls during the frost-free period. Summer is hot; maximum temperature could reach up to +40°C. In general, the winters are *mild to moderately cold.* The average temperature in January is -5°C, average absolute minimum air temperature is -20-25°C, absolute minimum temperature is –32°C. The growing season lasts 216 days (from April 1 to November 4), and the frost - free period lasts 258 days in average.

2.2. Objects of Study

The following deciduous oaks species were involved in the introduction test: *Q. castaneifolia* C. A. Mey., *Q. dentata* Thunb., *Q. frainetto* Ten., *Q. hartwissiana* Stev., *Q. iberica* Stev. ex M. Bieb., *Q. libani* G. Olivier, *Q. macranthera* Fisch. & C.A. Mey. ex Hohen., *Q. macrocarpa* Michx., *Q. macrocarpa subsp. oliviformis* (F. Michx.) A. Camus, *Q. pedunculiflora* K. Koch, *Q. robur var. praecox* Czern., *Q. robur var. tardiflora* Czern., *Q. robur 'Fastigiata'*, *Q. robur 'Fennessii'*, *Q. rubra* L., *Q. serrata* Murray and *Q. variabilis* Blume.

Table 1. Distribution of sampled plant species by calendar dates of budding in 2017

Sample	Date of bud burst	Number of samples
Tardiflora1	14.04	1
Tardiflora2	15.04	1
Praecox1- Praecox5	05.04	5
T1S11.04	11.04	1
T1S12.04- T3S12.04	12.04	3
T1S14.04- T4S14.04	14.04	4
T1S17.04- T7S17.04	17.04	7
T1S24.04	24.04	1
T1S02.05, T2S02.05	02.05	2
T1S08.05- T8S08.05	08.05	8

Six-year-old oak seedlings of *Q. macrocarpa*, *Q. robur* and *Q. rubra* were selected as the objects of the hyperspectral study. All seedlings were grown in the same soil and climatic conditions at the introduction nursery of the Botanical garden, SFedU. Five plants were randomly selected for each variant of experiment.

Two samples of *Q. robur var tardiflora* from the artificial population of *Q. robur* grown on the territory of the Botanical garden were isolated for genetic analysis. The first of two these samples that was planted in 1888, was designated as Tardiflora1. The second sample planted in 1975, was designated as Tardiflora2. Five samples of *Q. robur var. praecox* were randomly selected from an artificial oak tree population planted in 1930 (Praecox 1 - Praecox5).

In addition, 26 specimens of three-year-old seedlings, that have been grown from seeds of the Tardiflora1 plants, were selected for analysis and designated as T1S11.04 or T5S17.04, where the first letter denotes the phenological form (Tardiflora) and the second letter S f denotes seed generation followed by the date of onset of blooming in 2017 (Table 1).

2.3. Research Methods

2.3.1. Introduction Test

Oak plants were evaluated for their ecological and biological properties – winter hardiness, drought resistance, seed reproduction, resistance to diseases and pests during the introduction test. For this purpose, an Ogorodnikov`s (1993) point scale was used. The observations were performed since 1970.

Scale of plant winter hardiness assessment is as follows:

1) *Not hardy plant:* being without protection, it freezes to the surface of the soil or snow, poorly recovered, exists mainly in a vegetative state, short-lived.
2) *Slightly hardy plant:* one - and two-year shoots and flower buds freeze or shrink periodically, it is freezes to the surface of the soil or snow in extreme winters, it is able to recover and blooms occasionally
3) *Medium-hardy plant:* part of the annual growth and flower buds periodically freeze or skeletal branches are damaged, but the plant is well restored, blooms and bears fruits

4) *Hardy plant:* only the tops of individual shoots or (in extreme winters) part of the flower buds are damaged
5) *Highly resistant Plant:* no damage is observed.

Visual assessment of winter hardiness was conducted twice a year during the season: at the beginning of the growing season (end of April – beginning of May) and in mid-summer, when it is possible to determine the degree of regeneration of lost parts. The final conclusion on winter hardiness of plants was made after critical winters.

Scale of assessment of drought resistance of plants is as follows:

1) *Not drought-resistant plant*: under the influence of drought, growth is suppressed, leaves and shoots dry up, grow only when watered, but suffer from air drought and high temperature at that.
2) *Slightly drought-resistant plant*: growth is weak, leaf burns, underdevelopment of seeds and buds, plants need regular watering.
3) *Medium-drought-resistant plant*: it develops satisfactorily in the normal years, in the dry years its growth rate is changed, leaves are partially damaged, and periodic watering is required.
4) *Drought-resistant plant*: it grows well and develops without watering, it tolerate drought without damaging the above-ground organs (premature lowering of part of the leaves is possible), buds and seeds develop normally, the plant grows well, blooms and bears fruit occur in the year following the drought.
5) *Highly drought-resistant plant*: it successfully develops without watering, including on very dry and warmed soils.

Scale assessment of plant resistance to diseases and pests:

1) Frequent and very severe lesions that cover most of the plant body and inhibit its growth and development.
2) Periodic severe damage that significantly weakens growth and development;

3) Periodic moderate damage affecting mainly the vegetative organs.
4) Weak damage, covering no more than a fourth part of the plant in a small number of individuals, not significantly affecting the development.
5) Single events of damage or no damage

Observations of effects induced in plants by diseases and pests were carried out regularly, simultaneously with the phenological observations.

The rating scale of seed reproduction:

Rating scale of seed reproduction:

1) Plants do not bloom when they reach adulthood.
2) Plants bloom poorly, do not bear fruits, seeds or seeds do not germinate.
3) Plants bloom moderately, but quantity of seeds is lowered, or seeds have a low germination ability, plants can be propagated vegetatively
4) Plants bloom and bear fruit well, sometimes abundantly, seeds have a high germination ability, but self-seeding in the rainfed conditions is impossible.

Phenological observations and data processing were performed using a standard method (Zaitsev, 1981). Phenological cycle of plants was compared to the matrix of optimal timing of phenological phases of woody plants grown in Rostov-on-don (Kozlovsky et al., 2014).

2.3.2. Hyperspectral Studies

Spectral characteristics were studied using Cubert UHD 185 frame hyperspectral camera (Bareth et al., 2015; Aasen et al., 2015). Hyperspectral surveys were performed at the open air, from the South-Eastern side of studied area, in the cloudless daylight period from 11.00 to 13.00. The obtained data were presented in one panchromatic (1000 × 1000 pixels in size), and in 125 hyperspectral images (50 × 50 pixels in size). Spatial resolution of the obtained hyperspectral data was up to 35 mm^2. Imaging of each object was performed 5 times and the obtained data were averaged. Hyperspectral survey was performed 22.08.2019, 05.09.2019 and 13.09.2019.

Hyperspectral survey data was processed using "CubePilot" software with the subsequent analysis in the "R" environment for statistical calculations.

2.3.3. Genetic Studies

Extraction of DNA was performed from leaves, which had previously been disinfected and treated with a weak solution of sodium hypochlorite. DNA extraction was performed with a sorbent method using a commercial kit "Sorb-GMO-B" (Syntol, Russia). DNA concentration was determined by Qubit fluorometer 3.0 (Invitrogen, Russia) according to standard methods. DNA concentrations were adjusted to 5 ng/μL.

ISSR-primers were selected using the approach based on the published data (from UBC Primer Set #9) (Zietkiewicz et al., 1994; Godwin et al., 1997; Lopez-Aljornaet al., 2007; Coutinho et al., 2014). Earlier, five the most effective oligonucleotide primers were highlighted with different annealing temperature (Chokheliet al., 2016, 2017, 2018) (Table 2), used in this study.

Table 2. Characteristics of ISSR-primers

Primer name	Primer sequence (5'→3')	Annealing temperature (°C)	Length of fragments (bp)	Number of fragments for the total sample
UBC 811	$(GA)_8C$	53	200-1150	30
UBC 835	$(AG)_8YC$	53.5	250-1250	34
UBC 841	$(GA)_8YC$	52.5	325-1500	36
UBC 857	$(AC)_8YG$	54	225-1400	34
UBC 880	$(GGAG)_4$	49.5	250-1650	36
Total				170

Y - isanypyrimidine (Cytosine (C) or Thymine (T)).

Computer analysis of DNA polymorphism was performed using POPGENE32 software (Yeh et al., 1997) to determine the heterozygosity (H_e) (Nei, 1987) absolute number of alleles (n_a) and effective number of alleles (n_e) (Kimura and Crow, 1964). Based on matrices, binary features were calculated using matrix of genetic differences (Nei, 1972). In the matrix of genetic differences, unweighted pair-group method (UPGMA) was constructed using dendrograms reflecting the degree of relatedness of the studied populations according to the ISSR-spectra by means of Dendroscope and POPGENE32 software (Nei and Li, 1979).

3. Results and Discussion

3.1. Ecological, Physiological and Phenological Characteristics of Plants

Results of evaluation of the main ecological physiological and phenological characteristics obtained in this work, as well as the known botanical characteristics of *Quercus* genus species are presented in the following annotated list:

Q. castaneifolia C. A. Mey. Seeds were obtained from Kiev (Ukraine) in 1978. Sample seeds fell to the ground in 1993. The trees were up to 10 m high, winter hardiness 3, drought resistance 4, resistance to pests and diseases 4, seed reproduction 1. Blooms IV–V, fruitless. Homeland – North Caucasus, Iran.

Q. dentata Thunb. Seeds were obtained from Kiev (Ukraine) in 1977, from Bishkek (Kyrgyzstan) in 1989. Trees were up to 15 m high, winter hardiness 3, drought resistance 4, resistance to pests and diseases 4, seed reproduction 2. Blooms V, fruits IX-X. Homeland – Dalniy Vostok (Russian Far East), Japan, Korea, Northwest China.

Q. frainetto Ten. Seeds were obtained from Bishkek (Kyrgyzstan) in 1989. Tree were up to 15 m high, winter hardiness 3, drought resistance 3, resistance to pests and diseases 4, seed reproduction 1. Blooms V, fruitless. Homeland – Balkans, Turkey, Italy.

Q. hartwissiana Stev. Seeds were obtained from Tashkent (Uzbekistan) in 1982. Tree heights were up to 20 m, winter hardiness 5, drought resistance 4, resistance to pests and diseases 4, seed reproduction 3. Blooms V, fruits IX-X. Homeland – Asia Minor.

Q. iberica Stev. ex M. Bieb. Seeds were obtained from Tashkent (Uzbekistan) in 1982. Tree heights were 20 m, winter hardiness 4, drought resistance 4, resistance to pests and diseases 4, seed reproduction 3. Blooms V, fruits IX-X. Homeland – Asia Minor.

Q. libani G. Olivier. Seeds were obtained from Kiev (Ukraine) in 1977, Bishkek (Kyrgyzstan) in 1989. Tree heights were up to 6 m, winter hardiness 3, drought resistance 5, resistance to pests and diseases 4, seed reproduction 2. Blooms V, fruits X. Homeland – Asia Minor.

Q. macranthera Fisch. & C. A. Mey. ex Hohen. Seeds were obtained from Kiev (Ukraine) in 1977, from Tashkent (Uzbekistan) in 1986. Origin of the sample seeds planted in 1936 is unknown. Tree heights were up to 15 m, winter hardiness 4, drought resistance 4, resistance to pests and diseases 4, seed reproduction 3. Blooms V, fruits X. Homeland – Caucasus, North Iran.

Q. macrocarpa Michx. Origin of the seeds is unknown. Samples were stored in the collection since 1936. Tree heights are 28 m, winter hardiness 5, drought resistance 4, resistance to pests and diseases 4, seed reproduction 3. Blooms V, fruits X. Decorative period longevity is 80 years. Homeland – North America.

Q. macrocarpa subsp. oliviformis (F. Michx.) A. Camus. The origin of the seeds is not known. Samples in the collection since 1936. Tree height is 20 m, winter hardiness 5, drought resistance 4, resistance to pests and diseases 4, seed reproduction 3. Blooms V, fruits X. homeland – North America.

Q. pedunculiflora K. Koch. Seeds were obtained from Kiev (Ukraine) in 1981. Sample seeds fell to the ground in 1991. Trees were up to 20 m high, winter hardiness 4, drought resistance 4, resistance to pests and diseases 4, seed reproduction 1. V blooms, no fruit. Homeland – Caucasus, Asia Minor.

Q. robur var. praecox Czern. Seeds were obtained in Rostov-on-don (Russia) in 1936. Tree height is up to 26 m, winter hardiness 5, drought resistance 4, resistance to pests and diseases 3, seed reproduction 4. Blooms IV-V, fruits X. homeland – Europe, North Africa, Asia Minor.

Q. robur var. tardiflora Czern. The origin of the seeds is unknown. Sample seeds stored in the collection since 1888. Tree heights were 26 m, winter hardiness 5, drought resistance 4, resistance to pests and diseases 4, seed reproduction 4. Blooms IV-V, fruits X. Homeland-Europe, North. Africa, Asia Minor.

Q. robur 'Fastigiata'. Seeds were obtained by Bila Tserkva (Ukraine) in 1968, the origin of the seeds from which the sample tree was planted (in 1936) is unknown. Tree heights were up to 20 m, winter hardiness 5, drought resistance 4, resistance to pests and diseases 3, seed reproduction 3. Blooms IV-V, fruits X. Decorative period longevity is 75 years.

Q. robur 'Fennessii. Seeds were obtained from Bishkek (Kyrgyzstan) in 1988. Trees were up to 5 m high, winter hardiness 5, drought resistance 4, resistance to pests and diseases 4, seed reproduction 2. Blooms IV-V, fruits X.

Q. rubra L. Seeds were obtained from Bishkek (Kyrgyzstan) in 1988, the origin of the seeds from which the sample tree was planted (in 1936) is unknown. Tree heights were up to 15 m, winter hardiness 5, drought resistance 3, resistance to pests and diseases 4, seed reproduction 2. Blooms IV-V, fruits X. Homeland – North America.

Q. serrata Murray. Seeds were obtained from Bishkek (Kyrgyzstan) in 1988. the Sample seeds fell to ground in 1993. Trees were up to 2 m high, winter hardiness 4, drought resistance 4, resistance to pests and diseases 4. Homeland – Japan, China.

Q. variabilis Blume. Seeds were obtained from Bishkek (Kyrgyzstan) in 1988. Sample fell to ground in 1993. Tree heights were up to 2 m, winter hardiness 3, drought resistance 4, resistance to pests and diseases 4. Homeland – North China.

Introduction tests showed that the following forms of oaks have the highest resistance to winter stress factors: *Q. macrocarpa, Q. hartwissiana, Q. robur* (*Q. robur var. praecox* и *Q. robur var. tardiflora*), *Q. rubra.* Medium winter hardness (most plants were damaged at the immature stage of ontogenesis) has been found for: *Q. frainetto, Q. variabilis, Q. castaneifolia, Q. dentata, Q. libani.*

Q. libani was the most drought-resistant trees. The following species were of good drought resistance: *Q. castaneifolia, Q. variabilis, Q. hartwissiana, Q. iberica, Q. macrocarpa, Q. macrocarpa subsp. oliviformis.* Low drought-resistance was found for Q. frainetto, Q. rubra, Q. serrata, Q. variabilis that demonstrated growth arrest, chlorosis, chlorosis, leaf burns and fall of unripe fruits.

Regular and abundant fruiting was observed of *Q. robur* (*Q. robur var. praecox*). The break between harvest years does not exceed one or two years. *Q. hartwissiana Q. macrocarpa, Q. macranthera, Q. iberica, Q. macranthera* have long fruiting break (from two to five years). Their productivity was low. Other species bear fruit rarely and not abundantly or were just bloomed.

Q. robur (*Q. robur var. praecox*) plants was most significantly damaged by pest and diseases. In particular, it was damaged by oak leaf wrappers *(Tortrix viridana* L. and *T. loefflingeana* L.*)*, gall wasp *(Andricus curvator* Hart., *A. ostrea* Hart., *A. foecundatrix* Hart., *Diplolepis quercus folli* L.*)*, the mottled umber *(Erannis defoliaria* Cl.*)* and acorn weevil *(Curculio glandium* Marsham*)*. *Q. robur* frequently suffered from lesions of powdery mildew. *Q. libani*, *Q. macrocarpa, Q. macrocarpa subsp. oliviformis, Q. castaneifolia, Q. dentate* and *Q. rubra.* were resistant to pests and diseases, including powdery mildew.

The following list of phenological characteristics describes differences found for two phenological varieties of *Q. robur var. tardiflora* Czern. and *Q. robur var. praecox* Czern. which were isolated in 1858 by V. M. Chernyaev (1858):

1) *Q. robur var. tardiflora develops buds, blooms,* begins to grow shoots 2-3 weeks later than in *Q. r. var. praecox.*
2) *Q. robur var. tardiflora* is less susceptible to pests and diseases;

3) *Q. robur var. tardiflora* it has a taller and more regular crown shape, and a smooth trunk. The business quality of this species wood is higher than that of *Q. r. var. praecox.*
4) *Q. robur var. praecox* is more drought-resistant than *Q. r. var. tardiflora*;
5) Dead leaves from *Q. r. var. tardiflora* do not fall in autumn and remain attached to the shoots the end of winter or early spring.

Among the listed above differences between the two varieties, differences in the timing of the onset of the phenological phases observed in the first half of growing season (budding, beginning of flowering, beginning of growth of shoots) should be considered as a main diagnostic feature. There are two samples of *Q. robur var. tardiflora* revealed in the artificial population of oak petiolate in the Botanical garden of SFedU – the first of the 1888 and the second of the 1975 (first years of planting).-The first sample is the most interesting from the point of view of decorative qualities and life longevity. In contrast to *Q. robur var. praecox*, which enter the senile stage being 60-70 years old, it has no signs of the senile stage.

Currently, this sample *Q. robur var. tardiflora* has a height of 26 m with a trunk diameter of 108 cm. It differs from *Q. robur var. tardiflora variety* described by Chernyaev (1858) in that it sheds leaves in October that increases its decorative qualities.

Q. robur var. tardiflora is close to *Q. robur var. praecox* in terms of stability. In particular, the late onset of phenological phases of *Q. robur var. praecox* affects its stability in the following way:

- the sample can avoid the first generation of oak leaf wrapper (*Tortrix viridana* L.) and other leaf-eating insects due to later leaf-opening times;
- oak requires cross-pollination for good fruiting, that is lowered for this form due to differences in the timing of flowering with other samples of the artificial population. As a result, it blooms regularly, but bears fruits at intervals of two or three years. Seeds are damaged

by seed-eaters by 10-20%. Not damaged seeds have a laboratory germination rate of up to 100%, field germination up to 80%.

- later beginning time of the growing season (later bud break) allow the sample to avoid exposure to late spring frosts.

Calendar dates of occurrence of the main phenological phases of *Q. robur var. precox* and *Q. robur var. tardiflora* compared with the averaged optimal calendar dates for woody plants grown in Rostov region are presented in the Table 3.

Table 3. Timing of phenological phases of *Q. robur* varieties grown in Rostov region

Phenological phases	Timing of the onset of phenological phases, calendar date		
	Q. robur var. praecox	*Q. robur var. tardiflora*	Optimal for Rostov-on-Don
Bud break	14.IV±2.1	21.IV±3.4	1.IV – 25.IV
Beginning of shoot growth	15.IV±2.3	24.IV±2.7	4.IV – 30.IV
Opening of leaves	18.IV±2.1	1.V±2.2	7.IV – 1.V
Appearance of buds	18.IV±2.9	29.V±3.5	3.IV – 15.V
Beginning of flowering	24.IV±2.7	5.V±2.7	14.IV – 1.VI
Beginning of mass flowering	26.IV±2.5	6.V±2.6	16.IV – 5.VI
Full leafing	28.IV±2.1	9.V±2.1	16.IV – 12.V
End of mass flowering	30.IV±2.5	9.V±3.1	20.IV – 15.VI
End of flowering	2.V±2.7	7.V±2.2	24.IV – 23.VI
End of shoot growth	20.V±4.5	29.V±8.2	12.V – 29.VII
Secondary growth of shoots	15.VI±8.2	30.VI±14.6	26.V – 27.VII
Completion of aging of shoots	24.VII±6.5	7.VIII±8.2	24.VI – 29.VIII
Mass seed maturation	4.IX±5.7	16.IX±2.6	20.VII – 6.X
Mass fruit fall	13.IX±8.9	21.IX±6.4	22.VII – 7.XI
Mass autumn leaf staining	5.X±3.4	13.X±8.5	14.IX – 22.X
Mass leaf fall	12.X±1.3	28.X±4.1	2.X – 1.XI

Main phenological phases of *Q. r. var. tardiflora* occur 10-15 days later compared with *Q. r. var. praecox*, and dates of its occurrence are shifted left at the time range of the optimal dates (Table 3). Vegetation longevity (period from budding to mass leaf fall) of this exote is of great importance in conditions of continental climate. All woody plants are distinguished by their vegetation longevity to the following groups: plants with a short vegetation period of less than 170 days; plants with an optimal vegetation period - from 170 to 206 days; plants with a long growing season – more than 206 days. The average vegetation period of woody plants is 188±0.4 days (Kozlovsky et al., 2014). From such a classification, *Q. r. var. precox* and *Q. r. var. tardiflora* have optimal vegetation periods lasting 182 and 190 days, respectively.

The phenological rhythm of the selected sample contributes to its high resistance to the climate stress factors of the region. Sample of *Q. r. var tardiflora is* propagated by autumn seeding. Its phenological features (as a parent plant) appeared to be well repeated in seed generations.

Table 4. Timing of the onset of phenological phases of *Q. macrocarpa* compared with *Q. rubra*

Phenological phases	The timing of the onset of phenological phases, calendar date		
	Q. rubra	*Q. macrocarpa*	Optimal for Rostov-on-Don
Bud break	17.IV	16.IV	01.IV – 25.IV
Beginning of shoot growth	21.IV	23.IV	04.IV – 30.IV
Opening of leaves	21.IV	18.IV	03.IV – 15.V
Appearance of buds	23.IV	19.IV	07.IV – 01.V
Beginning of flowering	29.IV	26.IV	14.IV – 01.VI
Beginning of mass flowering	01.V	29.IV	16.IV – 05.VI
Full leafing	03.V	03.V	20.IV – 15.VI
End of mass flowering	03.V	02.V	16.IV – 12.V
End of flowering	06.V	04.V	24.IV – 23.VI
End of shoot growth	21.V	22.V	12.V – 29.VII
Secondary shoot growth	28.V	15.VI	26.V – 27.VII
Completion of aging of shoots	26.VII	28.VII	24.VI – 29.VII
Mass seed maturation	20.IX	19.IX	20.VII – 06.X
Mass fruit	10.X	07.X	06.IX – 14.X
Mass autumn leaf staining	17.X	20.X	22.VII – 07.XI
Mass leaf fall	20.X	20.X	22.IX – 24.X

The phenological development cycles of two North American oaks – *Q. rubra* and *Q. macrocarpa* do not differ in the location of introduction (except of "secondary shoots growth" phase), that indicates a close phylogenetic relationship between these species. Their phenological phases occur 5-10 days later than in the case of the indigenous species *Q. robur* (*Q. r. var. praecox*). Phenological cycles of *Q. macrocarpa* and *Q. rubra* correspond to the optimal phenological cycle of woody plants inhabiting Rostov region (table 4). This explains high winter hardiness of the studied North American species. At the same time, *Q. macrocarpa* is a drought-resistant species, whereas *Q. rubra* does not tolerate drought well. In addition, *Q. rubra* needs acidic soils and does not tolerate salinization.

3.2. Hyperspectral Data Analysis

List of the 81st spectral index was compiled using the literature data as follows: "Boochs," "Boochs2," "CARI," "Carter2," "Carter3," "Carter4," "Carter5," "Carter6," "CI," "CI2," "ClAInt," "CRI1," "CRI2," "CRI3," "CRI4," "D1," "D2," "Datt," "Datt2," "Datt3," "Datt4," "Datt5," "Datt6," "DD," "DDn," "DPI," "DWSI4," "EGFN," "EGFR," "EVI," "GI," "Gitelson," "Gitelson2," "GMI1," "GMI2," "Green NDVI," "Maccioni," "MCARI," "MCARI2," "MPRI," "MSAVI," "mSR2," "MTCI," "MTVI," "NDVI," "NDVI2," "NDVI3," "NPCI," "OSAVI," "OSAVI2," "PARS," "PRI," "PRI_norm," "PRI*CI2," "PSRI," "PSSR," "PSND," "RDVI," "REP_Li," "SAVI," "SPVI," "SR," "SR1," "SR2," "SR3," "SR4," "SR5," "SR6," "SR8," "Sum_Dr1," "Sum_Dr2," "TCARI," "TCARI/OSAVI," "TCARI2," "TCARI2/OSAVI2," "TGI," "TVI," "Vogelmann," "Vogelmann2," "Vogelmann3," "Vogelmann4." For the further calculations, we selected that indexes whose values were distributed according to the normal law and demonstrated a relatively low coefficient of variation (coefficient of variation CV≤10%). Those indexes were REP_Li (1) и Cl (2) in the experiment (Table 5, Figure 2.).

$$REP\_Li = 700 + 40((R_{re} - R_{700})/(R_{740} - R_{700}),$$
$$R_{re} = (R_{670} - R_{780})/2 \text{ (Guyot and Baret, 1988)} \quad (1)$$

$$CI = R_{675} \text{ x } R_{690}/R^2{}_{683} \text{ (Zarco-Tejada et al., 2003)} \quad (2)$$

Table 5. Values of the coefficient of variation of the REP_Li and SL indexes for different species of oaks

Species	Coefficient of variation (CV), %	
	REP_Li	Cl
Q. robur	0.32%	6.52%
Q. rubra	0.91%	6.71%
Q. macrocarpa	0.50%	6.90%

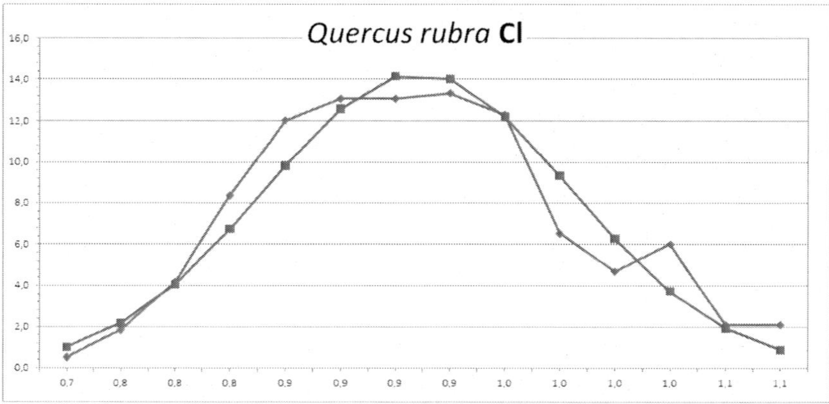

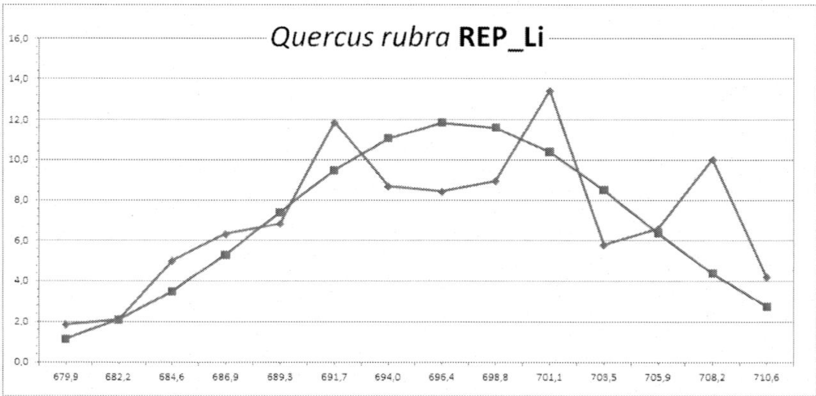

Figure 2. Empirical and theoretical distribution of the values of the REP_Li and Cl indexes measured for Q. *rubra* (as an example).

Analysis of the statistical complex of the trivariant experiment (Q. *macrocarpa*; *Q. robur*; *Q. rubra*) for the Cl index performed by the method of one-factor variance analysis showed that:

- factorial (intergroup) variance is equal to 5.2918, random (intergroup) variance 0,0048;
- influence of the factor was 68% that is statistically significant at P = 99,9%.

Results of the multiple group comparison averages the statistical complex using the Scheffe method are presented in Table 6.

Table 6. Comparison of group averages of Cl index in statistical trivariant complex from *Q. rubra, Q. macrocarpa and Q. robur*

Cl (1 week | 3 week | 4 week)			
	Q. rubra	*Q. macrocarpa*	*Q. robur*
Q. rubra		99.9* | 99.9 | 99.9	99.9 | 99.9 | 99.9
Q. macrocarpa	99.9 | 99.9 | 99.9		<95 | 99.9 | 99.9
Q. robur	99.9 | 99.9 | 99.9	<95 | 99,9 | 99,9	
REP_Li (1 week | 3 week | 4 week)			
	Q. rubra	*Q. macrocarpa*	*Q. robur*
Q. rubra		99.9 | 99.9 | 99.9	99.9 | 99.9 | 99.9
Q. macrocarpa	99.9 | 99.9 | 99.9		99.9 | <95 | 99.9
Q. robur	99.9 | 99.9 | 99.9	99.9 | <95 | 99.9	

* confidence probabilities (%) for differences of group averages of compared oak pairs are specified.

Significant differences in the value of the Cl index with a confidence probability P=99.9% were found for the pairs from the complex "*Q. macrocarpa* - *Q. rubra*" and "*Q. robur* - *Q. rubra*." Difference between "*Q. macrocarpa* - *Q. robur*" has not been proved at the minimal confidence probability allowed in the experiment (P=95%). Analysis of the statistical complex of trivariant experiment (*Q. macrocarpa; Q. robur; Q. rubra*) for the REP_Li index performed using the method of one-factor variance showed that:

- factorial (intergroup) variance is 42156.2, random (intergroup) variance is 30.9;
- influence of the factor was 72.4±0.1% that was statistically significant at P = 99.9%.

In the complex of "*Q. macrocarpa; Q. robur; Q. rubra,*" significant differences in the REP_Li index value with the confidence probability P=99.9% were proved in the pairs "*Q. macrocarpa - Q. rubra,*" "*Q. robur - Q. rubra*" and "*Q. macrocarpa - Q. robur*" (Table 6).

Distribution of REP_Li and Cl densities of *Q. rubra*, *Q. macrocarpa*, and *Q. robur* in the first, third, and fourth weeks of the survey is shown in Figures 3 and 4.

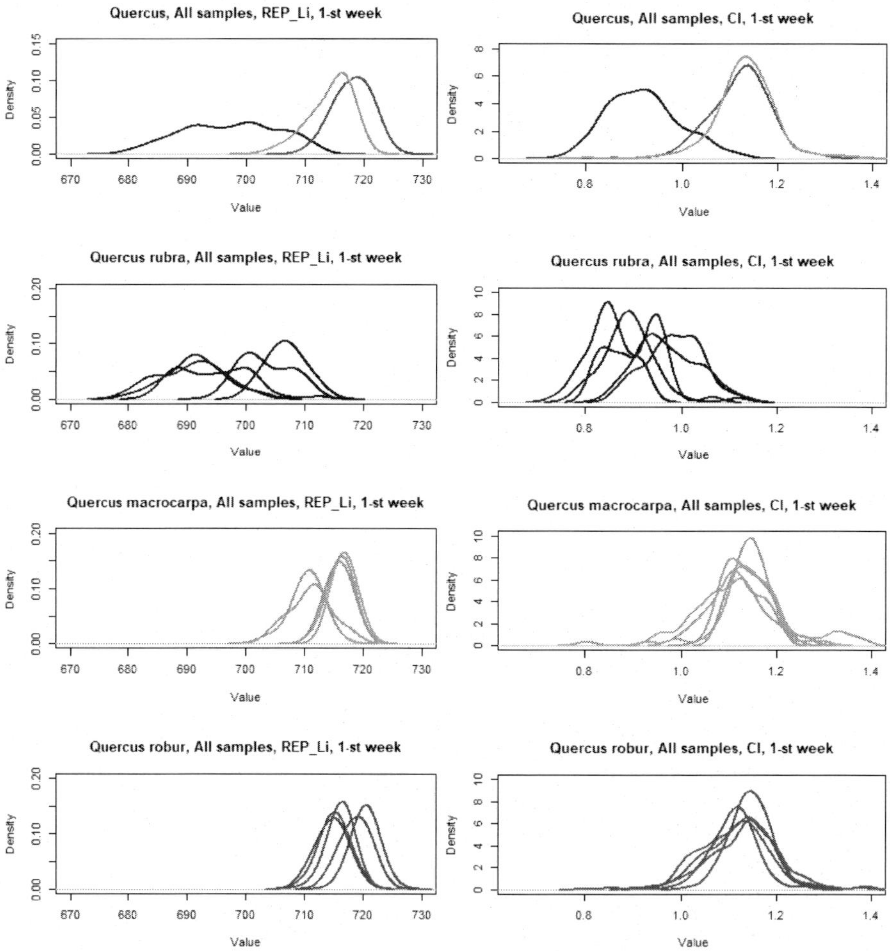

Figure 3. Distribution of REP_Li and Cl densities of *Q. rubra*, *Q. macrocarpa*, and *Q. robur* in the first week.

3.3. Results of Genetic Research

A study on the genetic indicators of the phenological forms of Q. *robur var. tardiflora* and *Q. robur var. praecox*, as well as on 26 samples of seed generation of *Q. robur var. tardiflora* were performed. The results showed that the observed number of alleles (na) in the artificial pedunculate oak population of the Botanical garden was 2.0000, and the effective number of alleles (ne) was 1.4467±0.3071 (mean±SD), the heterozygosity (He) was 0.2780±0.1484 (mean±SD).

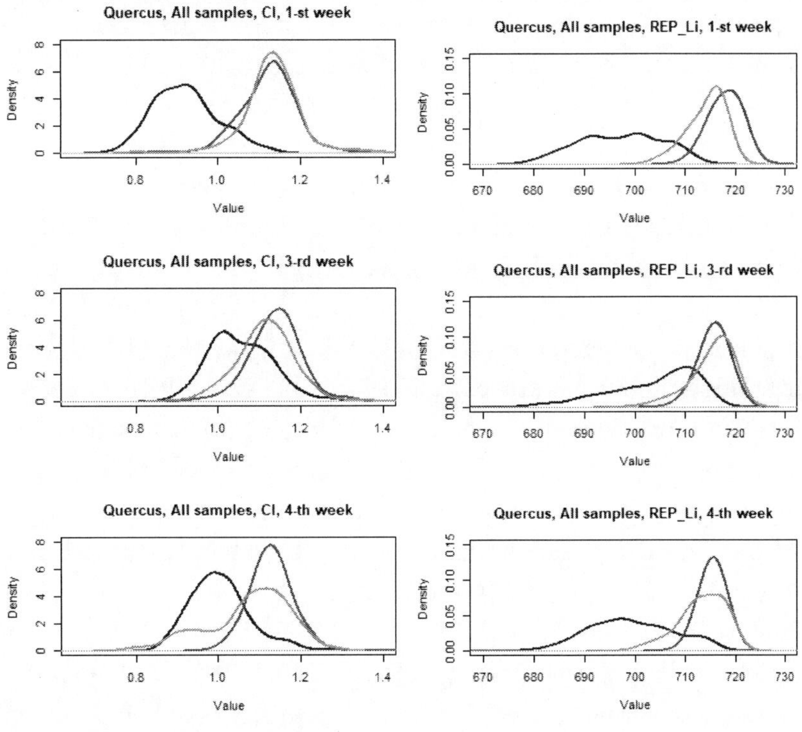

Figure 4. Dynamics of changes in the distribution densities of the REP_Li and Cl of *Q. rubra, Q. macrocarpa,* and *Q. robur.*

In addition, the effective number of alleles (ne) and the average heterozygosity (He) were calculated (using PopGen32 software) separately for samples of *Q. robur var. tardiflora*, including seed generation and samples of *Q. robur var. praecox*. As a result, it was found that the genetic indicators of the group *Q. robur var. tardiflora* (ne=1.3963, He=0.2534) are slightly lower compared with the group *Q. robur var praecox* (ne=1.4489, He=0.2578).

It is interesting, that the values of the main indicators of genetic polymorphism of the early and late forms of pedunculate oaks in the artificial population of the Rostov region are higher than in the planted populations of the Republic of Belarus (He=0.257 and He=0.231, He=0.253 and He=0.243, respectively).

The genetic similarity between these populations was in the range of 0.51 to 0.95 (table 7). UPGMA method has separated the samples into four clades (see dendrogram in Figure 5). Sample T1S11.04 lies outside clades. Both samples were of *Q. robur var. tardiflora* together with all five samples of *Q. robur var. praecox* have formed a common clade.

The maximum genetic similarity between phenological forms was 0.87, and the minimum was 0.60. Genetic similarity between Tardiflora1 and Tardiflora2 was I=0.80. The largest genetic similarity between Tardiflora1 and Rgaesoh1 was I=0.70. The smallest genetic similarity between Tardiflora2 and Praecox2, Tardiflora2 and Praecox5 was I=0.60. The largest genetic similarity between Praecox1 and Praecox4 was I=0.87. The smallest genetic similarity between Praecox2 and Praecox5 was I=0.76.

The other three clades were formed by seed generation samples obtained from Tardiflora1. The maximum genetic similarity between samples of seed generation of the phenological form Tardiflora1 is 0.95, and the minimum is 0.51. Calculated genetic similarity of Nei (I) (table 7) showed that the closest samples were T3S08.05 and T2S12.04 (I=0.95), T5S17.04 and T3S17.04 (I=0.95), T6S17.04 and T7S17.04 (I=0.95).

The most close samples were T2S12.04 and T3S17.04 (I=0.51). The corresponding dendrogram was constructed on the basis of values of genetic distances Nei (D) (Figure 5). Interestingly, that clade # 2 was formed by all samples of the seed generation of the late (May 02 and 08) blooming form (Tardiflora1). Clade # 3 was formed by samples of the seed generation of the late form (Tardiflora1) that bloomed on March 17 and 24. Clade # 4 was formed by samples of the seed generation of the late form(Tardiflora1) that bloomed on March 12 and 14.

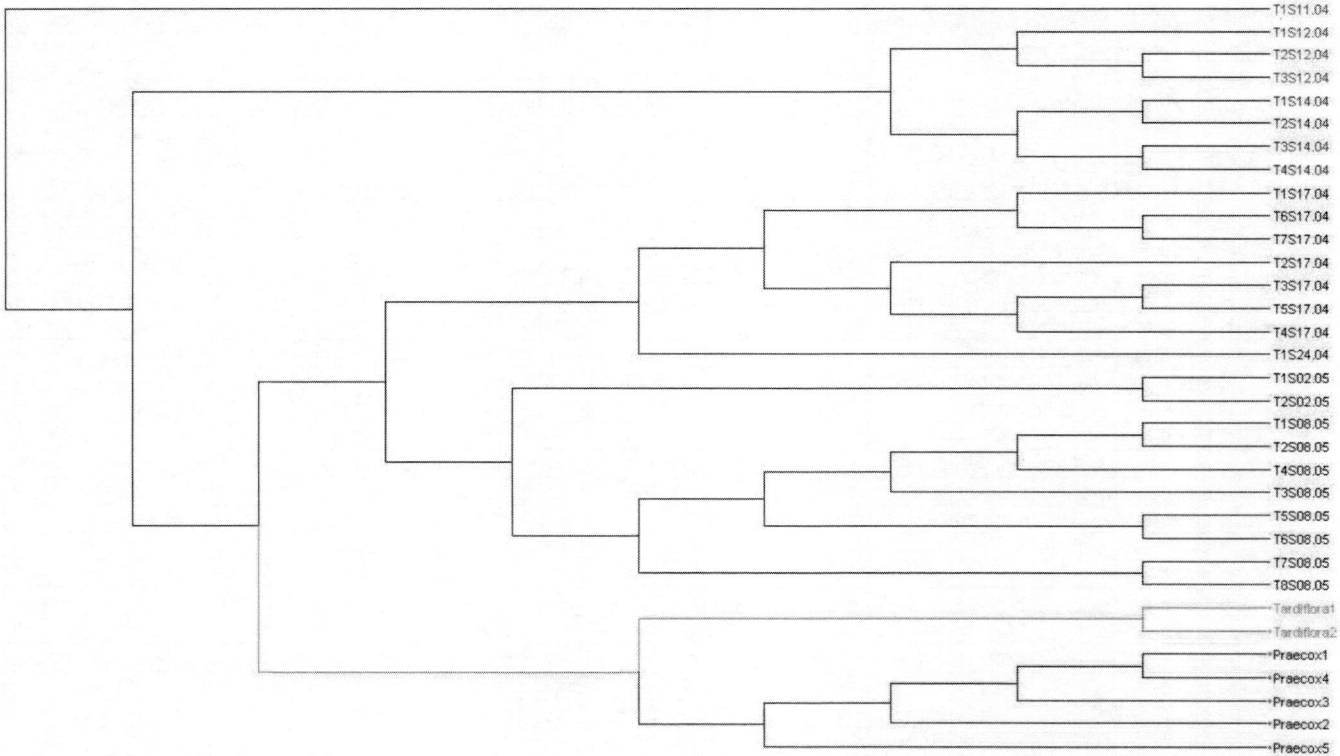

Figure 5. Dendrogram of genetic similarity of samples of pedunculate oak.

Table 7. Table of genetic similarity (I) and genetic distances (D) Nei

	1	2	3	4	5	6	7	8	9	10	11	12	13	14	15	16	17	18	19	20	21	22	23	24	25	26	27	28	29	30	31	32	33
1	***	0,55	0,59	0,58	0,56	0,65	0,61	0,63	0,59	0,57	0,56	0,52	0,56	0,55	0,56	0,64	0,64	0,65	0,65	0,65	0,62	0,64	0,61	0,62	0,56	0,55	0,64	0,61	0,58	0,53	0,58	0,57	0,59
2	0,60	***	0,78	0,77	0,58	0,68	0,61	0,62	0,54	0,58	0,55	0,52	0,56	0,58	0,56	0,56	0,60	0,62	0,60	0,64	0,59	0,63	0,58	0,59	0,56	0,55	0,59	0,56	0,54	0,52	0,55	0,55	0,57
3	0,53	0,25	***	0,81	0,62	0,69	0,65	0,66	0,55	0,57	0,51	0,53	0,52	0,58	0,57	0,59	0,63	0,63	0,64	0,64	0,60	0,65	0,58	0,61	0,58	0,59	0,60	0,56	0,55	0,53	0,58	0,57	0,58
4	0,55	0,26	0,21	***	0,70	0,72	0,70	0,70	0,62	0,63	0,58	0,61	0,63	0,62	0,62	0,61	0,62	0,68	0,64	0,68	0,60	0,66	0,61	0,61	0,59	0,57	0,59	0,56	0,57	0,54	0,59	0,57	0,55
5	0,58	0,55	0,48	0,36	***	0,82	0,75	0,72	0,65	0,60	0,57	0,58	0,58	0,59	0,61	0,58	0,65	0,69	0,65	0,68	0,64	0,68	0,65	0,64	0,62	0,64	0,64	0,54	0,71	0,64	0,65	0,67	0,63
6	0,44	0,39	0,37	0,33	0,20	***	0,79	0,75	0,69	0,62	0,61	0,60	0,62	0,62	0,63	0,65	0,69	0,71	0,71	0,75	0,68	0,71	0,66	0,65	0,68	0,61	0,67	0,60	0,69	0,64	0,68	0,69	0,65
7	0,50	0,49	0,43	0,36	0,28	0,23	***	0,82	0,60	0,59	0,59	0,58	0,59	0,58	0,59	0,58	0,65	0,69	0,69	0,71	0,64	0,65	0,60	0,66	0,68	0,60	0,68	0,63	0,62	0,56	0,59	0,61	0,59
8	0,46	0,47	0,41	0,36	0,33	0,29	0,19	***	0,58	0,56	0,56	0,56	0,58	0,55	0,59	0,61	0,60	0,66	0,65	0,66	0,63	0,61	0,60	0,62	0,62	0,55	0,63	0,58	0,59	0,56	0,56	0,58	0,55
9	0,52	0,61	0,60	0,48	0,44	0,37	0,51	0,55	***	0,81	0,81	0,81	0,79	0,84	0,84	0,74	0,69	0,74	0,67	0,72	0,66	0,70	0,69	0,69	0,68	0,65	0,65	0,56	0,69	0,69	0,74	0,69	0,66
10	0,56	0,55	0,56	0,46	0,51	0,48	0,53	0,57	0,21	***	0,88	0,86	0,87	0,82	0,81	0,75	0,66	0,67	0,61	0,64	0,62	0,66	0,66	0,67	0,70	0,67	0,63	0,56	0,66	0,66	0,72	0,68	0,63
11	0,57	0,60	0,68	0,55	0,56	0,49	0,52	0,58	0,22	0,13	***	0,89	0,95	0,81	0,79	0,75	0,66	0,66	0,64	0,66	0,69	0,67	0,70	0,70	0,73	0,68	0,68	0,61	0,64	0,65	0,71	0,64	0,62
12	0,66	0,65	0,64	0,49	0,54	0,51	0,54	0,58	0,22	0,15	0,11	***	0,90	0,86	0,82	0,76	0,66	0,68	0,65	0,68	0,66	0,67	0,69	0,68	0,72	0,65	0,62	0,56	0,62	0,64	0,68	0,63	0,60
13	0,58	0,57	0,65	0,46	0,55	0,48	0,53	0,55	0,24	0,14	0,05	0,11	***	0,84	0,82	0,75	0,65	0,65	0,64	0,66	0,68	0,66	0,67	0,66	0,71	0,64	0,64	0,59	0,64	0,63	0,69	0,66	0,62
14	0,59	0,54	0,55	0,47	0,52	0,47	0,54	0,60	0,17	0,20	0,21	0,15	0,17	***	0,95	0,72	0,71	0,72	0,69	0,71	0,68	0,69	0,70	0,71	0,68	0,70	0,65	0,60	0,65	0,61	0,69	0,65	0,65
15	0,58	0,57	0,56	0,48	0,49	0,46	0,53	0,53	0,18	0,21	0,23	0,20	0,19	0,05	***	0,73	0,71	0,73	0,67	0,71	0,66	0,66	0,69	0,69	0,66	0,71	0,66	0,61	0,67	0,64	0,70	0,67	0,66
16	0,44	0,57	0,52	0,50	0,55	0,43	0,55	0,49	0,30	0,28	0,29	0,28	0,28	0,32	0,32	***	0,74	0,69	0,68	0,71	0,66	0,68	0,71	0,69	0,71	0,66	0,68	0,63	0,64	0,65	0,66	0,65	0,61
17	0,44	0,51	0,46	0,48	0,44	0,37	0,44	0,51	0,37	0,42	0,41	0,41	0,44	0,34	0,35	0,30	***	0,81	0,74	0,78	0,70	0,74	0,72	0,71	0,66	0,68	0,66	0,59	0,64	0,58	0,64	0,61	0,58
18	0,43	0,47	0,46	0,39	0,37	0,34	0,37	0,42	0,30	0,40	0,41	0,39	0,44	0,32	0,32	0,37	0,21	***	0,80	0,84	0,78	0,77	0,76	0,76	0,71	0,69	0,69	0,58	0,67	0,64	0,69	0,64	0,64
19	0,43	0,51	0,44	0,44	0,44	0,34	0,37	0,44	0,40	0,49	0,44	0,43	0,45	0,37	0,40	0,38	0,30	0,22	***	0,95	0,86	0,86	0,80	0,82	0,72	0,72	0,68	0,58	0,64	0,57	0,65	0,61	0,71
20	0,43	0,45	0,44	0,39	0,38	0,29	0,35	0,42	0,33	0,45	0,41	0,39	0,42	0,34	0,35	0,35	0,25	0,18	0,05	***	0,84	0,85	0,82	0,82	0,74	0,74	0,70	0,61	0,65	0,59	0,68	0,62	0,69
21	0,47	0,52	0,51	0,51	0,44	0,38	0,44	0,46	0,41	0,48	0,37	0,42	0,39	0,38	0,41	0,41	0,36	0,25	0,15	0,17	***	0,82	0,83	0,81	0,78	0,76	0,69	0,59	0,66	0,59	0,66	0,61	0,64
22	0,45	0,46	0,44	0,42	0,39	0,35	0,43	0,50	0,36	0,41	0,40	0,40	0,41	0,37	0,41	0,39	0,31	0,26	0,15	0,16	0,19	***	0,85	0,85	0,78	0,77	0,71	0,59	0,63	0,58	0,67	0,63	0,71
23	0,50	0,55	0,54	0,50	0,44	0,41	0,51	0,51	0,37	0,42	0,36	0,37	0,40	0,36	0,37	0,35	0,33	0,27	0,22	0,19	0,19	0,16	***	0,91	0,81	0,80	0,75	0,59	0,62	0,58	0,65	0,60	0,65
24	0,48	0,53	0,50	0,50	0,45	0,43	0,42	0,47	0,37	0,40	0,36	0,39	0,42	0,34	0,37	0,37	0,35	0,27	0,19	0,19	0,22	0,16	0,10	***	0,82	0,81	0,76	0,62	0,62	0,57	0,65	0,59	0,68
25	0,57	0,58	0,55	0,53	0,48	0,38	0,39	0,48	0,39	0,36	0,32	0,33	0,34	0,38	0,41	0,34	0,41	0,34	0,32	0,31	0,25	0,25	0,22	0,20	***	0,82	0,79	0,64	0,65	0,62	0,66	0,63	0,64
26	0,60	0,59	0,52	0,56	0,45	0,50	0,51	0,59	0,44	0,40	0,39	0,43	0,45	0,36	0,35	0,42	0,38	0,37	0,33	0,30	0,28	0,26	0,22	0,21	0,20	***	0,77	0,62	0,61	0,58	0,64	0,59	0,62
27	0,45	0,52	0,51	0,53	0,44	0,40	0,39	0,46	0,43	0,46	0,38	0,47	0,44	0,44	0,41	0,39	0,41	0,37	0,39	0,36	0,37	0,35	0,29	0,28	0,24	0,26	***	0,80	0,70	0,65	0,66	0,64	0,66
28	0,49	0,58	0,57	0,57	0,62	0,51	0,46	0,54	0,58	0,58	0,49	0,57	0,52	0,51	0,50	0,46	0,52	0,54	0,54	0,50	0,53	0,53	0,52	0,48	0,45	0,48	0,22	***	0,66	0,60	0,64	0,63	0,60

	1	2	3	4	5	6	7	8	9	10	11	12	13	14	15	16	17	18	19	20	21	22	23	24	25	26	27	28	29	30	31	32	33
29	0,54	0,61	0,60	0,56	0,35	0,37	0,47	0,53	0,37	0,42	0,44	0,48	0,45	0,43	0,40	0,45	0,45	0,40	0,45	0,44	0,41	0,46	0,47	0,47	0,43	0,49	0,36	0,41	***	0,81	0,84	0,87	0,79
30	0,64	0,65	0,64	0,61	0,44	0,45	0,58	0,58	0,37	0,41	0,44	0,45	0,46	0,49	0,44	0,43	0,54	0,44	0,56	0,52	0,53	0,55	0,54	0,56	0,47	0,54	0,44	0,51	0,22	***	0,84	0,82	0,76
31	0,55	0,60	0,55	0,53	0,43	0,38	0,52	0,58	0,31	0,32	0,35	0,38	0,37	0,37	0,36	0,41	0,44	0,37	0,43	0,39	0,42	0,40	0,43	0,43	0,42	0,44	0,42	0,45	0,17	0,18	***	0,86	0,81
32	0,56	0,59	0,56	0,56	0,40	0,37	0,49	0,55	0,37	0,38	0,44	0,46	0,42	0,43	0,40	0,44	0,49	0,45	0,49	0,47	0,50	0,46	0,51	0,53	0,46	0,53	0,44	0,46	0,14	0,20	0,15		0,84
33	0,53	0,56	0,55	0,59	0,46	0,44	0,52	0,60	0,41	0,46	0,47	0,51	0,48	0,44	0,41	0,50	0,54	0,44	0,34	0,37	0,45	0,35	0,43	0,39	0,45	0,48	0,42	0,51	0,23	0,27	0,21	0,17	***

*** genetic similarity above the diagonal, genetic distances below the diagonal.

1-T1S11.04, 2-T1S12.04, 3-T2S12.04, 4-T3S12.04, 5-T1S14.04, 6-T2S14.04, 7-T3S14.04, 8-T4S14.04, 9-T1S17.04, 10-T2S17.04, 11-T3S17.04, 12-T4S17.04, 13-T5S17.04, 14-T6S17.04, 15-T7S17.04, 16-T1S24.04, 17-T1S02.05, 18-T2S02.05, 19-T1S08.05, 20-T2S08.05, 21-T3S08.05, 22-T4S08.05, 23-T5S08.05, 24-T6S08.05, 25-T7S08.05, 26-T8S08.05, 27-Tardiflora1, 28-Tardiflora2, 29-Praecox1, 30-Praecox2, 31-Praecox3, 32-Praecox4, 33-Praecox5.

CONCLUSION

Long term introduction tests performed in Botanical Garden of Southern University showed that the most promising oak species for cultivation in the Rostov region are *Q. macrocarpa, Q. hartwissiana, Q. iberica,* and *Q. robur var. tardiflora* Czern. Being resistant for the climate stress factors, diseases and pests, they grow quickly, have a long life span and high decorative qualities.

Our hyperspectral remote sensing studies on vegetation indices REP_Li and Cl demonstrated significant difference between oak species *Q. macrocarpa*, *Q. robur* and *Q. rubra* that suggest to the possibility of developing a method for remote identification of plant objects using unmanned aerial vehicles.

Plant samples of *Q. robur var. tardiflora* (Tardiflora1 and Tardiflora2), despite their different origin, have high genetic similarity (80%). It is also high (Imin=0.60, Imax=0.70) between *Q. robur var. tardiflora* from *Q. robur var. praecox.* Samples (varieties) of both phenological forms a common clade. They differ significantly from the samples of seed generation of Tardiflora1 (Imin=0.51, Imax=0.95), that may be caused by the self-pollination *Q. robur var. tardiflora,* due to the time discrepancy in the flowering phase between *Q. robur var. tardiflora* from *Q. robur var. praecox.*

ACKNOWLEDGMENTS

The research was financial supported by the project of the Ministry of Science and Higher Education of the Russian Federation within the framework of the state task (Southern Federal University, FENW-2020-0028).

REFERENCES

Aasen, H., Burkart, A., Bolten, A., Bareth, G., 2015. Generating 3D hyperspectral information with lightweight UAV snapshot cameras for vegetation monitoring: from camera calibration to quality assurance. *ISPRS J. Photogramm. Remote Sens.* 108 (5), 245–259.

Bareth, G., Aasen, H., Bendig, J., Gnyp, M. L., Bolten, A., Jung, A., Michels, R., Soukkamäki, J., 2015. Low-weight and UAV-based hyperspectral full-frame cameras for monitoring crops: spectral comparison with portable spectroradiometer measurements. *Photogr. – Fernerkundung – Geoinf.* 1, 69–79.

Caralunga V. V. Tragediya rossijskih dubrav // *Izvestiya vysshih uchebnyh zavedenij "Lesnoj zhurnal."* – 2005. – V. 6. – p. 23–30. [Tragedy of Russian oak groves // *News of higher educational institutions "Forests journal"*]

Chernyaev V. M. *O lesah Ukrainy*. [*About Ukraine forests*] M., 1858.–54 s.

Choheli V. A., Makarenko M. S., Usatov A. V., Markin N. V., Varduni T.V. Optimizaciya ISSR-analiza genoma duba chereshchatogo (Quercus robur L.) // *Nauchno-prakticheskaya konferenciya s mezhdunarodnym uchastiem "Genetika – fundamental'naya osnova innovacij v medicine i selekcii*." [Optimization of ISSR analysis of English oak (Quercus robur L.) // *Scientific and practical conference with international participation "Genetics – a fundamental basis of innovational medicine and selection*] – 2017. – 2-4 noyabrya, Rostov-na-Donu. – s. 40–42.

Chokheli V., Kagan D., Rajput V., Kozlovsky B., Sereda M., Shmaraeva A., Khibuhina T., Fedyaeva V., Shishlova Z., Dmitriev P., Varduny T., Kapralova O., Usatov A. The study of genetic variability of cenopopulations of pedunculate oak (*Quercus robur*) in the Rostov region with the use of ISSR - markers // *Int. J. Agric. Biol.* – 2018. – V. 20 (11). pp. 2544–2548. https://doi.org/10.17957/ IJAB/15.0802.

Chokheli Vasiliy, Kozlovsky Boris, Sereda Mikhail, Lysenko Vladimir, Fesenko Igor, Varduny Tatiana, Kapralova Olga, Bondarenko Elena. Preliminary Comparative Analysis of Phenological Varieties of Quercus robur by ISSR-Markers // *Journal of Botany.* – 2016. – V. 2016. – Article ID 7910451. – 7 p., doi:10.1155/2016/7910451.

Coutinho J. P., Carvalho A. and Lima-Brito J. Taxonomic and ecological discrimination of *Fagaceae* species based on ITS PCR-RFLP // *AoB PLANTS.* – 2014. – V. 7.

Efimov YU. P. *Fenologicheskie formy duba chereshchatogo v usloviyah central'noj lesostepi i ih lesohozyajstvennoe znachenie.* Avtoreferat dissertacii na soiskanie uchenoj stepeni kandidata sel'skohozyajstvennyh nauk. [*Phenological forms of English oak in conditions of Russian central forest-steppe zone and their economic importance.* Dissertation abstract for the degree of candidate of agricultural sciences]. Voronezh. – 1967.

Godwin I. D., Aitken E. A. B., Smith L.W. Application of inter simple sequence repeat (ISSR) markers to plant genetics // *Electrophoresis.* – 1997. – V. 18. – pp. 1524 –1528. https://doi.org/10.1002/elps.11501 80906.

Guyot, G., & Baret, F. (1988). Utilisation de la haute resolu-tion spectrale pour suivre l'etat des couverts vegetaux. *Spectral Signatures of Objects in Remote Sensing,* 287, 279–286. European Space Agency, Nordwijk.

Kagan D. I. i dr. Geneticheskij analiz i uchet plodonosheniya fenologicheskih form duba chereshchatogo // *Sb. nauch. tr. / In-t lesa NAN Belarusi.* – Gomel'. – 2011. – Vyp. 71: Problemy lesovedeniya i lesovodstva. – pp. 219–230. [Genetic analysis and evaluation of fruiting of English oak. // *Proceedings of Forests Institute of Academy of Science of Belarus Republic*].

Kagan D. I. *Populyacionno-geneticheskaya struktura duba chereshchatogo v lesosemennyh plantaciyah i nasazhdeniyah Belorusskogo poles'ya.* Dissertaciya na soiskanie uchenoj stepeni kandidata biologicheskih nauk // Gomel'. – 2012. – 162 s. [*Population-genetic structure of English oak in forest seed plantings of Belarus marshy woodlands.* Dissertation abstract for the degree of candidate of biological sciences].

Kimura M. and Crow J. F. The number of alleles that can be maintained in a finite population // *Genetics.* – 1964. – V. 49. – pp. 725–738.

Kozlovskij B. L., Fedorinova O. I., Kuropyatnikov M. V. Novyj perspektivnyj dlya zelenogo stroitel'stva i lesorazvedeniya v Rostovskoj oblasti obrazec Quercus robur L. // *Politematicheskij setevoj elektronnyj nauchnyj zhurnal Kubanskogo gosudarstvennogo agrarnogo universiteta (Nauchnyj zhurnal KubGAU) [Elektronnyj resurs].* – [New perspectives for green Engineering and forest cultivation. *Multidisciplinary net electronic scientific journal of Kuban state University*] Krasnodar: KubGAU. – 2015. – V. 02(106). – pp. 580–591. – IDA [article ID]: 1061502036. – Rezhim dostupa [Access rules]: http://ej.kubagro.ru/2015/02/pdf/36.pdf.

Kozlovskij B. L., Fedorinova O. I., Kuropyatnikov M. V. Zakonomernosti sezonnogo razvitiya drevesnyh rastenij pri introdukcii v Rostove-na-Donu // *Politematicheskij setevoj elektronnyj nauchnyj zhurnal Kubanskogo gosudarstvennogo agrarnogo universiteta* (Nauchnyj zhurnal KubGAU) [Seasonal development patterns of wood plants introducing in Rostov-on-Don // *Multidisciplinary net electronic scientific journal of Kuban state agricultural University*] [Elektronnyj resurs Electronic sourse]. – Krasnodar: KubGAU, 2014. – №05(099). S. 244 – 255. – IDA [article ID]: 0991405017. –Rezhim dostupa: [Access rules] http://ej.kubagro.ru/2014/05/pdf/17.pdf.

Kozlovskij B. L., Ogorodnikova T. K., Kuropyatnikov M. V., Fedorinova O. I. *Assortiment drevesnyh rastenij dlya zelenogo stroitel'stva v Rostovskoj oblasti*. [*Assortment of wood plants for green engineering in Rostov-on-Don*] Rostov n/D.: Izd-vo YUFU SFedU, 2009. 416 s.

Milenin A. I. *Ekologicheskie osobennosti fenologicheskih raznovidnostej duba chereshchatogo (Quercus robur L.) v usloviyah CCHR*. Avtoreferat dissertacii na soiskanie uchenoj stepeni kandidata sel'skohozyajstvennyh nauk. [*Ecological features of phenological varieties of English oak (Quercus robur L.) in CCHR conditions*] Voronezh. – 1997. – 24 s.

Molchanov A. G. Intensivnost' fotosinteza fenologicheskih form duba chereshchatogo v usloviyah nedostatochnogo uvlazhneniya // *Lesovedenie*. [Intensity of photosynthesis in phonological forms of

English oak in drought conditions // *Forest science* – 2012. – V. 4. – pp. 31–38.

Molchanov A. G. Intensivnost' fotosinteza rano- i pozdno raspuskayushchihsya form duba chereshchatogo v usloviyah nedostatochnogo uvlazhneniya // *Strukturnye i funkcional'nye otkloneniya ot normal'nogo rosta i razvitiya rastenij: Materialy mezhdunarodnoj konferencii.* [Intensity of photosynthesis in phonological forms of English oak in drought conditions // *Structural and functional deviations from normal plant growth and development. Proceedings of international conference.*] Petrazovodsk: KarNC RAN. – 2011. – pp. 204–208.

Nei M. Analysis of gene diversity in subdivided populations // *Proceedings of the National Academy of Sciences of USA.* – 1973. – V. 70. – P. 3321–3323.

Nei M. and Li W. H. Mathematical model for studying genetic variation in terms of restriction endonucleases // *Proc. Natl. Acad. Sci. USA.* – 1979. – V. 76. – pp. 5269–5273.

Nei M. Genetic distance between populations // *Amer. Nat.* – 1972. – V. 106. – pp. 283–292.

Nei M. *Molecular Evolutionary Genetics*. Columbia University Press, New York, USA. – 1987. – 512 p.

Ogorodnikov A. Y. A. Metodika vizual'noj ocenki bioekologicheskih svojstv drevesnyh rastenij naselennyh punktov stepnoj zony // *Itogi introdukcii rastenij*. [A method of imaging of bioelectric properties of wood plants inhabiting the steppe zone settlements // *Progress in plant reproduction]* Rostov n/D.: Izd-vo Rost. un-ta, 1993. S. 50-58.

Pirko Ya. V. et al. Genetic features of the phenological forms of *Quercus robur* (*Fagaceae*) according to the analysis of the introns polymorphism of β-tubulin genes and microsatellite loci // *Ukrainian botanical journal.* – 2018. – V. 75(5). – pp. 489-500.

Rubcov V. V., ZHirenko N. G., Utkina I. A. Intensivnost' toka pasoki u raznyh fenologicheskih form duba v yuzhnoj lesostepi // *Lesovedenie.* [Xylem sup flow rate and oak phenological forms in conditions of forest-steppe zone. *Forest science.*] – 2007. – V. 5. – pp. 44–55.

Sankaran, S.; Khot, R.; Espinoza, Z.; Jarolmasjed, S.; Sathuvalli, R.; Vandemark, J.; Miklas, N.; Carter, H.; Pumphrey, O.; Knowles, R. Low-altitude, high-resolution aerial imaging systems for row and field crop phenotyping: A review. *Eur. J. Agron.* 2015, 70, 112–123.

Sil'chenko I. I. Fenologicheskie formy duba chereshchatogo (Quercus robur L.) v razlichnyh tipah landshaftov bryanskoj oblasti // *Vestnik Bryanskogo gosudarstvennogo universiteta.* [Phenological properties of English oak (*Quercus robur* L.) in conditions of different landscapes of Bryank region // *Bulletin of Bryansk state university*] – 2012. – V. 4. – pp. 158–161.

Veresin M. M. O fenologicheskih formah duba chereshchatogo i ispol'zovanii ih v lesorazvedenii // *Izvestiya vysshih uchebnyh zavedenij. Lesnoj zhurnal.* [About phenological forms of English oak (*Quercus robur* L.) and their use in the forest cultivation. *Proceedings of higher education institutions.*] – 1958. – V. 3. – pp. 3–15.

Yeh F. C., Yang R. C., Boyle T. B. J., Ye Z. H., Mao J. X. POPGENE, the *User-friendly Shareware for Population Genetic Analysis* // Molecular Biology and Biotechnology Center, University of Alberta, Edmonton, Alberta, Canada. – 1997.

Zajcev G. N. *Fenologiya drevesnyh rastenij.* [*Phenology of wood plants*] M.: Nauka, 1981. 120 s.

Zarcotejada, J.; Guilléncliment, L.; Hernándezclemente, R.; Catalina, A.; González, R.; Martín, P. Estimating leaf carotenoid content in vineyards using high resolution hyperspectral imagery acquired from an unmanned aerial vehicle (UAV). *Agric. For. Meteorol.* 2013, 171–172, 281–294.

Zietkiewicz E., Rafalski A., Labuda D. Genome fingerprinting by simple sequence repeat (SSR) - anchored polymerase chain reaction amplification // *Genomics.* – 1994. – V. 20. – pp. 176 –183.

Zozulin G. M. Lesa Nizhnego Dona. [*Forests of Lower Don*] Rostov n/D: Izd-vo RGU, 1992. 203 s.

In: Quercus: Classification, Ecology and Uses ISBN: 978-1-53618-026-8
Editor: Bent J. Steffensen

Chapter 5

MEXICAN OAKS AS FOUNDATION SPECIES: THE CASE OF *QUERCUS CRASSIPES* AND *Q. CASTANEA*

***Leticia Valencia-Cuevas*[1], *Elgar Castillo-Mendoza*[2], *Miriam Serrano-Muñoz*[3] and *Efraín Tovar-Sánchez*[4,*]**

[1]Facultad de Ciencias Biológicas,
Universidad Autónoma del Estado de Morelos,
Cuernavaca, Morelos, México
[2]Laboratorio de Investigaciones Ambientales,
Centro de Investigación en Biotecnología,
Universidad Autónoma del Estado de Morelos,
Cuernavaca, Morelos, México
[3]Universidad Autónoma del Estado de Morelos,
Cuernavaca, Morelos, México
[4]Centro de Investigación en Biodiversidad y Conservación,
Universidad Autónoma del Estado de Morelos,
Cuernavaca, Morelos, México

[*] Corresponding Author's Email: efrain_tovar@uaem.mx.

ABSTRACT

Theory ecology is applied to comprehend how genetic variation within tree dominant species may be an important driver of ecological processes. In this sense, studies performed suggest that the genetic diversity of the foundation species can have robust organizational influence at the population, community and ecosystem levels. Foundation species are a small subset of the total species in an ecosystem; it has been suggested that they should capture most of the variation of the community structure and ecosystem processes. The trees of forest ecosystems are excellent candidates to be considered as foundation species because their architectural, functional, and physiological characteristics define the structure of the forests and can influence the microclimate; their biomass contributes significantly to ecosystem processes. Species of the genus *Quercus* (Fagaceae) are one of the most important tree canopy groups, due to its diversity and dominance in temperate forests in particular. In ecological terms, oaks are important for their ecological functions, also their canopy function as habitats for a wide variety of biological groups, including epiphytic plants, mammals, birds, fungi and arthropods. In particular, the gall-forming wasp belonging to the Cynipidae family (Cynipini Tribe) associated with the genus *Quercus* are considered one of the most specialized groups of herbivorous insects. The galls produced by cynipids form a microcosm with great ecological activity, harboring a closely associated community of guest organisms (including non-gall-inducing cynipids, flies, moths and beetles) and parasitoids. The parasitoids associated with cynipids are specific to them. It has been reported that their communities may be affected by the genetic attributes of plants. Therefore, oaks and the insects associated with the galls oak canopy represent an excellent system to study the effect of genetic variation of foundation species on cynipid communities, their associated parasitoids, and secondary fauna. Mexico is considered one of the most important centers of diversification of the genus *Quercus*, with 161 species (30%). *Quercus crassipes* and *Q. castanea* are two species of red oak, which are a canopy dominant elements of different forest types that support a wide variety of associated species. Therefore, the aim of this chapter is to determine if *Q. crassipes* and *Q. castanea* can be recognized as foundation species. This review denotes that both species are dominant oaks of Mexican temperate forests. They have wide geographic distribution and both species act as important reservoirs of canopy arthropod fauna. Also, we have shown that genetic diversity in both oak species results in extended phenotypes that extends to the community and ecosystem levels, in particular, on the associated community structure of oak – gall – inducing wasp –parasitoids and secondary fauna associated with oak canopy.

Keywords: foundation species, temperate forest, arthropods, gall forming insects

INTRODUCTION

In forested ecosystems, the structural or functional attributes of some species creates and defines ecological communities or ecosystems (Ellison et al. 2005, 479). From this perspective, foundation species are those that define and structure communities providing resources and creating locally stable conditions for species besides participating in the modulation and stabilization of ecosystem processes (Dayton 1972, 85; Ellison et al. 2005, 479; Whitham et al. 2006, 511).

According to Ellison and colleagues (2005, 479; 2010, 168), species with the potential to act as founding species in their ecosystems are those that possess the following characteristics: 1) species locally abundant and regionally common, 2) species with structural or functional characteristics that create habitats for a large number of associated species, and 3) species that modulate core ecosystem processes, as energy flux and biogeochemical cycles. The trees of forest ecosystems are excellent candidates as foundation species because their functional, architectural and physiological characteristics define the structure of the forests with implications in the microclimate. Also, the biomass and chemical constitution of these species contribute significantly to ecosystem processes (Ellison et al. 2005, 479).

Considering Ellison and colleagues' proposal, some species of the *Quercus* genus can be considered excellent candidates as founding species. Oaks (Fagaceae: *Quercus*) are one of most important tree and shrub group in temperate forests and include 531 species (Govaerts and Frodin 1998, 96), widely distributed worldwide and are dominant components of the canopy. Recent forest inventory data from Mexico and the US have revealed that oaks have the highest species diversity and biomass of all woody genera in those two countries (Cavender-Bares 2016, 38). Specifically, Mexico is considered one of the centers of diversification for the *Quercus* genus. They number 161 species, located in the Lobatae section (red oaks; 76 species),

Quercus section (white oaks, 81 species) and Protobalanus section (intermediate oaks, four species) (Valencia 2004, 33). Also, they offer key environmental services such as energy production, water cycle regulation, carbon sequestration, air pollution removal, etc. (Rampant et al. 2011, 1; Nowak et al. 2013, 396). In addition, several studies have revealed a great diversity of associated flora and fauna associated with oak species (e.g., birds, squirrels, arthropods, ectomycorrhizal and saprophytic fungi, lichens, epiphytic plants, among others, Sáenz-Jimenez 2010, 299–334; Cavender-Bares 2016, 39; Melhado-Carboney 2016, 25–49).

Furthermore, oaks show a high genetic diversity (Tovar-Sánchez et al. 2008, 232; Valencia-Cuevas et al. 2014, 646; Müller and Gailing 2019, 785–786) due to the combination of life history features, such as long generation times, woody life form, outcrossing mating system, high fecundity, and pollination by wind (Hamrick et al. 1979; 39), the high frequency of hybridization events that occur in these species (Kremer et al. 2012, 584) and the range of geographical distribution (Hamrick et al. 1992, 111). Genetic diversity in oak populations promote a gamma of resources and conditions (González-Rodríguez et al. 2004; 427; Tovar-Sánchez and Oyama 2004, 1354–1359) that can be exploited by different flora and fauna communities associated with these species (Tovar-Sánchez et al. 2013, 168–176; Valencia-Cuevas et al. 2018, 4–8). In fact, some studies have suggested that genetic diversity influences not only the fitness of unique species but also that its benefits can extend to the associated communities through tropic interactions (Bailey et al. 2006, 78; Valencia-Cuevas et al. 2018, 7; Castillo-Mendoza 2019, 42–84).

The canopies of forest ecosystems (aerial part where all the foliage, buds, fine branches, associated flora and fauna, the suspended soil and the interstices [air] of a forest are combined, Valencia-Cuevas and Tovar-Sánchez 2015, 2) contain most of the photosynthetically active biomass (Lowman 2004, 454). They are therefore the main energy absorption sites in the biosphere (Basset et al. 2003, 183), as well as sites in which an intensive exchange of oxygen, water vapor, and carbon dioxide takes place (Lowman 2004, 454). They are also recognized as habitats that support a wide variety of associated flora and fauna (Nadkarni et al. 2004, 3–23).

One example are arthropods, characterized by their great abundance, diversity, and functional importance. It has been documented that this phylum contains the highest species diversity on earth (from 2.5 to 3.7 million of species, Hamilton et al. 2010, 42), where they play important ecological roles (pollinators, herbivores, predators, parasites, and detritivores) and are involved in processes that maintain ecosystem function (Schowalter and Ganio 2003, 315). Different studies have shown that the oak canopy has a high diversity of arthropods suggesting their importance as habitat for this group of organisms. Ecto and endophagous arthropods have been reported to exist in oak canopy (Tovar-Sánchez et al. 2003, 81–86; Tovar-Sánchez and Oyama 2006a, 705–711; Tovar-Sánchez and Oyama, 2006b, 373–379; Tack et al. 2010, 2660–2672; Castagneyrol et al. 2012, e44247; Valencia-Cuevas et al. 2018, 4–8).

Endophagous insects include gall-forming wasps (Hymenoptera: Cynipidae) which are species with specific tissue and organ specialization in their oak hosts (Stone et al. 2002, 656–657). Galls are defined as abnormal structures of parts of plant tissues or organs (Pujade-Villar 2013, 3) produced by an inducing agent. Approximately 15,000 species of organisms such as viruses, bacteria, algae, fungi, protozoa, rotifers, nematodes, mites and insects are capable of inducing gall formation (Pujade-Villar 2013, 2). However, most galls are induced by insects. The Cinipidae family is divided into two groups: gall inducers (Aylacini, Diplolepidini, Eschatocerini, Pediaspidini, Cynipini) and those associated with galls as inquilines (Synergini) (Abe et al. 2007, 179). Gall wasps are obliged to oviposit inside the tissues of their oak hosts, and the larvae develop in galls induced on specific plant tissues (Stone and Schönrogge 2003, 512–514), representing microhabitats that support communities of specialist inhabitants (Pujade-Villar 2013, 7–12). However, galls induced by cynipid wasps are structures that have extraordinary ecological value because their hosts are not limited to gall inductors but generally support very complex networks that consist of gall inductor, inquiline, parasitoid insects and secondary fauna (Pujade-Villar 2013, 7–12). In particular, gall wasps are frequently associated with a diverse parasitoid community (Stone et al. 2002, 653) included in six families of Chalcidoidea (Eulophidae, Eupelmidae, Eurytomidae,

Ormyridae, Pteromalidae and Torymidae) (Askew 1980, 820; Askew et al. 2013, 83–120), whose role is to regulate the insect host populations (producing from 40 to100% of their mortality) (Cooper and Riesker 2010, 794–795). Finally, secondary fauna is characterized by establishing itself in the mature gall phase. These organisms use the gall to obtain food or as a refuge. In these structures it is common to find a wide variety of arthropods among them: beetles, aphids, formicides, spiders, pseudoscorpions, etc. (Pujade-Villar 2013, 11–12).

Quercus castanea Née and *Q. crassipes* Humb et Bonpl. are two species of red oaks of wide geographic distribution in Mexico. *Q. castanea* trees can be recognized easily in the field by its obovate, oblanceolate leaves, underside veins conspicuously elevated and reticulated, margins with two to five/six short teeth, eight–12 secondary veins on each side of the midvein, graygreenish coloration and fasciculate sessile trichomes (Valencia 1994, 48–49). *Q. crassipes* poses deciduous, coriaceous, narrowly elliptic and lanceolate leaves, their surface is barely lustrous with green or grayish coloration, and the lower surface is tomentose, white grayish and with revoluted margins. It has 10–19 secondary veins on each side of the midvein and stellate stipitate trichomes (Romero-Rangel et al. 2002, 578–579). Both species are dominant in the canopy of temperate forests (Valencia-Cuevas et al. 2014, 643) and constitute the habitat of different species (Tovar-Sánchez and Oyama 2006a, 705–711; b, 373–379; Tovar-Sánchez et al. 2013, 168–176; Melhado-Carboney 2016, 25–49; Valencia-Cuevas et al. 2018, 4–8).

In this chapter we present information that supports the hypothesis that *Q. castanea* and *Q. crassipes* are two oak species that play the role of foundation species in Mexican temperate forests, because both present characteristics as proposed by Ellison et al. (2005). To illustrate, we focus on the canopy associated arthropods with a major emphasis on specific oak–gall inductor–parasitoid interactions and associated secondary fauna reported for the more common gall found in each species.

SPECIES LOCALLY ABUNDANT AND REGIONALLY COMMON

Quercus castanea includes trees from 5 to 15 m in height with a trunk diameter of 30–60 cm. It is located between 1,900 and 3,500 m a.s.l. and found in more than 15 Mexican states (Valencia 2004, 41). Figure 1a shows the distribution of *Q. castanea* in Mexico. This species is found frequently in perturbed areas with a xerophytic scrub type of vegetation, and it is also localized in mountain cloud forests (Bello-González and Labat 1984, 30). Altitudinally, *Q. castanea* is distributed between 1100 to 2600 m. This species is one of the most widely distributed species within the red oak group in Mexico. It occurs in the Sierra Madre Oriental, Sierra Madre Occidental, Central Plateau, Trans-Mexican Volcanic Belt (TVB), Depresión del Balsas and Sierra Madre del Sur (Valencia 1994, 41).

Q. crassipes includes trees from 3 to 20 m in height and 0.40–1 m in trunk diameter. Altitudinally, it is found from 1,900 to 3,500 m in different vegetation types: *Pinus-Quercus*, *Quercus-cupressus*, *Quercus* forests, xerophytic scrubs, in transition between grassland vegetation and mixed forest, mountain cloud forests, and perturbed vegetation (Romero-Rangel et al. 2002, 578–579). It is found in 12 Mexican states, mainly in the southeast part of the Sierra Madre Oriental, in the TVB, although it can also be found in the Sierra Madre del Sur (Valencia 2004, 42). Figure 1b shows the distribution of *Q. crassipes* in Mexico.

Both species are dominant of the canopy of the temperate forests where they inhabit. For example, the density of both oak species (individuals/hectare) has been documented in six template forests through the TVB in Mexico: Parque Nacional El Tepozteco [*Q. castanea* (56.7), *Q. crassipes* (44.1)], Parque Ecológico de la Ciudad de México [*Q. castanea* (149.0), *Q. crassipes* (135.2)], Parque Barranca de Tarango [*Q. castanea* (161.6), *Q. crassipes* (120.1),], Parque Las Peñas [*Q. castanea* (186.7), *Q. crassipes* (150.0)] (Valencia-Cuevas et al. 2014, 643).

Other evidence that supports the dominance of *Quercus crassipes* in Mexican forests is the work of Figueroa-Rangel and Olvera-Vargas (2000,

91–98) who document that this species was the dominant species in thirty-two permanent plots (500 m^2 each) in Cerro Grande, Sierra de Manantlán, Jalisco.

Structural or Functional Characteristics that Create Habitat for a Large Number of Associated Species

Structurally complex habitats provide a wide range of resources and conditions for the establishment of a greater diversity of species (Halaj et al. 2000, 139–140; Ishii et al. 2004, 342–355). When plant communities determine the physical structure of the environment, they have great influence on the structure of animal communities (Strong et al. 1984, 63–68). In the case of species that inhabit the forests, the structure and complexity of their habitat is determined by the abundance, architecture of the vegetal species (e.g., shape and size of leaves, buds, twigs) and taxonomic diversity of vegetal communities (Valencia-Cuevas and Tovar-Sánchez 2015, 22–23). This results in a more complex environment, where the availability of resources and conditions is higher (Novotny et al. 2006, 1115–1118). An attribute of the complexity of oak forests is the level of genetic diversity that species present, because it is known that genetic diversity is the raw material upon which natural selection reacts (Lowe et al. 2004, 52), which promotes the individual's repertoire of phenotypic responses of the organisms. In this sense, it has been suggested that tree populations with higher levels of genetic diversity have the potential to express greater variation in phenotypic attributes such as morphology (Tovar-Sánchez and Oyama 2004, 1354–1359), tree architecture (Bailey et al. 2004, 2100–2102), chemical characteristics (Cheng et al. 2011, 107–117; Caseys et al. 2015, 17), phenology (Hunter and Price 1992, 728), etc.

Studies applying different genetic marker types representing neutral or potentially adaptive genetic variation have revealed that oaks show high genetic diversity levels (Müller and Gailing 2019, 785–786). Authors have suggested that this pattern is due to their life history and ecological characteristics and the high frequency of hybridization events that occur in these species (Kremer et al. 2012, 584). In a review, Hamrick and colleagues (1992, 95–124) used published data of the genetic diversity of 332 woody taxa and reported that woody species with long generation times, outcrossing breeding systems, wind or animal-ingested seed dispersal and large geographic ranges, have more genetic diversity within species and populations than woody species with other combinations of traits. In general, oaks group have the characteristics proposed by Hamrick and colleagues, including *Q. castanea* and *Q. crassipes* which, in part, explain their genetic diversity levels. For example, Tovar-Sánchez et al. (2008, 232) using chloroplast microsatellites in pure populations of *Q. crassipes* showed mean values of 0.49 in the Shannon diversity index. In addition, studies that have evaluated the genetic diversity levels of *Q. castanea* using nuclear microsatellites through its geographical distribution in Mexico (Valencia-Cuevas et al. 2014, 646; Oyama et al. 2018, 4), show *He* values from 0.61 to 0.873. The genetic diversity levels documented by both species are high, according to a recent review, where Müller and Gailing (2019, 786) summarize current literature related to oak genetic diversity.

Hybridization phenomena in the genus *Quercus* occur with high frequency in natural conditions (Whittemore and Schaal 1991, 2540; Curtu et al. 2007, 2; Gailing and Curtu 2014, 5–6) and this is believed to be related to a poor development of sterility barriers between *Quercus* species which facilitate interspecific crosses with closely related species (Whittemore and Schaal 1991, 2540). This phenomenon has contributed to the promotion of complex patterns of foliar morphological variation presented by oaks and that has led to problems in species identification (Tovar-Sánchez and Oyama 2004, 1359; González-Rodríguez and Oyama 2005, 427). Also, it contributes to an increase of genetic diversity levels of involved populations in hybridization events as a result of new genetic combinations in hybrids (Valencia-Cuevas et al. 2015, 1092). Valencia (1994, 156–158) proposed that a group of red oaks conformed by *Q. affinis*, *Q. crassipes*, *Q. crassifolia*, *Q. laurina*, *Q. mexicana*, and *Q. rubramenta* may experience genetic exchange when they occur in sympatric/mixed stands in Mexican forests.

This last scenario has been corroborated by Tovar-Sánchez and Oyama (2004, 1352) for the *Q. crassipes* × *Q. crassifolia* complex. Also, in a recent study Valencia-Cuevas et al. (2015) reported hybridization events between *Q. castanea*, *Q. laurina*, and *Q. crassifolia*. Tovar-Sánchez and Oyama (2004, 1352–1363) documented hybridization between *Q. crassipes* and *Q. crassifolia* in seven mixed populations in Mexico, using RAPDs as molecular markers. Also, in this work it was revealed that these events promote the occurrence of hybrids that presented intermediate morphology between the parental species. In addition, Tovar-Sánchez and colleagues (2008, 232) using chloroplast microsatellites revealed that mixed populations of *Q. crassipes* × Q. *crassifolia* complex had the highest levels of genetic variation (Shannon diversity) in comparison with allopatric populations, suggesting that there is introgression in these stands. For *Q. castanea*, Valencia-Cuevas et al. (2014, 641–652) documented a natural gradient of genetic diversity, using nuclear and chloroplast microsatellites, as increase the richness species of red oak communities in the TVB. Later, Valencia-Cuevas et al. (2015, 1085–1097) evidenced that *Q. castanea* is a species involved in hybridization events with other red oaks: *Q. laurina*, *Q. crassipes*, *Q. crassifolia*, a condition that promotes an increase in its genetic diversity levels and in part explains the occurrence of individuals with atypical foliar morphological in sympatric stands. The increase in the genetic diversity levels of oaks as a result of hybridization events can also generate morphological, phenological, architectural and chemical changes (Fritz 1999, 382; Cheng et al. 2011, 107–117), as previously mentioned. These changes can be translated into a broader mosaic of resources and conditions, whose benefits can extend to the associated communities through tropic interactions (Bailey et al. 2006, 78; Valencia-Cuevas et al. 2018, 7; Castillo-Mendoza 2019, 42–84).

The canopy of forest ecosystems has been recognized as a habitat that supports a wide biodiversity (e.g., plants, vertebrates, invertebrates, and fungal). Within the invertebrate group, arthropods are characterized by their great abundance, diversity, and functional importance. Particularly in temperate forests, species of the genus *Quercus* (oaks) are one of the most important tree canopy groups, for its diversity and dominance, as previously documented. Also, some species have foundation species characteristics, wide geographical distribution and high levels of genetic diversity, attributes that suggest their importance as habitat for the arthropods and for the establishment, organization, and maintenance of their communities.

Table 1 shows the list of species associated to *Quercus castanea* and *Q. crassipes* canopy. Also, some studies that show the species biodiversity associated to *Q. castanea* and *Q. crassipes* canopy are mentioned below.

Mites is an arthropod group that has been reported in *Q. crassipes* and *Q. castanea* canopy. For example, Palacios-Vargas et al. (2003, 287–292) reported the canopy oribatid mites from five localities of Mexican oak species in Central Mexico. They found that *Q. crassipes* harbored the highest number of oribatid mites during the dry and rainy seasons, while *Q. castanea* had more mites during the rainy season. They reported that the most important mites in abundance and frequency were *Camisia* sp., *Phauloppia* sp. and *Scapheremaeus* sp. Acorn drilling insects have also been reported in *Q. crassipes* and *Q. castanea*. Tovar-Sánchez et al. (2012, 129–141) reported 11 species of acorn drilling insects belonging to the following three orders: Coleoptera, Lepidoptera, Hymenoptera, associated with the canopy of both species.

Furthermore, Maldonado-López et al. (2015, 5) reported that of ten oak species analyzed (*Q. deserticola*, *Q. obtusata*, *Q. castanea*, *Q. candicans*, *Q. scythophylla*, *Q. magnoliifolia*, *Q. glaucoides*, *Q. dysophylla* and *Q. laeta*) in Lake Cuitzeo basin, Michoacán state, Mexico, *Q. castanea* was the second species that supported most of the gall species richness (33 species) and abundance. Similar results were reported for Pascual-Alvarado et al. (468–469) who evaluated the morphospecies richness in 80 oak species in Mexico. They found that three red oaks [*Q. crassifolia* (11), *Q. coccolobifolia* (11) and *Q. castanea* (10)] had a considerable number of galls collected throughout their broad distribution in Mexico. Likewise, Barranco-León (2016, 91–96) found 17 families and 51 species of Lepidoptera larvae associated to canopy of *Q. castanea* in the state park "Flor del Bosque", Puebla, Mexico. Also, he reported that *Q. castanea* supported more species richness than *Q. glabrescens*. Taken together, these results support the importance of *Q. crassipes* and *Q. castanea* as the habitat of diverse communities.

In general, when the genetic diversity of foundation species has been considered two patterns are detected: (1) the diversity of communities is positively and significantly related to the genetic diversity of host species, and (2) genetically more similar hosts support more similar associated communities.

These patterns have been reported in *Q. crassipes* and *Q. castanea*. For example, Tovar-Sánchez and Oyama (2006b, 707–710) reported a significant positive relationship between the diversity of gall-forming insects and the genetic diversity of the host plant (*Q. crassipes*, *Q. castanea*, hybrid) at the population level in Mexico. Also, the authors found that the composition, abundance, and diversity of the community of endophagous insects (gall-forming wasps and leaf miner moths) were affected by the genetic status of the host tree. Specifically, *Q. crassipes* registered the highest species diversity. Similar results were reported for the ectophagous arthropods associated to this oak complex (Tovar-Sánchez and Oyama 2006b, 373–376) because *Q. crassipes* trees registered the highest densities and diversities of arthropod fauna. Similarly, Tovar-Sánchez et al. (2013, 168–170) found that the diversity and density of canopy arthropods associated with *Q. castanea* and *Q. crassipes* responded positively and significantly to the genetic diversity of the host tree at the individual level, in central Mexico. Additionally, Melhado-Carboney (2016, 25–49) reported a significant effect of genetic diversity of oak host taxa on the richness and diversity of lichen species in the *Q. crassipes* × *Q. crassifolia* complex.

Population genetic diversity and individuals of *Q. castanea* positively influenced endophagous (gall-forming wasps and leaf miner moths) in terms of density and species richness (Valencia-Cuevas et al. 2018, 5). Additionally, indirect effects of the individual genetic diversity of the oak host were observed, affecting the species richness and density of parasitoid insects. Both results suggest that genetic variation in *Q. castanea* is one important factor that influences the dynamics of tritrophic interactions that involve oaks, herbivores, and parasitoids (Valencia-Cuevas et al. 2018, 7).

Species that Modulate Core Ecosystem Processes, as Energy Flux and Biogeochemical Cycles

From an ecological perspective, oaks are an important component of temperate forests in the Northern Hemisphere in terms of diversity, dominance and abundance (Hipp et al. 2018, 439). For example, data from the US and Mexico reveal that oaks have the highest biomass and species diversity of forest tree genera in those two countries (Cavender-Bares 2016, 38). With 161 species in Mexico, the oaks comprise 29% of the total aboveground biomass in forests surveyed (Cavender-Bares 2019, 671). Therefore, oaks can be important contributors to ecosystem function in Mexican temperate forest systems (Cavender-Bares 2019, 671). For example, they provide habitat and food for other trophic levels, including numerous insects, birds, squirrels, etc. which in turn provide food for many other organisms, promoting networks formed by various trophic levels. Also, associations with ectomycorrhizal and other fungi influence microbial composition and impact nutrient cycling and ecosystem function (Nguyen et al. 2016, 4039–4041; Cheeke et al. 2017, 436–437).

In a study conducted in the Cuitzeo basin in Mexico, Chávez-Vergara et al. (2015, 807–810) reported that the reabsorption of foliar nutrients of *Q. castanea* litter has influence on the chemical composition of the forest floor and, consequently, in the rates of organic matter decomposition and soil nutrient availability, suggesting a strong footprint of species traits on microbial activities and soil nutrient transformations in oak deciduous forests. Later, Chávez-Vergara et al. (2016, 807–810) determined that the associations of *Q. castanea* with a diverse fungi community influence microbial composition and impact nutrient cycling.

Table 1. Fauna associated to *Quercus castanea* and *Q. crassipes* canopy

Insects (order/family)	Arachnida (order/family)	Acari (family)	Miryapoda (order/family)	Epiphytes (family/species)	Birds (family/species)	Hymenoptera (family/species)
Quercus castanea						
Blattodea: Coleoptera: Carabidae, Bostrichidae, Curculionidae, Elateridae, Leiodidae, Ptiliidae, Scarabaeidae, Staphylinidae, Tenebrionidae, **Collembola:** Trogossitidae, Entomobryidae, **Dermaptera: Diptera**: Ceratopogonidae, Chironomidae, Culicidae, Dolichopodidae, Sciaridae, Syrphidae, Tephritidae, Tipulidae, **Ephemeroptera:**, **Hemiptera**: Aleyrodidae, Issidae, Lygaeidae, Reduviidae, Rhopalidae, Veliidae, **Heteroptera**, **Hymenoptera:** Ceraphronidae, Diapriidae, Formicidae, Icheumonidae, **Lepidoptera:** Arctiidae, Crambidae, Elachistidae, Erebidae, Geometridae, Hesperiidae, Lasiocampidae, Limacodidae, Lycaenidae, Lymantriidae, Megalopygidae, Mimallonidae, Noctuidae, Notodontidae, Nymphalidae, Papilionidae, Psychidae, Pterophoridae, Pyralidae, Riodinidae, Saturniidae, Riodinidae, Tortricidae, **Orthoptera:** Blattellidae, Blattidae, Tettigoniidae, **Psocoptera:** Pseudocaecillidae, Psyllipsocidae, Liposcelididae, **Scolopendromorpha** Scolopendridae, **Thysanoptera**: Phlaeothripidae	**Aranae**: Amaurobidae, Anyphaenidae, Araneidae, Clubionidae, Dyctinidae, Gnaphosidae, Nesticidae, Oonopidae, Pholcidae, Plectreuridae, Salticidae, Scytodidae, Theridiidae, Theridiosomatidae **Scorpions**: Vaejovidae, **Opilionida: Pseudoscorpionida:** Olpiidae,	Camisiidae, Carabodidae, Ceratozetidae, Cymbaeremaeidae, Damaeidae, Galamnidae, Liodidae, Opppidae, Oribatulidae, Oripodidae, Scheloribatidae, Phenopelopidae, Tectocepheidae, Trhypochthoniidae	**Chordeumatida**: Caseyidae, **Geophilomorpha**: Geophilidae, **Lithobiomorpha**: Henicopidae, Lithobiidae.	Bromeliaceae: *Tillandsia carlos-hankii T. bourgaei T. prodigiosa T. magdougalli*	*Aphelocoma ultramarina Piranga flava Pheuticus melanocephalus Dendroica townsendi Parus wollweberi Dendroica nigrescens Regulus calendula*	**Cynipidae:** *Andricus hastata*, *A. jaculi*, *Amphibolips hidalgoensis*, *A. maturus*, *A. michoacanensis*, *A. nassa*, *A. tarasco.* ***Eurytomidae*** *Eurytoma mucronura*

Insects (order/family)	Arachnida (order/family)	Acari (family)	Miryapoda (order/family)	Epiphytes (family/species)	Birds (family/species)	Hymenoptera (family/species)
			Quercus crassipes			
Coleoptera:, **Collembola**:, **Diptera**:, **Hemiptera**: Aleyrodidae, Anthocoridae, Miridae, **Hymenoptera**: Platygastridae, Braconidae, Sphecidae, **Lepidoptera, Orthoptera, Psocoptera Thysanoptera**		Camisiidae, Carabodidae, Ceratozetidae, Cymbaeremaeidae, Damaeidae, Galamnidae, Liodidae, Opppidae, Oribatulidae, Oripodidae, Scheloribatidae, Phenopelopidae, Tectocepheidae, Trhypochthoniidae				**Cynipidae** *Amphibolips cibriani*, *A. hidalgoensis*, *A. nr hidalgoensis*, *Cynips hirsuta*, *Neuroterus* sp. **Synergus** *Synergus citriformis* **Calcidoidea** Eulophidae, Eurytomidae, Pteromalidae, Ormyridae, Torymidae

Source: ***Quercus castanea***: Kinsey 1937, 39–40; Gibson et al. 1997, 477–498; Palacios-Vargas et al. 2003, 290–291; Mondragón-Chaparro and Cruz-Ruíz 2008, 12, 20–22; Pujade-Villar 2009, 812; Nieves-Aldrey et al. 2001, 78-83, Nieves-Aldrey and Pascual 2012, 22-24; Tovar-Sánchez et al. 2013, 5-8; Lorea-Padilla et al. 2015, 48-49; Aguilar et al. 2016, 2–6; Castillejos-Lemus 2016, 21–22 and 32-33; Sánchez-Flores et al. 2017, 264; Valencia-Cuevas et al. 2018, 15, 22; Barranco-León 2016, 19, 28–29, 56–58, 81–96. ***Q. crassipes***: Kinsey 1937, 39–40; Palacios-Vargas et al. 2003, 290–291, Jiménez-Muñoz 2007, 26–27, Melika et al. 2011, 53–55, Tovar-Sánchez et al. 2013, 5–8; Guzmán-Mendoza et al. 2016, 572–575; Sánchez-Flores et al. 2017, 264; Flores-Mercado 2018, 40, 63–70; Pujade-Villar et al. 2018, 151–158; Flores-Mercado et al. 2019, 443–445.

On the other hand, cinipid-induced galls are considered as microecosystems (Pujade-Villar 2013, 5–12), because multiple organisms can interact inside them. For example, in the gall induction process, the inductor species (Cynipini), inquilines (Synergini and Ceroptresini) and their parasitoids [Chalcidoidea (Hymenoptera)] (Pujade-Villar 2013, 5–12; Serrano-Muñoz et al. 2016, 177) are groups associated to the gall. In the same way, there are other organisms that inhabit the galls once the inducers and the primary fauna (inquilines and parasitoids) have abandoned it. Among them we find spiders, aphids, coccids, hymenopter, etc. (Pujade-Villar 2013, 5–12). Thus, cynipid gall wasps and their associated natural communities can be a powerful system to evaluate the influence of foundation species in the ecosystem process, as the energy flow. An excellent model to illustrate it is the gall induced by *Amphibolips hidalgoensis* (Cynipidae) and its associated communities, which has been reported in both *Q. castanea* and *Q. crassipes* canopy.

A. hidalgoensis is a gall-inducing wasp that has been reported in *Q. crassipes*, *Q. crassifolia* (Nieves-Aldrey 2012, 8; Melika et al. 2011, 53–56) and *Q. castanea* (Valencia-Cuevas et al. 2018, 15–22) canopies. The asexual generation of this wasp induces the formation of unilocular, large and globose with a maximum diameter of 6.5 cm gall. When it is in development, the gall has a thin wall of a light green color; once the gall dries it is light brown, with smooth and glabrous surface, with parenchyma and vascular bundles that support the larval chamber.

In addition to the inducer insect (herbivore), as a member of primary fauna in *A. hidalgoensis* galls induced in the *Q. castanea* canopy, the presence of Synergus (Cynipidae) has been reported, an inquiline insect that has lost the ability to induce galls (Valencia-Cuevas et al. 2018, 4). These insects produce their own larval chambers with its own cortical tissue. However, they are capable of killing the inductor wasp, by competition for space and/or food, if their eggs are placed very close to the larval chamber of the inductor (Pénzes et al. 2012, 1–3). Also, six Chalcidoidea families (parasitoids): Eulophidae, Eupelmidae, Eurytomidae, Ormyridae, Pteromalidae and Torymidae have been reported associated to this gall in the *Q. castanea* canopy (Valencia-Cuevas et al. 2018, 10), while that for *Q.*

crassipes reports parasitoid insects of Eurytomidae and Torymidae families (Flores-Mercado et al. 2019, 446). As secondary fauna associated to *A. hidalgoensis* gall in *Q. castanea* members of six families included in the Hymenoptera oden (Valencia-Cuevas et al. 2018, 10) have been reported, while in *Q. crassipes* members of three insect orders were reported (Flores-Mercado et al. 2019, 446). In total, implicated are 18 and eight species associated with *A. hidalgoensis* galls in *Q. castanea* and *Q. crassipes*, respectively (Table 2, Figure 2). In general, both oak host species share families and genera of parasitoid insects; however, *Q. castanea* presented a greater number of families and genus. In contrast, the secondary fauna is different between oak host species, being more diverse in *Q. castanea*.

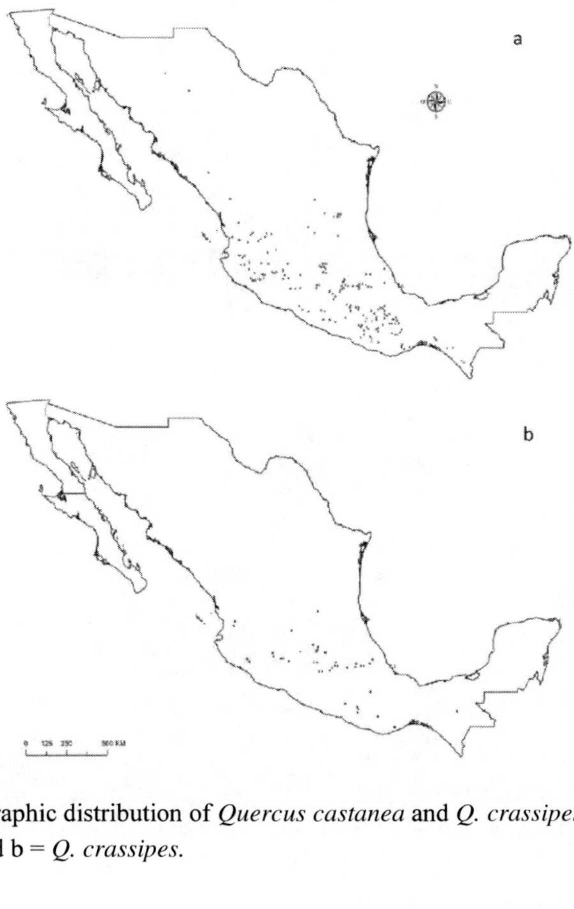

Figure 1. Geographic distribution of *Quercus castanea* and *Q. crassipes* in Mexico. a = *Q. castanea* and b = *Q. crassipes*.

Table 2. Arthropodofauna associated to *Amphibolips hidalgoensis* gall in *Quercus crassipes* (Flores-Mercado et al. 2019, 443–445) and *Q. castanea* (Valencia-Cuevas et al. 2018–22)

Quercus crassipes	*Quercus castanea*
Primary fauna Inductor (Cynipidae: Cynipini)	
Amphibolips hidalgoensis	*Amphibolips hidalgoensis*
Inquiline (Cynipidae: Synergini)	
Synergus citriformis	*Synergus* sp.
Parasitoids (Chalcidoidea)	
Eulophidae: *Eurytma* Eurytomidae: *Eurytoma* Torymidae: *Torymus*	Eupelmidae: *Brossema* sp. Eulophidae: *Cirruspilus* sp. *Clostocerus* sp. *Euplectrus* sp. Tetrastichinae sp. Eurytomidae: *Eurytoma* sp. Ormyridae: *Ormyrus* sp. Pteromelidae: Pteromelidae Torymidae: *Torymus* sp.
Secondary fauna	
Hemiptera: Anthocoridae Psocoptera Coleoptera	Hymenoptera: Apidea: Apidea sp.1 Bethylidae: Bethylidae sp.1 Braconidae: Braconidae sp.1 Elasmidae: *Elasmus* sp.1 Platygastridae: Scelionade sp. Morpho sp. 1 Sphecidae: Morpho sp.1

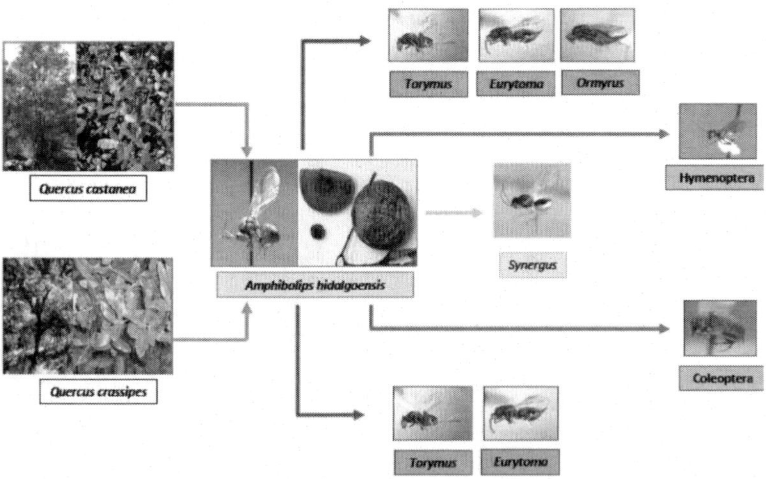

Figure 2. Arthropodofauna associated to *Amphibolips hidalgoensis* gall in *Q. castanea* (Valencia-Cuevas et al. 2018, 10) and *Q. crassipes* (Flores-Mercado et al. 2019, 446). **Primary fauna**, Inductor: *Amphibolips hidalgoensis* and gall (green); Inquiline, *Synergus* (yellow); parasitoids: *Torymus*, *Eurytoma* and *Ormyrus* (red); **Secondary fauna**. Coleoptera and Hymenoptera (blue).

CONCLUSION

This review denotes that *Q. crassipes* and *Q. castanea* are dominant oaks of Mexican temperate forests. They have wide geographic distribution and both species act as important reservoirs of canopy arthropod fauna. The studies show that the hybrid phenomenon plants increase the genetic diversity of the population in *Q. crassipes* and *Q. castanea* and may also play a major role in the structure of canopy community, increasing the species richness and diversity of arthropod community. In this review, we have shown that genetic diversity in both oak species results in extended phenotypes that extends to the community and ecosystem levels, in particular, in the endophagous insect. The results indicate that both oak species act as important reservoirs of canopy biodiversity. Therefore, we conclude that *Q. crassipes* and *Q. castanea* can be recognized as foundation species. Furthermore, we suggest that in terms of conservation is fundamental to conserve genetic diversity of foundation species.

REFERENCES

Abe, Yoshihisa, Melika, George & Stone, Graham N. (2007). "The diversity and phologeography of Cynipid gallwasps (Hymenoptera: Cynipidae) of the oriental and eastern Palearctic regions, and their associated communities." *Oriental Insects*, *41*, 169–212. doi: 10.1080/00305316.2007.10417504.

Aguilar-Peralta, Joan S., González-Rodríguez, Antonio, González-Esquivel, José, G. & Cuevas-Reyes, Pablo. (2016). "Acorn predation patterns in species of the genus *Quercus* (Fagaceae) along an altitudinal gradient in the Tequila Volcano, Jalisco." *Biológicas*, *18*, 1, 1–9.

Askew, Richard R. (1980). "The diversity of insect communities in leaf-mines and plant galls." *Journal of Animal Ecology*, *49*, 817–829. doi:10.2307/4229.

Askew, Richard R., Melika, George, Pujade-Villar, Juli, Schönrogge, Karsten, Stone, Graham N. & Nieves-Aldrey, José Luis. (2013). "Catalogue of parasitoids and inquilines in cinipid oak in the West Palaearctic." *Zootaxa*, *3643*, 001–133. doi: 10.11646/zootaxa.3643.1.1.

Bailey, Joshep K., Bangert, Randy K., Schweitzer, Jennifer A., Trotter R. Talbot III, Shuster Stephen M. & Whitham, Tomas G. (2004). "Fractal geometry is heritable in trees." *Evolution*, *58*, 2100–2102. doi:10.1111/j.0014-3820.2004.tb00493.x.

Bailey, Joshep K., Wooley, Stuart C., Lindroth Richard L. & Whitham, Tomas G. (2006). "Importance of species interactions to community heritability: a genetic basis to trophic level interactions." *Ecological Letters*, *9*, 78–85. doi:10.1111/j.1461-0248.2005.00844.x.

Barranco-León, María N. (2016). "*Factors that influence the diversity and distribution of lepidoptera in Parque Estatal Flor del Bosque*, *Puebla, Mexico*." PhD diss., Potosino Science and Technology Research Institute A. C.

Basset, Yves, Horlyck, V. & Wright, Joseph. (2003). "Forest canopies and their importance." In *Studying forest canopies from above: the international canopy crane network*, edited by Yves Basset, V. Horlyck, and J. Write, 27–33. Panama City: Smithsonian Tropical Research Institute.

Bello-González, Miguel A. & Labat, Jean-Noel. (1984). "Oaks (*Quercus*) of Michoacán State, Mexico." Collection Etudes Mesoaméricaines.

Caseys, Celine, Stritt Cristoph, Glauser Gaetan, Blanchard, Thierry & Lexer, Christian. (2015). "Effects of hybridization and evolutionary constraints on secondary metabolites: the genetic architecture of phenylpropanoids in European *Populus* species." *PLoS ONE*, *10*, e0128200. doi:10.1371/journal.pone.0128200.

Castagneyrol, Bastein, Lagache Lélia, Giffard Brice, Kremer, Antoine & Jactel, Hervé. (2012). "Genetic diversity increases insect herbivory on oak saplings." *PLoS One*, *7*(8), e44247. doi:10.1371/ journal.pone.0044247.

Castillejos-Lemus, Dohuglas E. (2016). "Diversity and genetic structure of *Amphibolips michoacaensis* Nieves-Aldrey & Maldonado 2012

Hymenoptera: Cynipidae) associated to *Quercus castanea* Née 1801 (Fagales: Fagaceae).” PhD. diss., National Autonomous University of Mexico.

Castillo-Mendoza E. (2019). “Effect of hybridization of the *Q. glabrescens* × *Q. rugosa* and *Q. glabrescens* × *Q. obtusata* complex on the community of gall-forming insects and their parasitoids.” PhD. diss. National Autonomous University of Mexico.

Cavender-Bares Jeannine. (2016). “Diversity, distribution and ecosystem services of the North American oaks.” *International Oaks*, *27*, 37–49.

Cavender-Bares Jeannine. (2019). “Diversification, adaptation, and community assembly of the American oaks (*Quercus*), a model clade for integrating ecology and evolution”. *New Phytologist*, *221*, 669–692 doi: 10.1111/nph.15450.

Chávez-Vergara, Bruno, González-Rodríguez, Antonio, Etchevers, Jorge D., Oyama, Ken & Garcia-Oliva, Felipe. (2015). “Foliar nutrient resorption constrains soil nutrient transformations under two native oak species in a temperate deciduous forest in Mexico.” *European Journal of Forest Research*, *134*, 803–817. doi:10.1007/s10342-015-0891-1.

Chávez-Vergara, Bruno, Rosales-Castillo, Andrei, Merino, Agustín, Vázquez-Marrufo, Gerardo, Oyama, Ken & García-Oliva, Felipe. (2016). “*Quercus* species control nutrients dynamics by determining the composition and activity of the forest floor fungal community.” *Soil Biology and Biochemistry*, *98*, 186–195. doi:10.1016/ j.soilbio.2016. 04.015.

Cheeke, Tania E., Phillips, Richard P., Brzostek, Edward R., Rosling, Anna, Bever James D. & Fransson, Petra. (2017). “Dominant mycorrhizal association of trees alters carbon and nutrient cycling by selecting for microbial groups with distinct enzyme function.” *New Phytologist*, *214*, 432–442. doi:10.1111/nph.14343.

Cheng, Dandan, Vrieling Klaas & Klinkhammer, Peter G. L. (2011). “The effect of hybridization on secondary metabolites and herbivore resistance: implications for the evolution of chemical diversity in plants.” *Phytochemestry Reviews*, *10*, 107–117. doi:10.1007/s11101-010-9194-9

Cooper, W. Rodney & Rieske, Lynne K. (2010). "Gall structure affects ecological associations of *Dryocosmus kuriphilus* (Hymenoptera: Cynipidae)." *Environmental Entomology*, *39*, 787–797. doi:10.1603/EN09382.

Curtu, Alexandru L., Gailing Oliver & Finkedley, Reiner. (2007). "Evidence for hybridization and introgression within a species-rich oak (*Quercus* spp.) community". *BMC Evolutionary Biology*, *7*, 218. doi:10.1186/1471-2148-7-218.

Dayton, Poul K. (1972). "Toward an understanding of community resilience and the potential effects of enrichments to the benthos at McMurdo Sound, Antarctica." In *Proceedings of the colloquium on conservation problems in Antarctica*, edited by B.C. Parker, 81–96. California: Allen Press.

Ellison, Aaron M., Bank, Michael S., Clinton, Barton D., Colburn, Elizabeth A., Elliott, Katherine, Ford, Chelcy R., Foster, David R., Kloeppel, Brian, D., Knoepp, Jennifer D., Lovett, Gary M., Mohan, Jaqueline, Orwig, David A., Rodenhouse, Nicholas L., Sobczak, William V., Stinson, Kristina A., Stone, Jeffrey K., Swan, Christopher M., Thompson, Jill, Von Holle, Betsy & Webster, Jackson R. (2005). "Loss of foundation species: consequences for the structure and dynamics of forested ecosystems." *Frontiers and Ecology and Environment*, *3*, 479–486.doi:10.1890/1540-9295(2005)003[0479:LOFSCF]2.0.CO;2.

Ellison, Aaron M., Barker-Plotkin, Audrey A., Foster, David, R. & Orwig, David A. (2010). "Experimentally testing the role of foundation species in forests: the Harvard Forest Hemlock Removal Experiment." *Methods in Ecology and Evolution*, *1*, 168–179. doi:10.1111/j.2041-210X.2010.00025.x.

Figueroa-Rangel, Blanca L. & Olvera-Vargas, Miguel. (2000). "Regeneration patterns in relation to canopy species composition and site variables in mixed oak forests in the Sierra de Manantlán Biosphere Reserve, Mexico." *Ecological Research*, *15*, 249–261. doi:10.1046/j.1440-1703.2000.00346.x.

Flores-Mercado, Noemi. (2018). "*Indigenous knowledge and entomofauna associated with some gall morphotypes of Quercus crassipes Humb &*

Bonpl. and Quercus conspersa Benth." Master diss., Mexico, University of Veracruz.

Flores-Mercado, Noemi, Rangel-Villafranco, Monica, Pujade-Villar, Juli. Cardenas-Camargo, Israel. & Chaires-Grijalva, Martha P. (2019). "Diversity of entomofauna in three gall morphotypes associated with *Quercus crassipes* Humb. & Bonpl. in the State of Mexico." *Revista entomologica Méxicana. Entomología forestal*, *6*, 442–447.

Fritz, Robert S. (1999). "Resistance of hybrid plants to herbivores: genes, environment, or both?" *Ecology*, *80*, 382–391. doi:10.1890/0012-9658(1999)080[0382:ROHPTH]2.0.CO;2.

Gailing, Oliver & . Curtu, Alexandru L. (2014). "Interspecific gene flow and maintenance of species integrity in oaks." *Annals of Forest Research*, *57*, 5–18. doi:10.15287/afr.2014.171.

Gibson, Gary. A. P., Huber, John T. & Woolley, James B. (1997). "*Annotated keys to the genera of Nearctic Chalcidoidea (Hymenoptera).*" Canada: NRC Research Pess.

Guzmán-Mendoza, Rafael, Castaño-Meneses, Gabriela & Nuñez-Palenius, Héctor G. (2016). "The diversity of ant communities (Hymenoptera: Formicidae) and their connections with other arthropods from three temperate forests of Central Mexico." *Revista de Biología Tropical*, *64*, 2, 571–585. doi:10.15517/rbt.v64i2.18979.

González-Rodríguez, Antonio & Oyama, Ken. (2005). "Leaf morphometric variation in *Quercus affinis* and *Q. laurina* (Fagaceae), two hybridizing Mexican red oaks." *Botanical Journal of Linnean Society*, *147*, 427–435. doi:10.1111/j.1095-8339.2004.00394.x.

Govaerts, Rafaël & Frodin, David G. (1998). "*World checklist and bibliography of Fagales (Betulaceae, Corylaceae, Fagaceae and Ticodendraceae)*." Kew, UK: Royal Botanic Gardens.

Halaj, Juraj, Ross, Darell W. & Moldenke, Andrew R. (2000). "Importance of habitat structure to the arthropod food-web in Douglas-fir canopies." *Oikos*, *90*, 139–152. doi:10.1034/j.1600-0706.2000.900114.x.

Hamilton, Andrew J., Basset, Yves, Benke, Kurt K., Grimbacher, Peter S., Miller, Scott E., Novotny, Vojtech Samuelson, Allan G., Stork, Nigel E., Weiblen, George D. & Jean, Yen, D. L. (2010). "Quantifying uncertainty in estimation of tropical arthropod species richness." *American Naturalist*, *176*, 90–95. doi:10.1126/science.1191058.

Hamrick, J. L., Godt Mary Jo, W. & Sherman-Broyles, Susan L. (1992). "Factors influencing levels of genetic diversity in woody plant species." *New Forests*, *6*, 95—124. doi:10.1007/978-94-011-2815-5_7.

Hamrick, James L., Linhart, Yan B. & Mitton, Jeffry B. (1979). "Relationships between life history characteristics and electrophoretically detectable genetic variation in plants." *Annual Review of Ecology and Systematics*, *10*, 173–200. doi:10.1146/annurev.es.10.110179.001133.

Hipp, Andrew L., Manos, Paul S., González-Rodríguez, Antonio, Hahn, Marlene, Kaproth Matthew, McVay John D., Valencia Avalos, Susana & Cavender-Bares, Jeannine. (2018). "The American oaks exhibit convergent evolution in biogeography, climatic niche, and diversification rates." *New Phytologist*, *217*, 439–452. doi.org/ 10.1111/nph.14773.

Hunter, Mark D. & Price, Peter W. (1992). "Playing chutes and ladders: heterogeneity and the relative roles of bottom-up and top-down forces in natural communities." *Ecology*, *73*, 724–732. doi.org/10.2307/1940152.

Kinsey, Alfred C. (1937). "New Mexican gall wasps (Hymenoptera, Cynipidae) II." *Revue de Entomologia*, *7*, 428–471.

Ishii, Hiroaki T., Tanabe, Shin-ichi & Hiura, Tsutom. (2004). "Exploring the relationships among canopy structure, stand productivity, and biodiversity of temperature forest ecosystems." *Forest Science*, *50*(3), 342–355. doi:10.1093/forestscience/50.3.342.

Jiménez-Muñoz, Edith. (2007). "*Herbivoría de bellotas de Q. crassipes Humb & Bonpl (Fagaceae) en un bosque fragmentado en el estado de Hidalgo*." Dissertation, Autonomous University of Hidalgo State.

Kremer, Antoine, Abbott, Albert G., Carlson, John E., Manos, Paul S., Plomion, Christophe, Sisco, Paul, Staton, Margaret E., Ueno, Saneyoshi & Vendramin, Giovanni G. (2012). "Genomics of Fagaceae". *Tree Genetics & Genomes*, *8*, 583– 610. doi:10.1007/ s11295-012-0498-3.

Lorea-Padilla, Francisco J., López-Barbosa, Edmundo C., González-Rodríguez, Antonio & Cuevas-Reyes, Pablo. (2015). "Spatial variation of the canopy arthropod community associated to *Quercus castanea* along a moisture gradient". *Boletín de la sociedad mexicana entomológica.*, *1*, 47–52.

Lowe, Andrew, Harris, Stephen & Ashton, Paul. (2004). *Ecological genetics: design, analysis and application.* UK: Blackwell.

Lowman, Margaret D. (2004). "Tarzan or Jane? A short history of canopy biology." In *Forest Canopies*, edited by Lowman, Margaret D. and Rinker, H.B. 453–464. San Diego, California: Elsevier Publishers.

Maldonado-López, Yuritzi, Cuevas-Reyes, Pablo, Stone, Graham N., Nieves-Aldrey, José Luis & Oyama, Ken. (2015). "Gall wasp community response to fragmentation of oak tree species: importance of fragment size and isolated trees." *Ecosphere*, *6*, 1–15. doi.org/10.1890/ES14-00355.1.

Melhado-Carboney, Jennie Charlotte. (2016). "Effect of hybridization of the *Quercus crassifolia* × *Q. crassipes* (Fagaceae) complex on the associated lichenic community." Master diss., Autonomous University of Morelos State.

Melika, George, Equihua-Martínez, Armando, Estrada-Venegas, Edith G., Cibrián-Tovar, David, Cibrián-Llanderal, Vicor D. & Pujade-Villar, Juli. (2011). "New *Amphibolips* gallwasp species from Mexico (Hymenoptera: Cynipidae)." *Zootaxa.*, *3105*, 47–59. doi: 10.11646/zootaxa.3105.1.2.

Mondragón-Chaparro, Demetria & Cruz-Ruiz, Gabriel I. (2008). "Seasonal variation of the macro-arthropod community associated to *Tillandsia carlos-hankii* (Bromeliaceae) in an oak-pine forest in Oaxaca, Mexico." *Brenesia*, *70*, 11–22.

Müller, M. & Gailing, Oliver. (2019). "Abiotic genetic adaptation in the Fagaceae." *Plant Biology*, *21*, 783–795. doi:10.1111/plb.13008.

Nadkarni, Nalini M., Parker, Geoffrey G., Rinker H. Bruce & Jarzen, David M. (2004). "The nature of forest canopies." In *Forest canopies*, edited by Lowman, Margaret D. and Rinker, H.B. 3–23. San Diego, California: Elsevier Publishers.

Nguyen, Nhu H., Williams, Laura, Vincent, John B., Stefanski, Artur, Cavender-Bares, Jeaninne, Messier, Christian, Paquette, Alaine, Gravel, Dominique, Reich, Peter B. & Kennedy, Peter G. (2016). "Ectomycorrhizal fungal diversity and saprotrophic fungal diversity are linked to different tree community attributes in a field-based tree experiment." *Molecular Ecology*, *25*, 4032–4046. doi:10.1111/mec.13719.

Nieves-Aldrey, José Luis. (2001). "Hymenoptera, Cynipidae." In *Fauna Ibérica*, vol. 16, edited by M.A. Ramos, J. Alba, X. Bellés, J. Gosálbez, A. Guerra, E. Macpherson, F. Martín, J. Serrano, and J. Templado. 636. Madrid: National Musium of Natural Science.

Nieves-Aldrey, José Luis, Pascual, Enrique, Maldonado-López, Yuritzi, Medianero, E. & Oyama, Ken. (2012). "Revision of the *Amphibolips* species of Mexico excluding the "niger complex" Kinsey (Hymenoptera: Cynipidae), with description of seven new species." *Zootaxa*, *3545*, 1–40. doi: 10.11646/zootaxa.3545.1.1.

Novotny, Vojtech, Drozd, Pavel, Miller, Scott E., Kulfan, Miroslav, Janda, Milan, Basset, Yves & Weiblens, George D. (2006). "Why are there so many species of herbivorous insects in tropical rainforest." *Science*, *313*, 1115–1118. doi:10.1126/science.1129237.

Nowak, David J., Hirabayashi, Satosi, Bodine, Allison & Hoehn, Robert. (2013). "Modeled PM2.5 removal by trees in ten U.S. cities and associated health effects." *Environmental Pollution*, *178*, 395–402.

Oyama, Ken, Ramírez-Toro, Wilfrido, Peñaloza-Ramírez, Juan M., Pérez-Pedraza, Alberto E., Torres-Miranda, César A., Ruiz-Sánchez, Eduardo & González-Rodríguez, Antonio. (2018). "High genetic diversity and connectivity among populations of *Quercus candicans*, *Quercus crassifolia*, and *Quercus castanea* in a heterogeneous landscape in Mexico." *Tropical Conservation Science*, *11*, 1–14. doi:10.1177/1940082918766195.

Palacios-Vargas, José G., Iglesias, Ricardo & Castaño-Meneses, Gabriela. (2003). "Mites from Mexican canopies." *Insect Science and its Application*, *23*, 287–292. doi:10.1017/S1742758400012340.

Pascual-Alvarado, Enrique, Nieves-Aldrey, José Luis, Castillejos-Lemus, Douhglas E., Cuevas-Reyes Pablo & Oyama, Ken. (2017). “Diversity of galls induced by wasps (Hymenoptera: Cynipidae, Cynipini) associated with oaks (Fagaceae: *Quercus*) in Mexico.” *Botanical Sciences*, *95* (3), 461–472. doi:10.17129/botsci.1215.

Pénzes, Zsolt, Chang-Ti, Tang. Bihari, Péter, Bozsó, Miklós, Schwéger, Szabina & Melika, George. (2012). “Oak associated inquilines (Hymenoptera: Cynipidae: Synergini).” *Tiscia Monograph. Series*., *11*, 1–66.

Pujade-Villar, July. (2013). “Oak galls: a miniature ecosystem that makes possible multidisciplinary studies.” In *Entomología Mexicana*. Vol. *12*. Tomo 1, edited by Equihua-Martínez, Armando, Estrada-Venegas, Edith. G., Acuña-Soto, Jesús. A., and Chaires-Grijalva, Martha P. 1–22. Sociedad Mexicana de Entomología y Colegio de Postgraduados.

Pujade-Villar, Juli, Barrera-Ruíz, Uriel M. & Cuesta-Porta, Víctor. (2018). “Description of *Amphibolips cibriani* Pujade-Villar n. sp. for Mexico (Hymenoptera: Cynipidae: Cynipini).” *Dugesiana*, *52*, 2, 151–158. doi:10.32870/dugesiana.v25i2.7051.

Pujade-Villar, Juli, Equihua-Martínez, Armando, Estrada-Venegas, Edith G. & Chagoyán-García, Cristina. (2009). “State of knowledge of Cinipini (Hymenoptera: Cynipidae) in Mexico: study perspectives.” *Neotropical Entomology*, *38*, 809–821. doi:10.1590/S1519-566X2009000600015.

Rampant, Patricia F., Lesur, Isabelle, Boussardon, Clément, Bitton, Frédérique, Martin-Magniette, Marie-Laure, Bodénès, Catherine, Le Provost, Grégoire, Bergès, Hélène, Fluch, Sylvia, Kremer, Antonie & Plomion, Christophe. (2011). “Analysis of BAC end sequences in oak, a keystone forest tree species, providing insight into the composition of its genome.” *BMC Genomics*, *12*, 292. doi:10.1186/1471-2164-12-292.

Romero-Rangel, Silvia, Rojas-Zenteno, Carlos E. & Aguilar, María L. (2002). “The *Quercus* (Fagaceae) genus in Mexico State.” *Annals of the Missouri Botanical Garden*, *89*, 551–593.

Sáenz-Jiménez, Fausto. (2010). “An approximation to the fauna associated with oak forests of Guantiva-La Rusia-Iguaque corridor (Boyacá-Santander, Colombia).” *Colombia Forestal*, *13*(2), 299–334.

Sánchez-Flores, Oscar A., García-Martínez, Oswaldo & Sánchez-Peña, Sergio R. (2017). "The Whitefly *Tetraleurodes perileuca* (Cockerell) (Hemiptera: Aleyrodidae) in Mexico and the First Record of its Parasitoid *Amitus granulosus* MacGown and Nebeker (Hymenoptera: Platygastridae)." *Journal of The Kansas Entomological Society*., *90*(3), 262–264. doi.org/10.2317/0022-8567-90.3.262.

Schowalter, Timmoty D. & Ganio, Lisa M. (2003). "Diel, seasonal and disturbance-induced variation in invertebrate assemblages." In *Arthropods of Tropical Forests* edited by Basset, Yves, Novotny, Vojtec, Miller, Scott E., Kitching, Roger L. 315– 28. Cambridge, UK: Cambridge Univ. Press

Serrano-Muñoz, Miriam, Villegas-Guzmán, Gabriel A., Callejas-Chavero, Alicia, José R. Lomeli-Flores, Barrera-Ruíz, Uriel M., Pujade, July & Ferrer-Suay, Mar. (2016). "Hymenopterans associated with *Andricus quercuslanigera* galls (Hymenoptera: Cynipidae, Chalcidoidea) from sierra de Guadalupe, State of Mexico." *Entomología Mexicana*, 177–182.

Stone, Graham N., Schönrogge, Karsten, Atkinson, Rachel J., Bellido, David & Pujade-Villar, Juli. (2002). "The population biology of oak gall wasps (Hymenoptera: Cynipidae)." *Annual Review of Entomology*, *47*, 633–68. doi:10.1146/ annurev.ento.47.091201.145247.

Stone, Graham N. & Schönrogge, Karsten. (2003). "The adaptive significance of insect gall morphology." *Trends in Ecology and Evolution*, *18*, 512–522. doi:10.1016/s0169-5347(03)00247-7.

Strong, Donald R., Lawton, John H. & Southwood, Richard. (1984). *Insects on plants. Community patterns and mechanisms*. Cambridge, Massachusetts: Harvard University Press.

Tack, Ayco J., Ovaskainen, Otso, Pulkkinen, Pertti & Roslin, Tomas. (2010). "Spatial location dominates over host plant genotype in structuring an herbivore community". *Ecology*, *91*, 2660–2672. doi:10.1890/09-1027.1.

Tovar-Sánchez, Efraín, Cano-Santana, Zenón & Oyama, Ken. (2003). "Canopy arthropod communities on Mexican oaks at sites with different disturbance regimes." *Biological Conservation*, *115*, 79–87. doi:10.1016/S0006-3207(03)00096-X.

Tovar-Sánchez, Efraín & Oyama, Ken. (2004). "Natural hybridization and hybrid zones between *Quercus crassifolia* and *Quercus crassipes* (Fagaceae) in Mexico: morphological and molecular evidence." *American Journal of Botany*, *91*, 1352–1363. doi:10.3732/ ajb.91.9.1352.

Tovar-Sánchez, Efraín & Oyama, Ken. (2006a). "Community structure of canopy arthropods associated in *Quercus crassifolia* × *Quercus crassipes* complex." *Oikos*, *112*, 370– 381. doi:10.1111/j.0030-1299.2006.14029.x.

Tovar-Sánchez, Efraín & Oyama, Ken. (2006b). "Effect of hybridization of the *Quercus crassifolia* × *Quercus crassipes* complex on the community structure on endophagous insects." *Oecologia*, *147*, 702–713. doi:10.1007/s00442-005-0328-5.

Tovar-Sánchez, Efraín, Mussali-Galante, Patricia, Esteban-Jiménez, Rocío, Piñero, Daniel, Arias, Dulce M., Dorado, Oscar & Oyama, Ken. (2008). "Chloroplast DNA polymorphism reveals geographic structure and introgression in *Quercus crassifolia* × *Quercus crassipes* hybrid complex in Mexico." *Botany*, *86*, 228–239. doi:10.1139/B07-128.

Tovar-Sánchez, Efraín, Valencia-Cuevas, Leticia, Castillo-Mendoza, Elgar, Mussali-Galante, Patricia, Pérez-Ruiz, Rigoberto, V. & Ana Mendoza. (2013). "Association between individual genetic diversity of two oak host species and canopy arthropod community structure." *European Journal of Forest Research*, *132*, 165–179 doi:10.1007/ s10342-012-0665-y.

Tovar-Sánchez, Efraín, Valencia-Cuevas, Leticia, Mussali-Galante, Patricia, Ramírez-Rodríguez, Rolando & Castillo-Mendoza, Elgar. (2015). "Effect of host-plant genetic diversity on oak canopy arthropod community structure in central Mexico." *Revista Chilena de Historia Natural*, *88*, 1–12. doi:10.1186/s40693-015-0042-6.

Valencia-Cuevas, Leticia & Tovar-Sánchez, Efraín. (2015). "Oak canopy arthropod communities: which factors shape its structure?" *Revista*

Chilena de Historia Natural, *88*(1), 1–22. doi:10.1186/s40693-015-0045-3.

Valencia-Cuevas, Leticia, Piñero, Daniel, Mussali-Galante, Patricia, Valencia-Avalos, Susana & Tovar-Sánchez, Efraín. (2014). "Effect of a red oak species gradient on genetic structure and diversity of *Quercus castanea* (Fagaceae) in Mexico." *Tree Genetics & Genomes*, *10*, 641–652. doi:10.1007/s11295-014-0710-8.

Valencia-Cuevas, Leticia, Mussali-Galante, Patricia, Piñero, Daniel, Castillo-Mendoza, Elgar, Rangel-Altamirano, Guadalupe & Tovar-Sánchez, Efraín. (2015). "Hybridization of *Quercus castanea* (Fagaceae) across a red oak species gradient in Mexico." *Plant Systematics and Evolution*, *301*, 1085–1097. doi:10.1007/s00606-014-1151-4.

Valencia-Cuevas, Leticia, Mussali-Galante, Patricia, Cano-Santana, Zenón, Pujade-Villar, July, Equihua-Martínez, Armando & Tovar-Sánchez, Efraín. (2018). "Genetic variation in foundation species governs the dynamics of trophic interactions." *Current Zoology*, *64*, 13–22. doi:10.1093/cz/zox015.

Valencia, Susana. (1994). "Contribution to the taxonomic delimitation of three species of the genus *Quercus* subgenus: Erythrobalanus: *Q. laurina* Humboldt et Bonpland, *Q. affinis* Scheidweiler and *Q. ghiesbregthii* Martens et Galeotti." Master diss. National Autonomous University of Mexico.

Valencia, Susana. (2004). "Diversity of *Quercus* genus (Fagaceae) in Mexico." *Boletín de la Sociedad Botanica de México*, *7*, 33–53. doi:10.17129/botsci.1692.

Whitham, Tomas G., Bailey, Joseph K., Scheweitzer, Jennifer A., Shuster, Stephen M., Bangert, Randy K., LeRoy, Carry J., Lonsdorf, Eric V., Allan, Gery J., DiFazio, Stephen P., Potts, Brad M., Fischer, Dylan C., Gehring, Catherine A., Lindroth, Richard L., Marks, Jane C., Hart, Stephen C., Wimp, Gina M. & Wooley, Wooley, Stuart C. (2006). "A framework for community and ecosystem genetics: form genes to ecosystems." *Nature*, *7*, 510–523. doi:10.1038/nrg1877.

Whittemore, A. T. & Schaal, B. A. (1991). "Interspecific gene flow in sympatric oaks." *Proceedings of the National Academy of Sciences, USA*, *88*, 2540–2544. doi:10.1073/pnas.88.6.2540.

Wimp, Gina M., Young, William P., Woolbright, Scott A., Martinsen, Gregory D., Keim, Paul & Whitham, Tomas G. (2004). "Conserving plant genetic diversity for dependent animal communities." *Ecology Letters*, *7*, 776–780. doi:10.1111/j.1461-0248.2004.00635.x.

In: Quercus: Classification, Ecology and Uses ISBN: 978-1-53618-026-8
Editor: Bent J. Steffensen

Chapter 6

PHYTOCHEMICAL POTENTIAL AND BIOLOGICAL ACTIVITIES OF EXTRACTS FROM OAK LEAVES (*QUERCUS* SPP.)

Martha Rocío Moreno-Jiménez[1], Jessica Alejandra Reyna-Rojas[1], Nuria Elizabeth Rocha-Guzmán[1], José Alberto Gallegos-Infante[1], Claudia Ivette Gamboa-Gómez[2], Verónica Cervantes-Cardoza[1] and Rubén Francisco González-Laredo[1,*]

[1]UPIDET, TecNM/Instituto Tecnológico de Durango, Durango, México

[2]Biomedical Research Unit-Mexican Social Security Institute (IMSS), Durango, Dgo., México

[*] Corresponding Author's Email: rubenfgl@itdurango.edu.mx.

ABSTRACT

In recent years, interest in the study of bioactive compounds extracted from natural sources has increased. Customarily, oak leaves (*Quercus* spp) have been used as food for livestock and as traditional medicine thanks to their nutritional and phytochemical composition, respectively. Different extracts of oak species have demonstrated antioxidant, anti-inflammatory, antimutagenic, anticancer, antitopoisomerase, gastroprotective, hepatoprotective and genotoxic potentials, in addition to promoting the enzymatic inhibition of cholinesterase and α-glucosidase.

These biological effects have been directly related to the phytochemical richness of oak leaves, which can be affected by environmental factors and ethnobotanical origins. The encapsulation of oak infusions and its fermentation with the Kombucha fungus have proved to be an alternative for the improvement of organoleptic and functional properties of the drink. In addition, emerging technologies such as PEF (pulsed electric fields) have shown to enhance their phenolic composition and their biological activity. The most widely explored phytochemical group in oak leaves is polyphenols, of which condensed and hydrolysable tannins have been reported in high concentrations in different *Quercus* species. This chapter includes a summary of the findings reported over the last years in different oak species, including a description of their phytochemical composition and their biological effects.

Keywords: bioactivity, extracts, infusions, leaves, non-timber uses, tannins

INTRODUCTION

Oaks (*Quercus* spp) represent the second most important group of temperate-cold climate in Mexico. In previous years, some authors have shown that oak forests (5.5%) and pine-oak forests (13.7%) represented around 10% of the country's surface.

Three variants of the oak orders have been described: *Erythrobalanus* (also known as red oak), *Leucobalanus* (i.e., white oak or Lobatae) and *Protobalanus* (i.e., golden cup oaks or intermediate oaks). White oaks grow from low slopes of mountainous massifs or hills in semi-arid areas up to stream margins in canyons with humid forests. These are distribuited between the sea level and 3,500 m and include mainly the following species:

Quercus repanda Humb. Et Bonpl., *Q. frutex* Trel., *Q. macdougalii.*, *Q. glabrescens* Benth., *Q. oleoides*, *Q. magnoliifolia*, *Q. glaucescens* Humb. et Bonpl., *Q. dumosa* Nutt, *Q. liebmanii*, *Q. magnoliifolia*, *Q. muehlenbergii* Englem., *Q. obtusata*, *Q. fusiformis* Small., *Q. insignis* and *Q. macdougallii*. On the other hand, red oaks prefer humid sites, being frequent at the bottom of canyons or the medium-high slopes with forested areas of temperate-subhumid climates. Its distribution ranges from 150 to 3,100 m and includes: *Quercus ocoteifolia* Liebm, *Q. laurina*, *Q. crassipes* Humb. et Bonpl., *Q. miquihuanensis* Nixon et C. H. Müll., *Q. hintoniorum*, *Q. salicifolia* Née, *Q. conspersa* Benth., *Q. aristata*, *Q. elliptica* Née, *Q. macvaughii* R. Spellenberg, *Q. duratifolia* C. H. Müll., *Q. salicifolia*, *Q. hirtiifolia*, *Q. mullerii* M. Martínez, *Q. elliptica*, *Q. castanea*, *Q. candicans*, and *Q. acutifolia* Née. By last, the distribution of golden oaks is between 200 and 2,800 meters above sea level. Species like *Q. palmeri* Engl., and *Q. chrysolepis* Liebm., are at altitudinal intervals of 1000 and 2000 m, respectively.

Approximately 600 to 900 species have been described so far and divided into two diversity centers, one of them in Southeast Asia with around 125 species, and the second in Mexico with around 135–150 species. Eighty-six of this species are considered endemic to Central and Southern Mexico. These figures have been updated, describing 109 endemic species of Mexican oaks, which represent 2.2% of the total input of woody resources of which 47% belongs to the *Quercus* section, 61% to *Lobatae* and 1% to *Protobalanus*. Among the species widely distributed in Mexican territory, that is, present in more than 15 States are: *Q. candicans* Née, *Q. castanea* Née, *Q. crassifolia* Humb. et Bonpl., *Q. laeta* Liebm., *Q. microphylla* Née, *Q. obtusata* Humb. et Bonpl., and *Q. rugosa* Née.

Oak distribution is wide in the Mexican Republic, mainly in the temperate zones of the Northern Hemisphere, temperate forests, covered oak forests ($9x10^6$ Ha of the territory), and in mountainous areas. The majority of Mexican oak species (over 95%) are located between 1,200 and 2,800 m, and their presence is confirmed along the country, except in the States of Yucatán, Quintana Roo and Campeche. The state of Coahuila has 31 oak species, of which white species predominate in 67% of the total *Quercus* population. Likewise, they have been registered as endemic species *Q. carmenensis*, *Q. coahuilensis*, *Q. graveii*, *Q. hintoniorum*, *Q. invaginata* and *Q. saltillensis*. Durango State has 41 species of *Quercus*, 22 whites and 19 red ones (Luna-José et al., 2003).

Two ways of reproduction of oak species have been described: sexual reproduction, and clonal growth. Sexual reproduction occurs from the production of acorns, which can be regular, sporadic or massive after long periods of non-reproduction (mast sowing or normal masting), while clonal growth occurs through the suction of roots. Some species of oak are reproduced by both clonal diffusion and sexual reproduction, such as *Q. eduardii* and *Q. potosina,* species that grow in protected areas of the Sierra Fria of Aguascalientes, Mexico (Alfonso-Corrado et al., 2007). In a demographic study by same authors, lasting four years, it was observed that both species produced clonal offspring annually during the study time, and that they reproduced sexually only once, which has suggested that even low extractions of adult individuals from the area can negatively affect the whole population.

NON TIMBER USES OF OAK SPECIES

Although oaks are used mainly to obtain wood, cellulose (54%), lumber (30%), firewood (7%), and coal (6%) are also current utilizations. However, non-timber uses for the different species of oaks have been observed over the years. According to FAO, when talking about non-timber forest products, it generally refers to *"biological origin goods other than wood, derived from forests, wooded land and trees located outside forests,"* which includes products of roots, stems, bark, leaves, gills, flowers and fruits, excluding the use of wood, chips, charcoal and firewood. Fifty-five non-woody oak species have been used by various ethnic groups in Mexico, such as the Tepehuas, Totonacos, Mixes, Mixtecos, Purépechas, Zapotecos, Otomíes, Mazahuas, Tarahumaras, Tepehuanes and Náhuatls (Luna-José et al., 2003). States such as Tlaxcala, Morelos, Mexico City, Nayarit, Sonora, Guerrero, Michoacán, Durango, Estado de México, Chiapas, Chihuahua,

Hidalgo, Jalisco, Oaxaca, Puebla, Veracruz and Nuevo Leon have registered non-timber uses for some species of *Quercus*, which have been destined mainly to medicinal uses (96% of the total species), alimentary, as fodder (22.5%), and artisanal (16.4%). In addition, the tannins present in oak species have been intended to the tanning of animal skins. For medicinal purposes almost all organs from the plant are used: bark, leaves, flowers, roots and gills; for nutritional purposes, acorns, buds of leaves, flowers, leaves and gills; as forage or livestock feed, acorns and foliage of plants; and for the development of crafts, the acorns (Luna-José et al., 2003).

Generally, acorn, bark and leaves from oaks have shown greater potential for non-timber applications. Luna-José et al., (2003) have described 31 illnesses, of which most are digestive tract and skin diseases, which are traditionally treated with oak species, mainly bark and leaves. The decoction of bark from species such as *Q. robur* and *Q. petrea* are commonly used in Bulgaria for its astringent properties and for the treatment of dermatitis (Leporatti and Ivancheva, 2003). The astringent, tonic and antiseptic properties of the oak bark have been described and related to the presence of tannic acid compounds. Somewhere else in Italy, additionally to the bark, the fruit (acorns) of *Q. robur* and *Q. petrea* have shown astringent, anti-inflammatory, mild antiseptic and anti-hemorrhagic effects (Leporatti and Ivancheva, 2003). Acorns store important nutrients such as carbohydrates and lipids, so they are used mainly as food. As an example, acorns of *Q. emoryi* contain 35% oil, 50% starch and 4% sucrose. Particularly, oak leaves represent an organ that is easily obtainable and differentiable among other species for their morphological characteristics, and therefore is widely used for its therapeutic properties.

Traditional Uses of Oak Leaves (Medicinal)

Several species such as *Q. faginea* and *Q. ilex* are consumed orally as infusions of their leaves. The first one has shown expectorant and mucolytic effects, and the second one is used in cases of hypotension (González et al., 2010). Likewise, infusions of another oak leaves have been studied for their antioxidant, anti-inflammatory and anticancer activity, in addition to the genotoxic effect in HeLa cells, antimicrobial, antitopoisomerase and gastroprotective effects.

Infusions of *Q. resinosa* have shown different biological effects (see Table 1), which have been directly related to their phenolic composition. In contrast, the high concentrations of gallic acid and catechins in the beverage cause a decrease in their sensory acceptability due to their dark coloration and astringent taste (Rocha-Guzmán et al., 2012). The latter phenomenon is due to the formation of an insoluble precipitate in the saliva due to the interaction of polyphenols with the proteins rich in proline from saliva (Vázquez-Cabral et al., 2014). To improve or mask the astringency of this drink, efforts have been made ranging from encapsulation (Rocha-Guzmán et al., 2010; Gallegos-Infante et al., 2013) to fermentation of the beverage (Vázquez-Cabral et al., 2014; Vázquez-Cabral et al., 2016a).

In this way, the fermentation (@ 7 days, 10% of sugar, 10% of starting culture, and 2.5% of inoculum at 25°C) of infusions of *Q. resinosa* leaves with the *Kombucha* consortium activated changes on the sensorial characteristics of the drink as pH, color, flavor (e.g., decrease in astringency and bitterness), the phenolic composition and therefore in the antioxidant capacity (DPPH scavenging assay) (Vázquez-Cabral et al., 2014). Kombucha analogues prepared by the fermentation of infusions of oak leaves (*Q. resinosa, Q. arizonica, Q. convallata*) showed anti-inflammatory activity by reducing the levels of pro-inflammatory cytokines IL-6 and TNF-alpha, and antioxidant by decreasing in the production of nitric oxide (NO) in induced macrophages with oxidative damage by H_2O_2 (Vazquez Cabral et al., 2017). Recently, non-thermal conservation technologies such as pulsed electric fields (PEF) have been used in Kombucha analogues of leaf infusions of *Q. obtusata*, using square waves at certain electric field intensity (37.3–53.4 kV/cm), time of PEF processing (445.3-1979.2 μs), an outlet temperature (18.31 ± 0.98°C), an input energy (21.2–136.5 kJ/L) and two feed flow rates (51.42 and 102.85 L/h), observing higher phenolic content and antioxidant capacity using the higher specific energies. In this study, the acetic acid bacteria showed to be more sensitive than the yeasts to the treatment with PEF (Vázquez Cabral et al., 2016b).

Two wall materials such as lactose-caseinate and maltodextrin have been used for the microencapsulation by drying of *Q. resinosa* infusions (Rocha-Guzmán et al., 2010; Gallegos-Infante et al., 2013). Lactose-caseinate capsules of micro and nanometric sizes have shown greater antioxidant capacity at low concentrations (Rocha-Guzmán et al., 2010). Sometime later, *Q. resinosa* infusions encapsulated by spray-drying showed the best performance and DPPH radical scavenging using maltodextrin (Gallegos-Infante et al., 2013). The microencapsulation is a method that may be used to protect phenolics from degradation by ambient factors such as oxygen, temperature and processing, preserving their biological effect and antioxidant activity (Macías-Cortés et al., 2020).

In addition to aqueous extracts, other solvents have been studied for the extraction and methods for the encapsulation of bioactive compounds. Meanwhile, the alcoholic extracts of leaves from different oak species have shown hepatoprotective, gastroprotective (Moharram et al., 2015), antioxidant (Moharram et al., 2015; Custódio et al., 2015; Sarwar et al., 2015; Uddin and Rauf, 2012), antimicrobial (Güllüce et al., 2004; Khurram et al., 2011; Khurram et al., 2012; Sakar et al., 2005, Sarwar et al., 2015; Şohretoğlu et al., 2007, Uddin and Rauf, 2012; Jamil et al., 2012), anticandidal (Khurram et al., 2011), and antifungal activities (Şohretoğlu et al., 2007) as well as the enzymatic inhibition of cholinesterase and α-glucosidase (Custódio et al., 2015). In the ethanol extracts only the antioxidant activity has been explored (Popovic et al., 2013) (Table 1).

Less conventional is the use of leaves and fruits of *Q. infectoria* in aqueous silver nitrate solutions for the synthesis of silver nanoparticles. It results to be a method able to reduce the silver ions in silver nanoparticles in 80 min to moderate conditions with antibacterial activity towards pathogenic bacteria of plants such as *Pectobacterium carotovorum, Ralstonia solanasearum, Erwina amylovora* and *Xanthomonas citri* (Chahardooli et al., 2014).

Table 1. Biological effects of herbal extracts of different oak species

Biological activity	Oak species	Extraction method	References
Antibacterial	*Quercus cerris*	MeOH 80%	Şohretoğlu et al., 2007
	Quercus coccifera	MeOH 80%	Şohretoğlu et al., 2007
	Quercus dilatata	Methanolic fractions	Jamil et al., 2012
	Quercus macranthera	MeOH 80%	Şohretoğlu et al., 2007
	Quercus pubescens	MeOH 80%	Şohretoğlu et al., 2007
Anticancer	*Quercus sideroxyla*	Infusion 1% (w/v)	Moreno-Jiménez et al., 2015
	Quercus resinosa	Infusión	Rocha-Guzmán et al., 2009
Anticandidal	*Quercus baloot*	MeOH 80%	Khurram et al., 2011 and 2012
Antifungical	*Quercus coccifera*	MeOH 80%	Şohretoğlu et al., 2007
Anti-inflammatory	*Quercus arizonica*	Infusion 1% (w /v) fermented with the fungus *Kombucha*	Vázquez-Cabral et al., 2017
	Quercus convallata	Infusion 1% (w /v) fermented with the fungus *Kombucha*	Vázquez-Cabral et al., 2017
	Quercus durifolia	Infusion 1% (w/v)	Moreno-Jiménez et al., 2015
	Quercus sideroxyla	Infusion 1% (w/v)	Moreno-Jiménez et al., 2015
	Quercus resinosa	Infusion 1% (w /v) fermented with the fungus *Kombucha*	Vázquez-Cabral et al., 2017
	Quercus sideroxyla	Infusion 1% (w/v)	Moreno-Jiménez et al., 2015
Antimicrobial	*Quercus aucheri*	MeOH 80%	Sakar et al., 2005
	Quercus grisea	Infusion	Sánchez-Burgos et al., 2013
	Quercus ilex	Soxhlet with MeOH	Güllüce et al., 2004
	Q. aucheri	Fractions of MeOH 80%	Sakar et al., 2005
	Quercus incana	MeOH	Sarwar et al., 2015
	Quercus resinosa	Infusion	Sánchez-Burgos et al., 2013
	Quercus robur	MeOH	Uddin and Rauf, 2012

Biological activity	Oak species	Extraction method	References
Antimutagenic	*Quercus brantii*	Hydroalcoholic extract by maceration	Mirzaei and Mirzaei, 2013
Antioxidant	*Quercus arizonica*	Infusion 1% (w /v) fermented with the fungus *Kombucha*	Vázquez-Cabral et al., 2017
	Quercus brantii	Hydroalcoholic extract by maceration	Mirzaei and Mirzaei, 2013
	Quercus brantii	Hydroalcoholic extract	Mirzaei et al., 2013
	Quercus convallata	Infusion 1% (w /v) fermented with the fungus *Kombucha*	Vázquez-Cabral et al., 2017
	Quercus durifolia	Infusion	Rivas-Arreola et al., 2010
	Quercus durifolia	Infusion 1% (w/v)	Rocha-Guzmán et al., 2012
	Quercus durifolia	Infusion 1% (w/v)	Moreno-Jiménez et al., 2015
	Quercus durifolia	Aqueous acetone (70%) followed by methanol (50%)	Gamboa-Gómez et al., 2012
	Quercus eduardii	Infusion	Rivas-Arreola et al., 2010
	Quercus eduardii	Infusion 1% (w/v)	Rocha-Guzmán et al., 2012
	Quercus eduardii	Infusion 1% (w/v)	Moreno-Jiménez et al., 2015
	Quercus eduardii	Aqueous acetone (70%) followed by methanol (50%)	Gamboa-Gómez et al., 2012
	Quercus grisea	Infusion	Sánchez-Burgos et al., 2013
	Quercus incana	MeOH	Sarwar et al., 2015
	Quercus laeta	Infusion	Sánchez-Burgos et al., 2013
	Quercus obtusata	Infusion	Sánchez-Burgos et al., 2013
	Quercus pétrea	EtOH 80%	Popovic et al., 2013
	Quercus resinosa	Infusión	Rocha-Guzmán et al., 2009
	Quercus resinosa	Microencapsulated infusion by spray drying	Rocha-Guzmán et al., 2010
	Quercus resinosa	Infusion 1% (w/v)	Rocha-Guzmán et al., 2012
	Quercus resinosa	Microencapsulated infusion by spray drying	Gallegos-Infante et al., 2013
	Quercus resinosa	Infusion	Sánchez-Burgos et al., 2013
	Quercus resinosa	Infusión 1% (w/v)	Vázquez-Cabral et al., 2014

Table 1. (Continued)

Biological activity	Oak species	Extraction method	References
	Quercus resinosa	Infusion 1% (w /v) fermented with the fungus *Kombucha*	Vázquez-Cabral et al., 2014
	Quercus resinosa	Infusion 1% (w /v) fermented with the fungus *Kombucha*	Vázquez-Cabral et al., 2017
	Quercus resinosa	Aqueous acetone (70%) followed by methanol (50%)	Gamboa-Gómez et al., 2012
	Quercus robur	MeOH	Uddin and Rauf, 2012
	Quercus robur	EtOH 80%	Popovic et al., 2013
	Quercus robur	MeOH 80%	Moharram et al., 2015
	Quercus sideroxyla	Infusion	Rivas-Arreola et al., 2010
	Quercus sideroxyla	Infusion	Rocha-Guzmán et al., 2012
	Quercus sideroxyla	Infusion 1% (w/v)	Moreno-Jiménez et al., 2015
	Quercus sideroxyla	Aqueous acetone (70%) followed by methanol (50%)	Gamboa-Gómez et al., 2012
	Quercus suber	Soxhlet with MeOH	Custódio et al., 2015
Antitopoisomerase	*Quercus resinosa*	Infusion	Sánchez-Burgos et al., 2013
Enzymatic inhibition of α-glucosidase	*Quercus gilva*	MeOH	Indrianingsih et al., 2015
	Quercus phillyraeoides	MeOH	Indrianingsih and Tachibana, 2016
	Quercus suber	Soxhlet with MeOH	Custódio et al., 2015
Enzymatic inhibition of cholinesterase	*Quercus suber*	Soxhlet with MeOH	Custódio et al., 2015
Gatroprotective	*Quercus grisea*	Infusion	Sánchez-Burgos et al., 2013
	Quercus laeta	Infusion	Sánchez-Burgos et al., 2013
	Quercus obtusata	Infusion	Sánchez-Burgos et al., 2013
	Quercus robur	MeOH 80%	Moharram et al., 2015

OAK LEAVES AS A SOURCE OF PHYTOCHEMICAL COMPOUNDS

The growth conditions of oak species influence the chemical composition of their organs including the leaves. It has been observed that the growth of trees of *Q. pubescens* and *Q. ilex* in conditions rich in CO_2, shows no presence of proteins and chlorophyll, unlike those grown under environmental conditions. Likewise, differences have been identified in the activities of antioxidant enzymes such as catalase, glutathione peroxidase and ascorbate peroxidase, as well as the sum of dehydroascorbate and ascorbate, which are diminished in leaves of trees grown in environments enriched with CO_2 compared to the environmental concentrations of CO_2 (Schwanz and Polle, 1998). In another study, it was observed that *Q. rubur* leaves grown under elevated CO_2 conditions decreased their total nitrogen concentrations, total phenolic compounds increased and the leaf was softer than normal. The administration of these leaves as food for worms of the winter moth, promoted that they have eaten more and produced more feces in comparison with the cultivated in normal environmental conditions. In elevated CO_2 conditions the development of the leaf was not modified, unlike its nutritional composition.

Among the biological compounds reported in leaves of different oak species are alkaloids, anthraquinones, cardiac glycosides, carotenoids, chlorophyll, cumarins, flavonoids, flobatannins, glycosides, hydrolysable tannins, phenols, proanthocyanidins, proline, proteins, reducing sugars, saponins, steroids, tannins, and terpenoids (Table 2).

The phytochemical groups most widely explored in oak leaves are the saponins and polyphenols. The name saponin is derived from the Latin *"sapo"*, which means *"soap"*, due to the formation of foam in aqueous solution. Its structure results from the union of nonpolar aglycones to one or more monosaccharide molecules (Sparg et al., 2004). Its glycosidic structure contains mostly oligosaccharides, arranged in a linear or branched way, and has main sugar residues as glucose, galactose, xylose, arabinose, rhamnose or glucuronic acid, glycosidically bound to an aglycone. Saponins are usually found in roots, tubers, leaves, blooms or seeds (Sparg et al., 2004).

Ethanolic extracts of *Q. laurifolia* Michx. have shown four triterpenic saponins: galloyl ester of quercilicoside-A, quercilicoside-A, β-D-glucopyranosyl ester of the 2α,3β,19α,24-tetrahydroxyolean-l2-ene-28-oic acid (i.e., sericoside), and β-D-glucopyranosyl ester of the 2α,3 β,23-trihydroxyolean-l2-ene-28-oic acid (i.e., arjunglucoside II); and eleven glycosides flavonoids: 3-O-isorhamnetin β-D-glucopyranoside, 3-O-kaempferol β-D-glucopyranoside (i.e., astragalin), 3-O-kaempferol 6-O-(trans-*p*-coumaroyl)- β-D-glucopyranoside (i.e., tiliroside), 3-O-kaempferol 6-O-galloyl-β-D-glucopyranoside, 3-O-quercetin 6-O-galloyl-β-D-glucopyranoside (i.e., tellimoside), 3-O-kaempferol 2,6-di-O-(trans-*p*-coumaroyl)-β-D-glucopyranoside, 3-O-kaempferol 2-0-(cis-*p*-coumaroyl)-6-O-(trans-*p*-coumaroyl)-β-D-glucopyranoside), 3-O-kaempferol 3,4-di-O-acetyl-2,6-di-O-(trans-*p*-coumaroy1)-β-D-glucopyranoside, 3-O-quercetin-β-D-galactopyranoside (i.e., hyperoside), 3-O-kaempferol 2,3-di-O-acetyl-4-O- (cis-p-coumaroyl) -6-O-(trans-*p*-coumaroyl) -β-D-glucopyranoside), and 3-O-quercetin α-L-arabinopyranoside (i.e., guajaverin) (Romussi and Caviglioli, 1993) (Figure 1).

Phenolic Composition of Different *Quercus* Species

Polyphenols are secondary metabolites naturally present in plants. These molecules are responsible for structural and protective functions in plants that act in defense mechanisms against pathogens, parasites and predators, and contribute to taste, color, astringency and bitterness. Their consumption has been suggested for its antioxidant potential and direct effects in the prevention of diseases related to oxidative stress (Dai and Mumper, 2010). These properties have led to the use of different technologies for the qualification and quantification of phenolic compounds in oak leaves.

Table 2. Phytochemical compounds reported in different *Quercus* species

Phytochemicals	*Quercus* species	References	Phytochemicals	*Quercus* species	References
Alkaloids	*Q. baloot*	Khurram et al., 2012	Proanthocyanidins	*Q. macranthera*	Şohretoğlu et al., 2007
	Q. dilatata	Jamil et al., 2012		*Q. petrea*	Popovic et al., 2013
	Q. durifolia	Sánchez-Burgos et al., 2013		*Q. pubescens*	Şohretoğlu et al., 2007
	Q. grisea	Sánchez-Burgos et al., 2013		*Q. robur*	Popovic et al., 2013
	Q. laeta	Sánchez-Burgos et al., 2013	Phenols	*Q. brantii*	Mirzaei and Mirzaei, 2013
	Q. obtusata	Sánchez-Burgos et al., 2013		*Q. petrea*	Popovic et al., 2013
	Q. resinosa	Sánchez-Burgos et al., 2013		*Q. robur*	Popovic et al., 2013
Anthraquinones	*Q. montana*	Subhashini et al., 2016	Proline	*Q. robur*	Popovic et al., 2013
	Q. dilatata	Jamil et al., 2012		*Q. petrea*	Popovic et al., 2013
	Q. griffithii	Subhashini et al., 2016	Proteins	*Q. suber*	Subhashini et al., 2016
	Q. macrocarpa	Subhashini et al., 2016		*Q. serrata*	Subhashini et al., 2016
Cardiac glycosides	*Q. dilatata*	Jamil et al., 2012	Reducing sugars	*Q. grifffithii*	Subhashini et al., 2016
	Q. durifolia	Sánchez-Burgos et al., 2013		*Q. macrocarpa*	Subhashini et al., 2016
	Q. grisea	Sánchez-Burgos et al., 2013		*Q. montana*	Subhashini et al., 2016
	Q. laeta	Sánchez-Burgos et al., 2013		*Q. robur*	Uddin and Rauf, 2012
	Q. resinosa	Sánchez-Burgos et al., 2013		*Q. serrata*	Subhashini et al., 2016
Carotenoids	*Q. robur*	Popovic et al., 2013		*Q. suber*	Subhashini et al., 2016
Coumarins	*Q. dilatata*	Jamil et al., 2012	Saponins	*Q. durifolia*	Sánchez-Burgos et al., 2013
Chlorophyll	*Q. robur*	Popovic et al., 2013		*Q. dilatata*	Jamil et al., 2012
	Q. petrea	Popovic et al., 2013		*Q. grifffithii*	Subhashini et al., 2016
Flavonoids	*Q. baloot*	Khurram et al., 2011		*Q. grisea*	Sánchez-Burgos et al., 2013
	Q. brantii	Mirzaei and Mirzaei, 2013		*Q. laeta*	Sánchéz-Burgos et al., 2013
	Q. cerris	Şohretoğlu et al., 2007		*Q. macrocarpa*	Subhashini et al., 2016
	Q. coccifera	Şohretoğlu et al., 2007		*Q. montana*	Subhashini et al., 2016
	Q. durifolia	Sánchez-Burgos et al., 2013		*Q. resinosa*	Sánchez-Burgos et al., 2013

Table 2. (Continued)

Phytochemicals	*Quercus* species	References	Phytochemicals	*Quercus* species	References
	Q. griffithii	Subhashini et al., 2016		*Q. robur*	Uddin and Rauf, 2012
	Q. grisea	Sánchez-Burgos et al., 2013		*Q. serrata*	Subhashini et al., 2016
	Q. laeta	Sánchez-Burgos et al., 2013		*Q. suber*	Subhashini et al., 2016
	Q. macranthera	Şohretoğlu et al., 2007	Steroids	*Q. baloot*	Khurram et al., 2011
	Q. macrocarpa	Subhashini et al., 2016		*Q. durifolia*	Sánchez-Burgos et al., 2013
	Q. montana	Subhashini et al., 2016		*Q. robur*	Uddin and Rauf, 2012
	Q. obtusata	Sánchez-Burgos et al., 2013	Tannins	*Q. baloot*	Khurram et al., 2011
	Q. petrea	Popovic et al., 2013		*Q. dilatata*	Jamil et al., 2012
	Q. pubescens	Şohretoğlu et al., 2007		*Q. durifolia*	Sánchez-Burgos et al., 2013
	Q. resinosa	Sánchez-Burgos et al., 2013		*Q. griffithii*	Subhashini et al., 2016
	Q. robur	Popovic et al., 2013		*Q. grisea*	Sánchez-Burgos et al., 2013
	Q. serrata	Subhashini et al., 2016		*Q. laeta*	Sánchez-Burgos et al., 2013
	Q. suber	Subhashini et al., 2016		*Q. macrocarpa*	Subhashini et al., 2016
Flobatannins	*Q. griffithii*	Subhashini et al., 2016		*Q. montana*	Subhashini et al., 2016
	Q. macrocarpa	Subhashini et al., 2016		*Q. obtusata*	Sánchez-Burgos et al., 2013
	Q. montana	Subhashini et al., 2016		*Q. petrea*	Popovic et al., 2013
	Q. serrata	Subhashini et al., 2016		*Q. resinosa*	Sánchez-Burgos et al., 2013
	Q. suber	Subhashini et al., 2016		*Q. robur*	Uddin and Rauf, 2012
Glycosides	*Q. griffithii*	Subhashini et al., 2016		*Q. robur*	Popovic et al., 2013
	Q. macrocarpa	Subhashini et al., 2016		*Q. serrata*	Subhashini et al., 2016
	Q. montana	Subhashini et al., 2016		*Q. suber*	Subhashini et al., 2016
	Q. serrata	Subhashini et al., 2016	Terpenoids	*Q. baloot*	Khurram et al., 2011
	Q. suber	Subhashini et al., 2016		*Q. baloot*	Khurram et al., 2012

Phytochemicals	*Quercus* species	References	Phytochemicals	*Quercus* species	References
Hydrolyzable tannins	*Q. cerris*	Şohretoğlu et al., 2007		*Q. dilatata*	Jamil et al., 2012
	Q. coccifera	Şohretoğlu et al., 2007		*Q. griffithii*	Subhashini et al., 2016
	Q. dilatata	Jamil et al., 2012		*Q. macrocarpa*	Subhashini et al., 2016
	Q. macranthera	Şohretoğlu et al., 2007		*Q. montana*	Subhashini et al., 2016
	Q. pubescens	Şohretoğlu et al., 2007		*Q. robur*	Uddin and Rauf, 2012
	Q. cerris	Şohretoğlu et al., 2007		*Q. serrata*	Subhashini et al., 2016
	Q. coccifera	Şohretoğlu et al., 2007		*Q. suber*	Subhashini et al., 2016

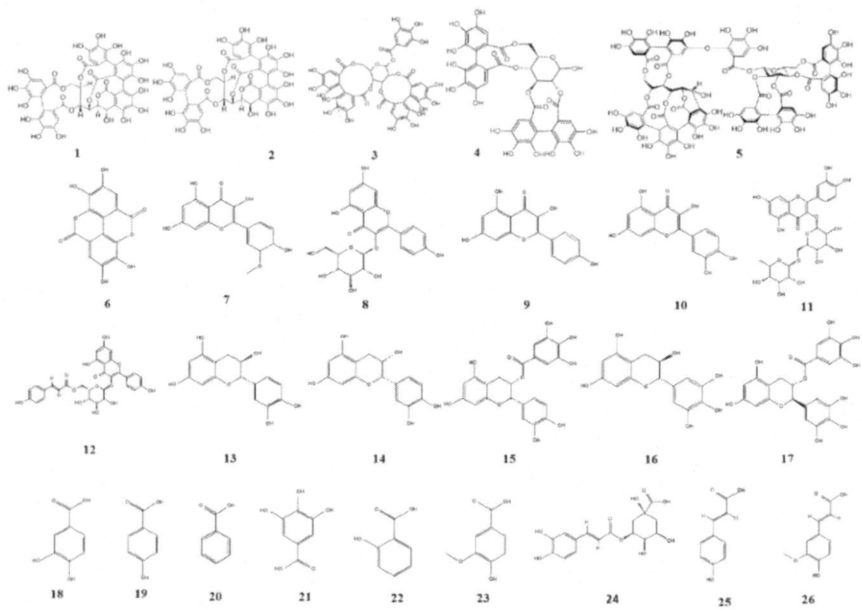

Figure 1. Estructure of representative phenolic compounds in *Quercus* leaves.
1 Castalagin; 2 Vescalagin; 3 Casuarictin: 4 Pedunculagin; 5 Cocciferin D2; 6 Ellagic acid; 7 Isorhamnetin; 8 Kaempferol-3-O-D-glucoside; 9 Kaempferol; 10 Quercetin; 11 Rutin; 12 Tiliroside; 13 Catechin; 14 Epicatechin; 15 Epicatechin gallate; 16 Gallocatechin; 17 Gallocatechin gallate; 18 3,4-dihydroxybenzoic acid; 19 4-hydroxybenzoic acid; 20 Benzoic acid; 21 Gallic acid; 22 Salicylic acid; 23 Vanillic acid; 24 Chlorogenic acid; 25 p-coumaric acid; 26 Ferulic acid.

Technologies as two-dimensional paper chromatography (2-DPC), thin layer chromatography (TLC), and UV spectroscopy, have allowed to observe the high presence of different flavonoids as apigenin, kaempferol, myricetin, naringenin, quercetin and rutin (Figure 1) in *Q. brantii* leaf, gall and bark (Noori et al., 2015). Plus, a spectrophotometric analysis of ethanol extracts (70% ethanol) of *Q. dentata* leaves allowed the elucidation of four acylated flavonoid glycosides (kaempferol 3-O-[2''-O-(trans-p-coumaroyl)-6''-O-actyl]-β-D-glucopyranoside, kaempferol 3-O-[2''-di-O-(trans-p-coumaroyl)-3'',4''-di-O-acetyl]-β-D-glucopyranoside, kaempferol 3-O-[2''-di-O-(trans-p-coumaroyl)-β-D-glucoside, 3-O-kaempferol-6-O-(trans-p-coumaroyl)-β-D-glucoside), and another five flavonoids (kaempferol, quercetin, 3-O-kaempferol-β-D-glucoside, 3-O-isorhamnetin-β-D-glucoside, 3-O-quercetin-β-D-glucoside) (Zhou et al., 2001).

In 2009, Karioti et al. applied an extraction scheme to deactivate the enzymes responsible for the degradation, oxidation and polymerization of proanthocyanidin. They observed proanthocyanidin glycosides (e.g., afzelechins), proanthocyanidin B3, prodelfinidin C, and small species as oxidation products of catechin (e.g., dehydrodicatechin A). Additionally, they found some flavonoids: quercetin, two flavonol glycosides (e.g., quercetin-3-O-glucopyranoside and isorhamnetin-3-O-glucopyranoside), four acylated glycosides of flavonol (e.g., kaempferol-3-O-(6'-galloyl)-glucopyranoside, quercetin-3-O-(6'-galiloyl)-glucopyranoside, tiliroside, and kaempferol-3-O-(2",6"-di-E-*p*-coumaroyl)-glucopyranoside) and ellagic acid (Figure 1) in the methanolic extracts from leaves of *Q. ilex* (Güllüce et al., 2004).

Likewise, chromatographic techniques have allowed the identification of phenolic compounds in methanol extracts (Moharram et al., 2015), and acetonic extracts (HPLC-DAD and HPLC-ESI-MS) to name a few, in leaves of different *Quercus* species (Salminen et al., 2004). This technology has made possible to identify the presence of certain chemical groups, mainly of phenolic nature, such as hydrolysable tannins, flavanols, flavonols, phenolic acids, etc. (Table 3).

Some studies have shown differences in the phenolic composition of oak leaves according to the subgenus to which they belong (*Erythrobalanus, Leucobalanus* and *Protobalanus*). In this way, through an analysis of size and shape, twelve species of oak from the Chihuahuan Desert were classified based on their phenolic composition, and the compounds with the greatest influence on the interspecific allometric variation of the phenolic compounds were determined. It was found that the subgenus of black oak (*Q. emory* and *Q. hypoleucoides*) shows lower phenolic content and diversity than the subgenus of white oak (*Q. arizonica*, *Q. gambelii*, *Q. grisea*, *Q. havardii*, *Q. muhlenbergii*, *Q. mohriana*, *Q. oblongifolia*, *Q. pungens Q. rugosa*, and *Q. turbinella*). Phenolic acids, ellagitannins, and flavonoid glycosides were explored as part of the chemical characterization of the species studied (Yarnes et al., 2006). Also, in a study with leaves of *Q. resinosa*, one of the most recently studied Mexican oak species for its biological effects, it was observed that this species belonging to the group of white oaks together with *Q. grisea*, *Q. arizonica* and *Q. convallata*, showed the highest content of hydrolysable tannins (494–2844 mg/L), and greater diversity and content of phenolic compounds compared to the red oak species (García-Villalba et al., 2013).

Alternatively, hybridization and seasonality are other parameters that affect the phytochemical composition of oak leaves. Leaves of the hybrid *Q. gambelli* x *Q. grisea*, have shown variations in the concentrations of ellagitannins and flavonoid glycosides according to the month of collection. Castalagin (Figure 1) has been observed in higher concentrations in September and in minor ones in June (Yarnes et al., 2006). Also, differences in phenolic concentrations of *Q. sideroxyla* leaves between spring and summer have been observed. The total phenolic content determined by the Folin-Ciocalteu method is diminished in spring and increased in summer (Arámbula-Salazar et al., 2010). For its part, in a study conducted on *Q. ilex* leaves during regrowth after a forest fire in summer and winter, has shown higher concentrations of hexosides of the flavonols, quercetin, kaempferol and rhamnetin in winter. Correspondingly, epicatechin was observed in similar concentrations in both seasons (Brossa et al., 2009). In this way, the climate has shown correlation with the chemical composition of oak leaves, as described by Passarinho et al., (2006), who observed in *Q. suber* the highest concentrations of quinic acid in spring and early summer and contrary to quercitol in autumn-winter season. This response has been related to the protective role of both compounds; the first against heat and high irradiation, and the second during the plant stress (Passarinho et al., 2006). In a recent study using green and freshly senescent leaves of *Q. rubra*, the influence of temperature (heating and non-heating) and rain precipitation (i.e., hot-dry and hot-humid environments) on the tannic composition was observed. Proanthocyanidins, the main group present in leaves of this species showed differences in the degree of polymerization, according to the environmental exposure. In dry and environmental conditions, tannin concentrations were higher, and they were poorly polymerized, while in wet conditions, tannin concentrations were lower but with a higher degree of polymerization (Top et al., 2017).

Table 3. Phenolic compounds reported in different extracts of *Quercus* leaves

Phenolics	*Quercus* species	References	Phenolics	*Quercus* species	References
HYDROLYZABLE TANNINS			FLAVANOLS		
Castalagin	*Q. arizonica*	García-Villalba et al., 2017	Catechin	*Q. arizonica*	Vázquez-Cabral et al., 2017
	Q. arizonica	Yernes et al., 2006		*Q. aucheri*	Sakar et al., 2005
	Q. convallata	García-Villalba et al., 2017		*Q. convallata*	Vázquez-Cabral et al., 2017
	Q. emoryi	Yernes et al., 2006		*Q. durifolia*	Rocha-Guzmán et al., 2012
	Q. gambelii	Yernes et al., 2006		*Q. durifolia*	Moreno-Jiménez et al., 2015
	Q. grisea	García-Villalba et al., 2017		*Q. durifolia*	Rivas-Arreola et al., 2010
	Q. grisea	Yernes et al., 2006		*Q. durifolia*	Sánchez-Burgos et al., 2013
	Q. havardii	Yernes et al., 2006		*Q. eduardii*	Rocha-Guzmán et al., 2012
	Q. hypoleucoides	Yernes et al., 2006		*Q. eduardii*	Moreno-Jiménez et al., 2015
	Q. mohriana	Yernes et al., 2006		*Q. eduardii*	Rivas-Arreola et al., 2010
	Q. muhlenbergii	Yernes et al., 2006		*Q. ilex*	Karioti et al., 2009
	Q. oblongifolia	Yernes et al., 2006		*Q. oleoides*	Moctezuma et al., 2014
	Q. oleoides	Moctezuma et al., 2014		*Q. resinosa*	Rocha-Guzmán et al., 2012
	Q. pungens	Yernes et al., 2006		*Q. resinosa*	Vázquez-Cabral et al., 2014
	Q. resinosa	García-Villalba et al., 2017		*Q. resinosa*	Vázquez-Cabral et al., 2017
	Q. robur	Salminen et al., 2004		*Q. sideroxila*	Rocha-Guzmán et al., 2012
	Q. rugosa	Yernes et al., 2006		*Q. sideroxila*	Moreno-Jimenez et al., 2015
	Q. turbinella	Yernes et al., 2006		*Q. sideroxila*	Rivas-Arreola et al., 2010

Table 3. (Continued)

Phenolics	*Quercus* species	References	Phenolics	*Quercus* species	References
Castavaloninic acid	*Q. arizonica*	Yernes et al., 2006		*Q. suber*	Custódio et al., 2015
	Q. gambelli	Yernes et al., 2006	Epicatechin	*Q. arizonica*	Vázquez-Cabral et al., 2017
	Q. grisea	Yernes et al., 2006		*Q. convallata*	Vázquez-Cabral et al., 2017
	Q. havardii	Yernes et al., 2006		*Q. durifolia*	Rocha-Guzmán et al., 2012
	Q. mohriana	Yernes et al., 2006		*Q. eduardii*	Rocha-Guzmán et al., 2012
	Q. muhlenbergii	Yernes et al., 2006		*Q. ilex*	Brossa et al., 2009
	Q. oblongifolia	Yernes et al., 2006		*Q. resinosa*	Rocha-Guzmán et al., 2012
	Q. pungens	Yernes et al., 2006		*Q. resinosa*	Vázquez-Cabral et al., 2014
	Q. rugosa	Yernes et al., 2006		*Q. resinosa*	Vázquez-Cabral et al., 2017
	Q. turbinella	Yernes et al., 2006		*Q. sideroxila*	Rocha-Guzmán et al., 2012
Casuarictin (Galoil-bis-HHDP-hexoside)	*Q. arizonica*	García-Villalba et al., 2017		*Q. sideroxila*	Moreno-Jiménez et al., 2015
	Q. grisea	García-Villalba et al., 2017		*Q. suber*	Custódio et al., 2015
	Q. resinosa	García-Villalba et al., 2017	Epicatechin gallate	*Q. arizonica*	Vázquez-Cabral et al., 2017
	Q. robur	Salminen et al., 2004		*Q. convallata*	Vázquez-Cabral et al., 2017
Cocciferin D2	*Q. arizonica*	Yernes et al., 2006		*Q. ilex*	Brossa et al., 2009
	Q. convallata	García-Villalba et al., 2017		*Q. resinosa*	Vázquez-Cabral et al., 2017
	Q. gambelii	Yernes et al., 2006	Gallocatechin	*Q, aucheri*	Sakar et al., 2005
	Q. grisea	Yernes et al., 2006		*Q. arizonica*	Vázquez-Cabral et al., 2017
	Q. havardii	Yernes et al., 2006		*Q. convallata*	Vázquez-Cabral et al., 2017
	Q. mohriana	Yernes et al., 2006		*Q. resinosa*	Vázquez-Cabral et al., 2017
	Q. muhlenbergii	Yernes et al., 2006	Gallocatechin gallate	*Q. arizonica*	Vázquez-Cabral et al., 2017

Phenolics	*Quercus* species	References	Phenolics	*Quercus* species	References
	Q. oblongifolia	Yernes et al., 2006		*Q. convallata*	Vázquez-Cabral et al., 2017
	Q. oleoides	Moctezuma et al., 2014		*Q. resinosa*	Vázquez-Cabral et al., 2017
	Q. oleoides	Moctezuma et al., 2014	PHENOLIC ACIDS		
	Q. pungens	Yernes et al., 2006	3-4-dihydroxybenzoic acid	*Q. arizonica*	Vázquez-Cabral et al., 2017
	Q. robur	Salminen et al., 2004		*Q. convallata*	Vázquez-Cabral et al., 2017
	Q. rugosa	Yernes et al., 2006		*Q. convallata*	Vázquez-Cabral et al., 2017
	Q. turbinella	Yernes et al., 2006		*Q. durifolia*	Vázquez-Cabral et al., 2017
Digaloil-HHDP-hexoside	*Q. arizonica*	García-Villalba et al., 2017		*Q. eduardii*	Vázquez-Cabral et al., 2017
	Q. convallata	García-Villalba et al., 2017		*Q. grisea*	Vázquez-Cabral et al., 2017
	Q. grisea	García-Villalba et al., 2017		*Q. resinosa*	García-Villalba et al., 2017
	Q. resinosa	García-Villalba et al., 2017		*Q. resinosa*	Vázquez-Cabral et al., 2017
	Q. sideroxila	García-Villalba et al., 2017		*Q. sideroxila*	Vázquez-Cabral et al., 2017
Mongolinin A	*Q. arizonica*	Yernes et al., 2006	4-hydroxybenzoic acid	*Q. resinosa*	Rocha-Guzmán et al., 2012
	Q. gambelli	Yernes et al., 2006		*Q. sideroxila*	García-Villalba et al., 2017
	Q. grisea	Yernes et al., 2006		*Q. sideroxila*	Rocha-Guzmán et al., 2012
	Q. havardii	Yernes et al., 2006	Benzoic acid	*Q. durifolia*	Rocha-Guzmán et al., 2012
	Q. mohriana	Yernes et al., 2006		*Q. eduardii*	Rocha-Guzmán et al., 2012
	Q. oblongifolia	Yernes et al., 2006		*Q. resinosa*	Rocha-Guzmán et al., 2012
	Q. oleoides	Yernes et al., 2006		*Q. resinosa*	Vázquez-Cabral et al., 2014
	Q. pungens	Yernes et al., 2006		*Q. sideroxila*	Rocha-Guzmán et al., 2012
	Q. rugosa	Yernes et al., 2006	Caffeic acid	*Q. arizonica*	García-Villalba et al., 2017
	Q. turbinella	Yernes et al., 2006		*Q. arizonica*	Vázquez-Cabral et al., 2017
Pedunculagin (Di-HHDP- hexoside)	*Q. arizonica*	Yernes et al., 2006		*Q. convallata*	Vázquez-Cabral et al., 2017
	Q. arizonica	García-Villalba et al., 2017		*Q. durifolia*	Rocha-Guzmán et al., 2012
	Q. convallata	García-Villalba et al., 2017		*Q. durifolia*	Moreno-Jiménez et al., 2015
	Q. gambelii	Yernes et al., 2006		*Q. eduardii*	Moreno-Jiménez et al., 2015

Table 3. (Continued)

Phenolics	*Quercus* species	References	Phenolics		References
	Q. grisea	Yernes et al., 2006		*Q. grisea*	García-Villalba et al., 2017
	Q. grisea	García-Villalba et al., 2017		*Quercus* species	García-Villalba et al., 2017
	Q. havardii	Yernes et al., 2006		*Q. resinosa*	Vázquez-Cabral et al., 2014
	Q. mohriana	Yernes et al., 2006		*Q. resinosa*	Vázquez-Cabral et al., 2017
	Q. muhlenbergii	Yernes et al., 2006		*Q. sideroxila*	Moreno-Jiménez et al., 2015
	Q. oblongifolia	Yernes et al., 2006	Chlorogenic acid	*Q. arizonica*	García-Villalba et al., 2017
	Q. oleoides	Moctezuma et al., 2014		*Q. arizonica*	Vázquez-Cabral et al., 2017
	Q. oleoides	Moctezuma et al., 2014		*Q. convallata*	García-Villalba et al., 2017
	Q. pungens	Yernes et al., 2006		*Q. convallata*	Vázquez-Cabral et al., 2017
	Q. resinosa	García-Villalba et al., 2017		*Q. durifolia*	García-Villalba et al., 2017
	Q. robur	Salminen et al., 2004		*Q. durifolia*	Moreno-Jiménez et al., 2015
	Q. rugosa	Yernes et al., 2006		*Q. eduardii*	García-Villalba et al., 2017
	Q. sideroxila	García-Villalba et al., 2017		*Q. emoryi*	Yernes et al., *2006*
	Q. turbinella	Yernes et al., 2006		*Q. grisea*	García-Villalba et al., 2017
Trigaloil-hexoside	*Q. arizonica*	García-Villalba et al., 2017		*Q. hypoleucoides*	Yernes et al., *2006*
	Q. grisea	García-Villalba et al., 2017		*Q. mohriana*	Yernes et al., *2006*
	Q. resinosa	García-Villalba et al., 2017		*Q. muhlenbergii*	Yernes et al., *2006*
Vescalagin	*Q. arizonica*	Yernes et al., 2006		*Q. oblongifolia*	Yernes et al., *2006*
	Q. arizonica	García-Villalba et al., 2017		*Q. pungens*	Yernes et al., *2006*
	Q. convallata	García-Villalba et al., 2017		*Q. resinosa*	García-Villalba et al., 2017
	Q. emoryi	Yernes et al., 2006		*Q. resinosa*	Vázquez-Cabral et al., 2014
	Q. gambelii	Yernes et al., 2006		*Q. resinosa*	Vázquez-Cabral et al., 2017
	Q. grisea	García-Villalba et al., 2017		*Q. suber*	Custódio et al., 2015

Phenolics	*Quercus* species	References	Phenolics	*Quercus* species	References
	Q. grisea	Yernes et al., 2006		*Q. turbinella*	Yernes et al., *2006*
	Q. havardii	Yernes et al., 2006		*Q.sideroxila*	García-Villalba et al., 2017
	Q. hypoleucoides	Yernes et al., 2006	Coumaric acid	*Q. arizonica*	García-Villalba et al., 2017
	Q. mohriana	Yernes et al., 2006		*Q. arizonica*	Vázquez-Cabral et al., 2017
	Q. muhlenbergii	Yernes et al., 2006		*Q. convallata*	Vázquez-Cabral et al., 2017
	Q. oblongifolia	Yernes et al., 2006		*Q. eduardii*	Moreno-Jiménez et al., 2015
	Q. oleoides	Moctezuma et al., 2014		*Q. grisea*	García-Villalba et al., 2017
	Q. pungens	Yernes et al., 2006		*Q. resinosa*	García-Villalba et al., 2017
	Q. resinosa	García-Villalba et al., 2017		*Q. resinosa*	Vázquez-Cabral et al., 2014
	Q. robur	Salminen et al., 2004		*Q. resinosa*	Vázquez-Cabral et al., 2017
	Q. rugosa	Yernes et al., 2006		*Q. sideroxila*	Moreno-Jimenez et al., 2015
	Q. turbinella	Yernes et al., 2006	Ellagic acid	*Q. arizonica*	García-Villalba et al., 2017
Vescavalonic acid	*Q. arizonica*	García-Villalba et al., 2017		*Q. convallata*	García-Villalba et al., 2017
	Q. convallata	García-Villalba et al., 2017		*Q. durifolia*	García-Villalba et al., 2017
	Q. grisea	García-Villalba et al., 2017		*Q. durifolia*	Rivas-Arreola et al., 2010
	Q. oleoides	Moctezuma et al., 2014		*Q. durifolia*	Moreno-Jiménez et al., 2015
	Q. resinosa	García-Villalba et al., 2017		*Q. eduardii*	García-Villalba et al., 2017
FLAVONOLS				*Q. eduardii*	Moreno-Jiménez et al., 2015
Isorhamnetin-cumaroil-hexoside	*Q. arizonica*	García-Villalba et al., 2017		*Q. eduardii*	Rivas-Arreola et al., 2010
	Q. convallata	García-Villalba et al., 2017		*Q. eduardii*	Rocha-Guzmán et al., 2012
	Q. eduardii	García-Villalba et al., 2017		*Q. grisea*	García-Villalba et al., 2017
	Q. grisea	García-Villalba et al., 2017		*Q. ilex*	Karioti et al., 2009
	Q. resinosa	García-Villalba et al., 2017		*Q. resinosa*	García-Villalba et al., 2017
	Q. sideroxila	García-Villalba et al., 2017		*Q. resinosa*	Vázquez-Cabral et al., 2014

Table 3. (Continued)

Phenolics	*Quercus* species	References	Phenolics	*Quercus* species	References
Isorhamnetin-rhamnoside-hexoside	*Q. arizonica*	García-Villalba et al., 2017		*Q. robur*	Moharram et al., 2015
	Q. durifolia	García-Villalba et al., 2017		*Q. sideroxila*	Moreno-Jiménez et al., 2015
	Q. eduardii	García-Villalba et al., 2017		*Q.sideroxila*	García-Villalba et al., 2017
	Q. grisea	García-Villalba et al., 2017		*Q.sideroxila*	Moreno-Jiménez et al., 2015
	Q. resinosa	García-Villalba et al., 2017		*Q.sideroxila*	Rivas-Arreola et al., 2010
	Q. sideroxila	García-Villalba et al., 2017	Ferulic acid	*Q. arizonica*	García-Villalba et al., 2017
Kaempferol-3-O-glycoside	*Q. convallata*	Vázquez-Cabral et al., 2017		*Q. arizonica*	Vázquez-Cabral et al., 2017
	Q. resinosa	Vázquez-Cabral et al., 2017		*Q. convallata*	Vázquez-Cabral et al., 2017
Kaempferol-acetyl-cumaroil-hexoside	*Q. arizonica*	García-Villalba et al., 2017		*Q. durifolia*	Moreno-Jiménez et al., 2015
	Q. convallata	García-Villalba et al., 2017		*Q. eduardii*	García-Villalba et al., 2017
	Q. eduardii	García-Villalba et al., 2017		*Q. eduardii*	Moreno-Jiménez et al., 2015
	Q. grisea	García-Villalba et al., 2017		*Q. grisea*	García-Villalba et al., 2017
	Q. resinosa	García-Villalba et al., 2017		*Q. resinosa*	Vázquez-Cabral et al., 2014
	Q. sideroxila	García-Villalba et al., 2017		*Q. resinosa*	Vázquez-Cabral et al., 2017
Kaempferol-acetyl- hexoside	*Q. arizonica*	García-Villalba et al., 2017		*Q. sideroxila*	Moreno-Jiménez et al., 2015
	Q. convallata	García-Villalba et al., 2017	Gallic acid	*Q. arizonica*	García-Villalba et al., 2017
	Q. grisea	García-Villalba et al., 2017		*Q. arizonica*	Vázquez-Cabral et al., 2017
	Q. resinosa	García-Villalba et al., 2017		*Q. convallata*	García-Villalba et al., 2017
Kaempferol	*Q. brantii*	Noori et al., 2015		*Q. convallata*	Vázquez-Cabral et al., 2017
Kaempferol-cumaroil- hexoside	*Q. arizonica*	García-Villalba et al., 2017		*Q. durifolia*	García-Villalba et al., 2017
	Q. convallata	García-Villalba et al., 2017		*Q. durifolia*	Moreno-Jiménez et al., 2015
	Q. durifolia	García-Villalba et al., 2017		*Q. durifolia*	Rivas-Arreola et al., 2010
	Q. eduardii	García-Villalba et al., 2017		*Q. durifolia*	Rocha-Guzmán et al., 2012
	Q. grisea	García-Villalba et al., 2017		*Q. eduardii*	García-Villalba et al., 2017

Phenolics	*Quercus* species	References	Phenolics	*Quercus* species	References
	Q. resinosa	García-Villalba et al., 2017		*Q. eduardii*	Moreno-Jiménez et al., 2015
	Q. sideroxila	García-Villalba et al., 2017		*Q. eduardii*	Rivas-Arreola et al., 2010
Kaempferol-galloyl- hexoside	*Q. arizonica*	García-Villalba et al., 2017		*Q. eduardii*	Rocha-Guzmán et al., 2012
	Q. convallata	García-Villalba et al., 2017		*Q. grisea*	García-Villalba et al., 2017
	Q. eduardii	García-Villalba et al., 2017		*Q. resinosa*	García-Villalba et al., 2017
	Q. grisea	García-Villalba et al., 2017		*Q. resinosa*	Rocha-Guzmán et al., 2012
	Q. resinosa	García-Villalba et al., 2017		*Q. resinosa*	Vázquez-Cabral et al., 2014
	Q. sideroxila	García-Villalba et al., 2017		*Q. resinosa*	Vázquez-Cabral et al., 2017
Kaempferol glycoside	*Q. arizonica*	Yernes et al., 2006		*Q. robur*	Moharram et al., 2015
	Q. durifolia	Rivas-Arreola et al., 2010		*Q. robur*	Salminen et al., 2004
	Q. gambelii	Yernes et al., 2006		*Q. suber*	Custódio et al., 2015
	Q. grisea	Yernes et al., 2006		*Q.sideroxila*	García-Villalba et al., 2017
	Q. havardii	Yernes et al., 2006		*Q.sideroxila*	Moreno-Jiménez et al., 2015
	Q. mohriana	Yernes et al., 2006		*Q.sideroxila*	Rivas-Arreola et al., 2010
	Q. muhlenbergii	Yernes et al., 2006		*Q.sideroxila*	Rocha-Guzmán et al., 2012
	Q. oblongifolia	Yernes et al., 2006			
	Q. pungens	Yernes et al., 2006	Salicylic acid	*Q. arizonica*	García-Villalba et al., 2017
	Q. robur	Salminen et al., 2004		*Q. arizonica*	Vazquez-Cabral et al., 2017
	Q. rugosa	Yernes et al., 2006		*Q. convallata*	García-Villalba et al., 2017
	Q. turbinella	Yernes et al., 2006		*Q. convallata*	Vazquez-Cabral et al., 2017
Kaempferol-malonyl- hexoside	*Q. convallata*	García-Villalba et al., 2017		*Q. durifolia*	García-Villalba et al., 2017
	Q. grisea	García-Villalba et al., 2017		*Q. eduardii*	Rocha-Guzmán et al., 2012
	Q. resinosa	García-Villalba et al., 2017		*Q. resinosa*	Rocha-Guzmán et al., 2012

Table 3. (Continued)

Phenolics	*Quercus* species	References	Phenolics	*Quercus* species	References
Quercetin	*Q. brantii*	Noori et al., 2015		*Q. sideroxila*	García-Villalba et al., 2017
	Q. ilex	Karioti et al., 2009	Vanillic acid	*Q. durifolia*	Moreno-Jiménez et al., 2015
	Q. robur	Salminen et al., 2004		*Q. eduardii*	Moreno-Jiménez et al., 2015
	Q. sideroxila	Rivas-Arreola et al., 2010		*Q. resinosa*	Rocha-Guzmán et al., 2012
	Q. suber	Custódio et al., 2015		*Q. resinosa*	Vázquez-Cabral et al., 2014
Quercetin-cumaroil- hexoside	*Q. arizonica*	García-Villalba et al., 2017		*Q. sideroxila*	Rocha-Guzmán et al., 2012
	Q. convallata	García-Villalba et al., 2017		*Q. sideroxila*	Moreno-Jiménez et al., 2015
	Q. durifolia	García-Villalba et al., 2017			
	Q. eduardii	García-Villalba et al., 2017			
	Q. grisea	García-Villalba et al., 2017			
	Q. resinosa	García-Villalba et al., 2017			
	Q. sideroxila	García-Villalba et al., 2017			
Quercetin-galloyl- hexoside	*Q. arizonica*	García-Villalba et al., 2017			
	Q. convallata	García-Villalba et al., 2017			
	Q. eduardii	García-Villalba et al., 2017			
	Q. resinosa	García-Villalba et al., 2017			
	Q. sideroxila	García-Villalba et al., 2017			
Quercetin glucuronide	*Q. arizonica*	Vázquez-Cabral et al., 2017			
	Q. convallata	Vázquez-Cabral et al., 2017			
	Q. resinosa	Vázquez-Cabral et al., 2017			
Quercetin- hexoside	*Q. durifolia*	García-Villalba et al., 2017			
	Q. grisea	García-Villalba et al., 2017			
	Q. resinosa	García-Villalba et al., 2017			
	Q. sideroxila	García-Villalba et al., 2017			

Phenolics	*Quercus* species	References	Phenolics	*Quercus* species	References
Rutin	*Q. arizonica*	Vázquez-Cabral et al., 2017			
	Q. brantii	Noori et al., 2015			
	Q. convallata	Vázquez-Cabral et al., 2017			
	Q. durifolia	García-Villalba et al., 2017			
	Q. eduardii	García-Villalba et al., 2017			
	Q. grisea	García-Villalba et al., 2017			
	Q. resinosa	Vázquez-Cabral et al., 2017			
	Q. resinosa	García-Villalba et al., 2017			
	Q. sideroxila	García-Villalba et al., 2017			
Tiliroside	*Q. ilex*	Karioti et al., 2009			

Tannins

The group of phenylpropanoids has been widely documented as part of the chemical composition of oak species, presenting high concentrations of condensed and hydrolysable tannins. Several studies support the fact that tannins are presented as a defense mechanism against herbivores within the so-called plant-insect interactions (Vaca-Sánchez et al., 2016).

Condensed tannins and hydrolysable tannins have been found in oak leaves, including *Q. robur*. Salminen et al. (2004) have proposed a biosynthetic route for the formation of hydrolysable tannins in oak leaves. This route begins with the binding of gallic acid and UDP-glucose to form 1-O-galloylglucose. Galloylations and position-specific steps allow the formation of 1,2,3,4,6-penta-O-galloylglucose, from which the first ellagitannin of the tellmangradin II via is formed. Then, after the additional oxidative coupling of two galloyl groups of tellmangradin II, casuarictin is formed (Figure 1). Pedunculagin is the precursor of vescalagin and castalagin, and it uses staquarin and casuarinin as intermediates. These authors assume that castavalonic acid is formed by galloylation in the hexahydroxydiphenoyl (HHDP) group of castalagin, and that cocciferin D2 results from the dimerization of pedunculagin and castavalonic acid.

In response to damage caused by herbivore mastication, which accounts for approximately 90% of the total leaf damage, it has been observed in *Q. oleoides* an increase in the production of ellagitannins (e.g., acutissimine B) and condensed tannins unities (i.e., catechin and catechin dimers). In addition, the concentrations of hydrolysable tannins such as mongolinine against damage by mineral herbivores, and mongolinin A and acutissimine B against the leaf miner have been increased. On the other hand, a negative relationship has been observed between the damage by insects that form galls and the response of vescalagin and acutissimin B (Moctezuma et al., 2014). Moreover, the purified tannins of leaves of *Q. ilex* (e.g., acutissimin B, pedunculagin, castalagin and phillyraeoidin A) have shown a high inhibition of lipid peroxidation, observing the best response in the presence of castalagin (53.9%), although this is not significantly related to the extract concentration (e.g., 10, 25 and 50 mg/mL). In addition, caffeic acid, tannic acid, catechin and quercetin present in leaves of *Q. ilex* significantly inhibited lipid peroxidation by TBARS measurements (Khennouf et al., 2010).

In a study where material from *Q. robur* trees was sampled during the summer season, the behavior of condensed and hydrolysable tannins showed to be dependent on the age of leaves. Hydrolysable tannins were found in leaves of all ages, although the young leaves have shown higher concentrations, while the mature leaves showed greater contents in proanthocyanidins. As part of its composition, the presence of flavonoid glycosides, ellagic acid and some hydrolysable tannins was also observed, reporting for the first time cocciferin D2 in *Q. robur* leaves. The increase in leaf phenolic content in summer has been attributed to soil texture and seasonal factors, including solar irradiation, the presence of predators and pathogens; likewise, the decrease in winter is also related to the moderation of mentioned factors (Salminen et al., 2004).

The condensed tannins are polymeric units of flavan-3-ol linked by carbon-carbon bonds and represent the second most abundant group of natural phenolic compounds after lignin. Its name is due to the fact that unlike hydrolysable tannins, during acid hydrolysis, the extension units become colored anthocyanins or insoluble flavanols; therefore, flavan-3-ols are the branch products in the pathway of anthocyanin biosynthesis. Although the greater abundance of condensed tannins in oak species has been observed in the bark, proanthocyanidins have been already described as part of the phenolic composition in leaves of many oak species. Variations in tannins concentrations in plants can be related to the genotype of the plant, environmental conditions and the state of tissue development (Barbehenn and Constabel, 2011).

Leaves of *Q. rubra*, belonging to the classification of red oaks, have shown to contain almost exclusively in its composition condensed tannins. The absence of hydrolysable tannins has shown a negative relationship between their phenolic composition and their oxidative activity, showing low concentrations of semiquinone radicals and browning rates of pedunculagin and pentagaloyl glucose compared to leaves of *Acer saccharum*, which presents higher concentrations of ellagitannins, condensed tannins and galloylglucose compounds.

Age of leaves is a parameter that has determined the presence and concentrations of different phenolic compounds. In the case of condensed tannins, it has been observed that the degree of polymerization of proanthocyanidins in *Q. incana* leaves is increased with the age of the leaves and therefore, their biological activity is modulated with respect to maturation (i.e., antioxidant capacity). Young oak leaves have shown concentrations of tannins (20–284 mg/g) that can double or even triple the amounts observed in mature leaves (8–125 mg/g) (Makkar and Singh, 1993). This characteristic supports the fact that mature oak leaves are used as livestock feed, since young leaves can cause intoxication to animals because of their high tannic acid content. In a study in which twelve brown Swiss bulls were fed with different concentrations of young leaves of *Q. pyrenaica*, the best fermentation of the food was observed at moderate doses of young oak leaves, while at high concentrations the microbiota was not able to detoxify them. With this, it has been concluded that the effect of young Pyrenean Oak leaves fed to cattle in ruminal fermentation *in vitro* depends on the administered dose of the incubated substrate (Doce et al., 2009). On the other hand, Mamber goats fed with leaves of *Q. calliprinos*, *Pistacia lentiscus* and *Cerafonia siliqua,* showed the ability to adapt to the nutritional and consumption environment of high concentration of tannins, which did not affect their serum metabolic indexes or cause intoxication.

Similarly, it is becoming important to understand the metabolism of hydrolysable tannins (i.e., ellagitannins) to urolithins in the cattle rumen (González-Barrio et al., 2012). This is because original oak leaf ellagitannins are transformed into urolithins (i.e., mostly isourolithins A and B) by the fecal and ruminal microbiota. However, it is relevant the difference from monogastric mammals, which have been reported only with urolithin A as the major metabolite formed. This variability may be related with the huge diversity in intestine microbiota and rumen that has been already detected.

The high concentration of tannins in leaves of *Q. incana* has influenced the enzymatic activities of bovine rumen, decreasing the activities of urease, carboxymethylcellulose, glutamate dehydrogenase and alanine aminotransferase, and increasing the activities of glutamate ammonia ligase. The most important affectations have been reported for micro-organisms closely linked to the solid matrix. The negative effects observed in ruminants due to the high tannic concentrations, have promoted the search for organic solvents that allow the detannification of *Q. incana* leaves. Potassium permanganate (0.03M) and potassium dichromate (0.02M) have shown percentages of 95% reduction in treatments performed under agitation for 20 min at 25°C. Likewise, it has been observed that the storage at 30°C of oak leaves in polyethylene bags sealed in the presence of 4% urea and 55% humidity, allows the inactivation of oak tannins (Makkar and Singh, 1993).

BIOLOGICAL ACTIVITIES OF OAK LEAVES

Anticancer Activity

The administration of *Q. sideroxyla* infusions to male Sprague-Dawley rats (1% w/v) induced with colon cancer by subcutaneous injection of 1,2-dimethylhydrazine (@ 21 mg/kg body weight) showed a decrease in the average number of tumors (0.9 ± 0.2 versus 2.6 ± 0.3), the multiplicity (1.2 ± 0.1 versus 2.0 ± 0.23) and the protein level of ß-catenin (2.2 times), in addition to decreasing the levels of the inflammatory markers COX-2 and IL-8 and modulating the expression of NF-κB in HT-29 cells (Moreno-Jiménez et al., 2015). It has been proposed that because of its phenolic composition the fresh leaves of *Q. resinosa* have anticancer activity *in vitro*, by decreasing the oxidative process and other DNA damage in human transformed cells (HeLa) (Rocha-Guzmán et al., 2009). Also, the high phenolic composition of *Q. brantii* leaves (136 mg gallic/g extract and 74.2 mg rutin/g extract) has been related with antioxidant activity showing a dose-dependence response and antimutagenic properties in hydroalcoholic extracts against sodium azide treatments (Mirzaei and Mirzaei, 2013).

Antimicrobial Activities

Several studies have focused on the antimicrobial (e.g., anticandical, antifungal, antibacterial) properties of crude hydroalcoholic extracts and fractions of extracts of leaves of *Q.* species, generally using the broth microdilution method. The ability to inhibit the growth of *Candida albicans* and bacteria of the genus *Brucella*, *Enterobacter*, *Escherichia*, *Neisseria*, *Pseudomonas* and *Bacillus* has been observed from methanolic extracts of *Q. ilex* leaves. However, these extracts do not show antifungal activity and ninety-six bacteria have shown to be resistant to those extracts (Güllüce et al., 2004). Bacteria such as *S. aureus*, *E. fecalia*, *E. coli*, *P. aeruginosa*, *C. albicans*, *C. krusei* and *C. parapsilosis* have shown greater inhibition by the ethyl acetate (EtOAc) extracts of the leaves from *Q. aucheri* in comparison to extracts with MeOH, *n*-BuOH and H_2O. The presence of two flavonoids: quercetin-3-O-α-L-arabinopyranoside and quercetin-3-O-D-galactopyranoside; and the precursors of tannins and a procyanidins (i.e., isolated as peracetates of (+)-catechin, (+)-gallocatechin and epicatechin-(4β→8)-catechin) in extracts of EtOAc may be involved in such antimicrobial activity (Sakar et al., 2005). On the other hand, aqueous extracts of *Q. resinosa* have shown greater antitopoisomerase, antimicrobial (i.e., against *C. albicans*, *K. pneumoniae*) and antioxidant activities by inhibiting the oxidation of LDL (IC_{50}: 0.220 mg/mL) compared to other species (e.g., *Q. grisea*, *Q. laeta* and *Q. obtusata*). This species has shown a phenolic and total flavonoid contents above the presented by other species studied, so that its phytochemical composition might be directly related to its biological effects (Sánchez-Burgos et al., 2013).

Ethanolic fractions of leaves from *Q. dilatata* have shown greater inhibition against strains of *E. coli*, *B. subtilus*, *S. aureus*, *M. luteus*, *B. bronchiseptica* and *S.setubal*, compared with fractions of *Chrozophora hierosolymitana* Spreng, *Chrysanthemum leucanthemum* L., and *Ephedra gerardiana* Wall, ex Stapf. This fraction has shown the most intense zone of bacterial inhibition (14.3–16.7 mm). The presence of alkaloids, tannins and saponins in the ethanol fraction of *Q. dilatata* could influence its antibacterial activity (Jamil et al., 2012). Likewise, the fractions with dichloromethane from hydroalcoholic extracts of *Q. baloot* in whose chemical composition flavonoids, terpenoids and steroids are included, have shown greater capacity for inhibition of *Candida albicans* ($R_f = 0.44$ and 0.50) (Khurram et al., 2011), *E. coli* ($R_f = 0.65$), *M. luteus* ($R_f = 0.42$ and 0.70), *B. subtilis* ($R_f = 0.42$ and 0.70) and *S. aureus* ($R_f = 0.70$) compared with fractions of *n*-hexane, ethyl acetate and *n*-butanol. The phytochemicals reported in this fraction included tannins, triterpenoids and flavonoids (Khurram et al., 2012).

Fractions with water, petroleum ether, EtOAc and *n*-butanol (BuOH) from methanolic extracts (80% methanol) from leaves of *Q. macranthera*, *Q. cerris*, *Q. pubescens* and *Q. coccifera* showed higher capacity of inhibition of fungal strains than bacteria. The majority of the fractions were effective against the *Candida* strains. The fractions of *n*-BuOH showed significant activity against *C. parapsilosis* (*Q. coccifera* MIC: 4.8 μg/mL) and *C. krusei* (*Q. macranthera*), while the EtOH fractions showed greater activity against *S. aureus* and *C. albicans.* In both fractions (*n*-BuOH and EtOH), the presence of some flavonoids was reported, including hydrolysable tannins and condensed tannins (Söhretoglu et al., 2007). Likewise, the *n*-BuOH fraction of *Q. incana* leaves has shown the presence of those polyphenol families, and has shown the highest activity against the bacterium *Micrococcus leuteus* (inhibition zone of 19 mm) and against the fungi *Aspergillus niger* (32 ± 0.55 mm) and *A. flavus* (28 ± 0.45 mm), also showing a good antioxidant activity (IC_{50}: 55.4 ± 0.21 μg/mL), though the fraction of ethyl acetate (IC_{50}: 23.21 ± 0.31 μg/mL) presented the best antioxidant response (Sarwar et al., 2015). Similarly, it has been shown that antimicrobial compounds in extracts of *Q. robur* moderately inhibit bacteria such as *Escherichia coli*, *Staphylococcus aureus*, *Klebsiella pneumonia*, *Staphylococcus epidermidis* and *Bacillus subtilis*, mainly in the EtOAc fraction (Uddin and Rauf, 2012).

Antioxidant Activity

Methanol extracts (80% methanol) from *Q. robur* leaves have been subjected to *in vitro* and *in vivo* studies. In determinations of antiradical power against DPPH· radicals, NO·, and $O_2^{\cdot}$, and the iron reducing capacity (FRAP), extracts from *Q. robur* leaves have shown greater antioxidant capacity than branches and acorns (Popovic et al., 2013). Likewise, extracts with hexane, chloroform, EtOAc and MeOH from the aerial parts of *Q. robur* have shown moderate antioxidant activity against DPPH radicals, and moderate antimicrobial activity against *Escherichia coli*, *Staphylococcus aureus*, *Klebsiella pneumoniae*, *Staphylococcus epidermis* and *Bacillus subtilis* (Uddin and Rauf, 2012).

Additionally, hydroalcoholic extracts from *Q. branttti* leaves showed higher hydroxyl scavenging activity (23-74%, IC_{50}: 1.57 ± 0.055 mg/mL) and superoxide radical inhibition (25-64%, IC_{50}: 2.1 ± 0.02 mg/mL) compared with *Juglans regia* extracts. On inhibition of chelating activity and inhibition of nitric oxide, the response was reversed, i.e., *Juglans regia*>*Q. branttti* (Mirzaei et al., 2013). Application of hydroalcoholic extracts of *Q. branttti* leaves include their utilization into the metal nanoparticles synthesis thanks to their strong composition of phenolic and flavonoid compounds (Korbekandi et al., 2015).

In *in vitro* determinations, leaf infusions of *Q. resinosa* have shown hypotensive activity by inhibition of the angiotensin-converting enzyme (ACE) (IC_{50}: 18ppm) where they displayed similar effect as the control (Captopril®, a commercial ACE inhibitor) (Rivas-Arreola et al., 2010), and antioxidant activity in ORAC and ABTS determinations, showing greater activity than white oak species such as *Q. sideroxila*, *Q. eduardii* and *Q. durifolia* (Rocha-Guzmán et al., 2012). Likewise, the alcoholic extracts of *Q. resinosa* have shown a greater antioxidant activity in comparison with other species (*Q. durifolia*, *Q. eduardii* and *Q. sideroxyla*) evaluated by inhibition of the DDPH radical (IC_{50}: 78.3 and 250.7μg/g), redox potential (147.0 and 201.6mV), inhibition of oxidation of LDL (95.6 and 97.2%) and inhibition of whitening with crocin (1.08 and 0.98g) in acetonic and methanolic extracts, respectively (Gamboa-Gómez et al., 2013).

Enzymatic Inhibition of α-Glucosidase

Aqueous and methanolic extracts of *Q. suber* leaves (10 mg/mL) have shown to significantly inhibit baker's yeast α-glucosidase by 97 and 89.1%, respectively, while 53.2% inhibition of mammalian α-glucosidase was observed in methanolic extracts. Saponins, phlobatannins, tannins, terpenoids, proteins, reducing sugars and glycosides have been described as part of the chemical composition of *Q. suber* leaves (Subhashini et al., 2016).

Additionally, compounds isolated from leaves of *Q. gilva* have shown capacity of inhibition of the enzyme α-glucosidase. The isolated tiliroside (IC_{50}: 28.36 ± 0.11 mmol/L) showed greater inhibitory activity than catechin (IC_{50}: 168.60 ± 5.15 mmol/L) and epicatechin (IC_{50}: 920.60 ± 10.10 mmol/L), which has more hydroxyl groups in its structure (Indrianingsih et al., 2015). The relationship between the decrease in hydroxyl groups and the decrease in the inhibitory activity of α-glucosidase has also been observed in leaves of *Q. phillyraeoides*, where fractions of condensed tannins showed greater inhibitory activity than β-sitosterol-D-glucoside (Indrianingsih and Tachibana, 2016). Therefore, it has been suggested that *Quercus* leaves are able to inhibit the carbohydrates digestive enzymes in the human diet and may be useful as an option for the control of glucose levels in patients with diabetes mellitus type II (Custódio et al., 2015).

Other Biological Effects

In addition to the main biological activities already described, other beneficial activities in oak leaves have been little explored. Methanolic extracts of *Q. suber* leaves have shown anticholinesterase activity (1 mg/mL) with a percentage inhibition of 79 (AChE) and 80 (BuChE) (Custódio et al., 2015). These inhibitors are widely used in the treatment of Alzheimer's diseases. Likewise, hepatoprotective and gastroprotective activity has been observed in *Q. robur* leaves in *in vivo* tests (Moharram et al., 2015). Also, Sánchez-Burgos et al. (2013) have explored the gastroprotective and antitopoisomerase potential of some oak species. This represents a field of opportunity for the exploration of bioactivities of extracts from oak leaves, as natural sources of compounds involved in the regulation of oxidative stress and prevention of diseases.

Finally, five compounds isolated from *Q. dentata* (Kaempferol 3-O-β-D-glucopyranoside, quercetin 3-O-β-D-glucopyranoside, kaempferol 3-O-(6"-trans-*p*-coumaroyl)-β-D-glucopyranoside, kaempferol 3-O-(2"-6"-di-trans-*p*-coumaroyl)-β-D-glucopyranoside, and kaempferol 3-O-(2",4"-di-acetyl-3"-cis-*p*-cumaroyl-6"-trans-*p*-coumaroyl)-β-D-glucopyranoside), have shown the ability to suppress the generation of superoxide induced by phorbol 12-myristate 13-acetate (PMA) while kaempferol 3-O- (2"-6"-di-trans-*p*-coumaroyl)-β-D-glucopyranoside suppresses superoxide generation induced by N-formylmethionyl-leucyl-phenylalanine (fMLP), and quercetin 3-O-β-D-glucopyranoside and kaempferol 3-O-(6"-trans-*p*-comaroyl)-β-D-glucopyranoside significantly suppresses the generation of superoxide induced by arachidonic acid (Meng et al., 2001).

Conclusion and Perspectives

Oak leaves have been used since old times as alternative medicine by different ethnic groups in Mexico. The phytochemical composition of leaves of different oak species has been widely explored, mainly the phenolic part, which has been related to different biological effects on pathologies related to oxidative stress. Although phenolic compounds represent an important part of the chemical footprint of leaves, other compounds also reported in leaves of the *Quercus* genus such as saponins, alkaloids and terpenoids have not been fully studied, despite being molecules with potential or proven biological effects. Likewise, other parts of the oak tree such as bark and wood have been widely studied for their phenolic and mainly tannic content, while publications focused on the study of their leaves are still limited. In this way, the exploration of the phenolic and non-phenolic phytochemical composition of oak leaves and their potential effect on preventing human pathologies represent a field of opportunity to develop alternative products as nutraceuticals obtained from these available natural resources.

References

Alfonso-Corrado, C., Clark-Tapia, R., and Mendoza, A. 2007. "Demography and management of two clonal oaks: *Quercus eduardii* and *Q. potosina* (Fagaceae) in central México." *Forest Ecology and Management* 251:129–141.

Arámbula-Salazar, J. A., Ibarra-Salinas, B. I., González-Laredo, R. F., Muñoz-Galindo, O. D., and Hernández-Vela, H. 2010. "Seasonal variation in the phenolic content of oak leaves *(Quercus sideroxyla)* in different soil textures." *Madera y Bosques* 16(3):49–59.

Barbehenn, R. V., and Constabel, C. P. 2011. "Tannins in plant-herbivore interactions." *Phytochemistry* 72:1551–1565.

Brossa, R., Casals, I., Pintó-Marijuan, M., and Fleck, I. 2009. "Leaf flavonoid content in *Quercus ilex* L. resprouts and its seasonal variation." *Trees* 23:401–408.

Chahardooli, M., Khodadadi, E., and Khodadadi, E. 2014. "Green synthesis of silver nanoparticles using oak leaf and fruit extracts (*Quercus*) and its antibacterial activity against plant pathogenic bacteria." *International Journal of Bioscience* 4:97–103.

Custódio, L., Patarra, J., Alberício, F., Neng, N. R., Florêncio-Nogueira, J. M., and Romano, A. 2015. "Phenolic composition, antioxidant potential and *in vitro* inhibitory activity of leaves and acorns of *Quercus suber* on key enzymes relevant for hyperglycemia and Alzheimer's disease." *Industrial Crops and Products* 64:45–51.

Dai, J., and Mumper, R. J. 2010. "Plant Phenolics: Extraction, analysis and their antioxidant and anticancer properties." *Molecules* 15:7313–7352.

Doce, R. R., Hervás, G., Belenguer, A., Toral, P. G., Giráldez, F. J., and Frutos, P. 2009. "Effect of the administration of young oak (*Quercus pyrenaica*) leaves to cattle on ruminal fermentation." *Animal Feed Science and Technology* 150:75–85.

Gallegos-Infante, J. A., Rocha-Guzmán, N. E., González-Laredo, R. F., Medina-Torres, L., Gomez-Aldapa, C. A., Ochoa-Martínez, L. A., Martínez-Sánchez, C. E., Hernández-Santos, B., and Rodríguez-Ramírez, J. 2013. "Physicochemical properties and antioxidant capacity of oak (*Quercus resinosa*) leaf infusions encapsulated by spray-drying." *Journal of Food Bioscience* 2:31–38.

Gamboa-Gómez, C. I., Hernández-Saavedra, D., Gallegos-Infante J. A., González-Laredo, R. F. Manzocco, L., and Rocha-Guzmán, N. E. 2013. "Polyphenolic content and antioxidant activity of leaf extracts from *Quercus durifolia*, *Quercus eduardii*, *Quercus sideroxyla* and *Quercus resinosa*." *Journal of Medicinal Plants Research* 7(35):2564–2573.

García-Villalba, R., Espín, J. C., Tomás-Barberán, F. A., and Rocha-Guzmán, N. E. 2017. "Comprehensive characterization by LC-DAD-MS/MS of the phenolic composition of seven *Quercus* leaf teas." *Journal of Food Composition and Analysis* 63:38–46.

González, Z. A., Garcia-Barriuso, M., and Amich, F. 2010. "Ethnobotanical study of medicinal plants traditionall used in the Arribes del Duero, Western Spain." *Journal of Ethnopharmacology* 131:343–355.

González-Barrio R., Truchado P., García-Villalba R., Hervás G., Frutos P., Espín J. C., and Tomás-Barberán F. A. (2012). "Metabolism of oak leaf ellagitannins and urolithin production in beef cattle." *Journal of Agricultural Food Chemistry* 28, 60(12):3068–3077.

Güllüce, M., Adıgüze, A., Ögütçü, H., Sengül, M., Karaman, I., and Sahin, F. 2004. "Antimicrobial effects of *Quercus ilex* L. extract." *Phytotheraphy Research* 18:208–211.

Indrianingsih, A. W., and Tachibana, S. 2016. "Bioactive constituents from the leaves of *Quercus phillyraeoides* A. Gray for -glucosidase inhibitor activity with concurrent antioxidant activity." *Food Science and Human Wellness* 5:85–94.

Indrianingsih, A. W., Tachibana, S., Dewi, R. T., and Itoh, K. 2015. "Antioxidant and a-glucosidase inhibitor activities of natural compounds isolated from *Quercus gilva* Blume leaves." *Asian Pacific Journal of Tropical Biomedicine* 5(9):748–755.

Jamil, M., Haq, I. U., Mirza, B., and Qayyum, M. 2012. "Isolation of antibacterial compounds from *Quercus dilatata* L. through bioassay guided fractionation." *Annals of Clinical Microbiology and Antimicrobials* 11:11.

Karioti, A., Bilia, A. R., Gabbiani, C., Messori, L., and Skaltsa, H. 2009. "Proanthocyanidin glycosides from the leaves of *Quercus ilex* L. (Fagaceae)." *Tetrahedron Letters* 50:1771–1776.

Khennouf, S., Amira, S., Arrar, L., and Baghiani, A. 2010. "Effect of some phenolic compounds and *Quercus* tannins on lipid peroxidation." *World Applied Sciences Journal* 8(9):1144–1149.

Khurram, M., Hameed, A., Khan, M. A., Amin, M. U. Hassan, M., Ullah, N., Manzoor, W., Najeeb, U., Qayum, A., Khan, F. A. and Bilal, M. 2012. "Antibacterial potentials of *Quercus baloot* Griff." *Journal of Medicinal Plants Research* 6(7):1244–1249.

Khurram, M., Hameed, A., Amin, M. U., Manzoor, W., Ullah, N., Chishti, K. A., Khayyam, S. U., Hassan, M., and Qayum, A. 2011. "Evaluation of anticandidal potential of *Quercus baloot* Griff. Using contact bioautography technique." *African Journal of Pharmacy and Pharmacology* 5(12):1538–1542.

Korbekandi, H., Chitsazi, M. R., Asghari, G., Najafi, R. B., Badii, A., and Iravani, S. 2015. "Green biosynthesis of silver nanoparticles using *Quercus brantii* (oak) leaves hydroalcoholic extract." *Pharmaceutical Biology* 53(6):807–812.

Leporatti, M. L., and Ivancheva, S. 2003. "Preliminary comparative analysis of medicinal plants used in the traditional medicine of Bulgaria and Italy." *Journal of Ethnopharmacology* 87:123–142.

Luna-José, A. L., Montalvo-Espinosa, L., and Rendón-Aguilar, B. 2003. Non-woody uses of oaks in Mexico. *Boletín de la Sociedad Botánica de México* 72:107–117.

Macías-Cortés, E., Gallegos-Infante, J. A., Rocha-Guzmán, N. E., Moreno-Jiménez, M. R., Medina-Torres, L., and González-Laredo, R. F. 2020. Microencapsulation of phenolic compounds: Technologies and novel polymers. *Revista Mexicana de Ingeniería Química 19*(2):491-521.

Makkar, H. P. S., and Singh, B. 1993. "Effect of storage and urea addition on detannification and in sacco dry matter digestibility of mature oak (*Quercus incana*) leaves." *Animal Feed Science and Technology* 41:247–259.

Meng, Z., Zhou, Y., Lu, J., Sugahara, J., Xu, S., and Kodama, H. 2001. "Effect of five flavonoid compounds isolated from *Quercus dentata* Thunb on superoxide generation in human neutrophils and phosphorylation of neutrophil proteins." *Clinica Chimica Acta* 306:97–102.

Mirzaei, A., Mirzaei, M., Khosravani, S. A., Salehpour, Z. 2013. "Radical scavenging potential of Iranian *Quercus brantii* and *Juglans Regia*." *Life Science Journal* 10(7s):1246–1250.

Mirzaei, N., and Mirzaei, A. 2013. "Antioxidant and antimutagenic property of Iranian oak and walnut plants." *International Journal of Biology, Pharmacy and Allied Science* 2(3):620–629.

Moctezuma, C., Hammerbacher, A., Heil, M., Gershenzon, J., Méndez-Alonzo, R., and Oyama, K. 2014. "Specific polyphenols and tannins are associated with defense against insect herbivores in the tropical oak *Quercus oleoides*." *Journal of Chemical Ecology* 40(5):458–467.

Moharram, F. A., Marzouk, M.S., Moneim-El Dib, R. A., El-Shenawy, S. M., Abdel-Rahman, R. F., and Ibrahim, R. R. 2015. "Hepatoprotective, gastroprotective, antioxidant activity and phenolic constituents of *Quercus robur* leaves." *Journal of Pharmaceutical Sciences and Research* 7(11):1055–1065.

Moreno-Jiménez, M. R., Trujillo-Esquivel, F., Gallegos-Corona, M. A., Reynoso-Camacho, R., González-Laredo, R. F., Gallegos-Infante, J. A., Rocha-Guzmán, N. E., and Ramos-Gómez, M. 2015. "Antioxidant, anti-inflammatory and anticarcinogenic activities of edible red oak (*Quercus* spp.) infusions in rat colon carcinogenesis induced by 1,2-dimethylhydrazine." *Food and Chemical Toxicology* 80:144–153.

Noori, M., Talebi, M., and Ahmadi, T. 2015. "Comparative studies of leaf, gall and bark flavonoids in collected *Quercus brantii* Lindl. (Fagaceae) from Lorestan Province, Iran." *International Journal of Plant Research* 5(2):42–49.

Passarinho, J. A. P., Lamosa, P., Baeta, J. P., Santos, H., and Ricardo, C. P. P. 2006. "Annual changes in the concentration of minerals and organic compounds of *Quercus suber* leaves." *Physiologia Plantarum* 127:100-110.

Popovic, B. M., Štajner, D., Zdero, R., Orlovic, S., and Galic, Z. 2013. "Antioxidant characterization of oak extracts combining spectrophotometric assays and chemometrics." *The Scientific World Journal.* ID 134656, 8 pp.

Rivas-Arreola, M. J., Rocha-Guzmán, N. E., Gallegos-Infante, J. A., González-Laredo, R. F., Bacon, J. R., Rong (Tsao) Cao, Proulx, A., and Intriago-Ortega, P. 2010. "Antioxidant activity of Oak (*Quercus*) leaves infusions against free radicals and their cardioprotective potential." *Pakistan Journal of Biological Sciences* 13(11):537–545.

Rocha-Guzmán, N. E., Gallegos-Infante, J. A., González-Laredo, R. F., Reynoso-Camacho, R., Ramos-Gómez, M., García-Gasca, T., Rodríguez-Muñoz, M. E., Guzmán-Maldonado, S. H., Medina-Torres, L., and Lujan-García, B. A. 2009. "Antioxidant activity and genotoxic effect on HeLa cells of phenolic compounds from infusions of *Quercus resinosa* leaves." *Food Chemistry* 115:1320–1325.

Rocha-Guzmán, N. E., Gallegos-Infante, J. A., González-Laredo, R. F., Harte, F., Medina-Torres, L., Ochoa-Martínez, L. A., and Soto-García, M. 2010. "Effect of high-pressure homogenization on the physical and antioxidant properties of *Quercus resinosa* infusions encapsulated by spray-drying." *Journal of Food Science* 75(5):57–61.

Rocha-Guzmán, N. E., Medina-Medrano, J. R., Gallegos-Infante, J. A., Gonzalez-Laredo, R. F., Ramos-Gómez, M., Reynoso-Camacho, R., Guzmán-Maldonado, H., and González-Herrera, S. M. 2012. "Chemical evaluation, antioxidant capacity, and consumer acceptance of several oak infusions." *Journal of Food Science* 77(2):162–166.

Romussi, G., Caviglioli, G., and Tomsasi, N. 1993. "Constituents of Cupuliferae, XVII: Triterpene saponins and flavonoids from *Quercus laurifolia* Michx." *Arch Pharm* 326(9):525–528.

Sakar, M. K., Söhretoglu, D., Özalp, M., Ekizoglu, M., Piacente, S., and Pizza, C. 2005. "Polyphenolic compounds and antimicrobial activity of *Quercus aucheri* leaves." *Turkish Journal of Chemistry* 29:555–559.

Salminen, J.-P., Roslin, T., Karonen, M., Sinkkonen, J., Pihlaja, K., and Pulkkinen, P. 2004. "Seasonal variation in the content of hydrolyzable tannins, flavonoid glycosides, and proanthocyanidins in oak leaves." *Journal of Chemical Ecology* 30(9):1693–1711.

Sánchez-Burgos, J. A., Ramírez-Mares, M. V., Larrosa, M. M., Gallegos-Infante, J. A., González-Laredo, R. F., Medina-Torres, L., and Rocha-Guzmán, N. E. 2013. "Antioxidant, antimicrobial, antitopoisomerase and gastroprotective effect of herbal infusions from four *Quercus* species." *Industrial Crops and Products* 42:57–62.

Sarwar, R., Farooq, U., Khan, U., Naz, S., Khan, S., Khan, A., Rauf, A., Bahadar, H., and Uddin, R. 2015. "Evaluation of antioxidant, free radical scavenging, and antimicrobial activity of *Quercus incana* Roxb." *Frontiers in Pharmacology* 6:1–7.

Schwanz, P., and Polle, A. 1998. "Antioxidative systems, pigment and protein contents in leaves of adult Mediterranean oak species (*Quercus pubescens* and *Q. ilex*) with lifetime exposure to elevated CO_2." *The New Phytologist* 140:411–423.

Şohretoğlu, D., Ekizoglu, M., Kiliç, E., and Sakar, M. K. 2007. "Antibacterial and antifungal activities of some *Quercus* species growing in Turkey." *FABAD Journal of Pharm Science* 32:127–130.

Sparg, S. G., Light, M. E., and van Staden, J. 2004. "Biological activities and distribution of plant saponins." *Journal of Ethnopharmacology* 94:219–243.

Subhashini, S., Maleeka-Begum, S. F., and Rajesh, G. 2016. "Phytochemical screening and TLC analysis of different *Quercus* species." *World Journal of Pharmacy and Pharmaceutical Sciences* 5(2):1220–1226.

Top, S. M., Preston, C. M., Dukes, J. S., and Tharayil, N. 2017. "Climate influences the content and chemical composition of foliar tannins in green and senesced tissues of *Quercus rubra*." *Frontiers in Plant Science* 8:423.

Uddin, G., and Rauf, A. 2012. "Phytochemical screening, antimicrobial and antioxidant activities of aerial parts of *Quercus robur* L." *Middle-East Journal of Medicinal Plants Research* 1(1):1–4.

Vaca-Sánchez, M. S., González-Rodríguez, A., Maldonado-López, Y., Fernandes, G. W., and Cuevas-Reyes, P. 2016. "Importancia de los taninos en especies del género *Quercus* como metabolitos secundarios asociados a defensa contra insectos herbívoros." ["Importance of tannins in species of the Quercus genus as secondary metabolites associated with defense against herbivorous insects."] *Biológicas.* 18(1): 10–20.

Vázquez-Cabral, B. D., Larrosa-Pérez, M., Gallegos-Infante, J. A., Moreno-Jiménez, M. R., González-Laredo, R. F., Rutiaga-Quiñones, J. G., Gamboa-Gómez, C. I., and Rocha-Guzmán, N. E. 2017. "Oak kombucha protects against oxidative stress and inflammatory processes." *Chemico-Biological Interactions* 272:1–9.

Vázquez-Cabral, B. D., Moreno-Jiménez, M. R., Rocha-Guzmán, N. E., Gallegos-Infante, J. A., González-Herrera, S. M., Gamboa-Gómez, C. I., and González-Laredo, R. F. 2016a. "Mexican oaks as a potential non-timber resource for kombucha beverages." *Revista Chapingo Serie Ciencias Forestales y del Ambiente* 22(1):73–86.

Vázquez-Cabral, D., Valdez-Fragoso, A., Rocha-Guzmán, N. E., Moreno-Jiménez, M. R., González-Laredo, R. F., Morales-Martínez, P. S., Rojas-Contreras, J. A., Mujica-Paz, H., and Gallegos-Infante, J. A. 2016b. "Effect of pulsed electric field (PEF)-treated kombucha analogues from *Quercus obtusata* infusions on bioactives and microorganisms." *Innovative Food Science and Emerging Technologies* 34:171–179.

Vázquez-Cabral, B. D., Rocha-Guzmán, N. E., Gallegos-Infante, J. A., González-Herrera, S. M., González-Laredo, R. F., Moreno-Jiménez, M. R., and Córdova-Moreno, I. T. S. 2014. "Chemical and sensory evaluation of a functional beverage obtained from infusions of oak leaves (*Quercus resinosa*) inoculated with the kombucha consortium under different processing conditions." *Nutrafoods* 13:169–178.

Yarnes, C. T., Boecklen, W. J., Tuominen, K., and Salminen, J.-P. 2006. "Defining phytochemical phenotypes: size and shape analysis of phenolic compounds in oaks (Fagaceae, *Quercus*) of the Chihuahuan desert." *Canadian Journal of Botanica* 84:1233–1248.

Zhou, Y. J., Xu, S. X., Che, Q. M., and Sun, Q. S. 2001. "Flavonoids from the leaves of *Quercus dentata*." *International Journal of Plant Science* 169(4):567–578.

In: Quercus: Classification, Ecology and Uses ISBN: 978-1-53618-026-8
Editor: Bent J. Steffensen

Chapter 7

Ecological Aspects of Oak Ecosystems with Dominance of *Quercus cerris* L. and *Q. frainetto* Ten. in Bulgaria

***Mariyana I. Lyubenova*[1,*] *and Violeta G. Dimitrova*[2]**
[1]Faculty of Biology, University of Sofia "St. Kliment Ohridski," Sofia, Bulgaria
[2]Dendrology Department, Unifersity of Forestry, Sofia, Bulgaria

Abstract

The chapter deals with considering the habitat and species diversity of oak forests in Bulgaria. The flora of studied oak forests, their structure and functionality - production, chemistry, intensity of the biological turnover, caloric content of the phytomass fractions, etc. are presented in detail. A summary of ecosystem services provided by oak ecosystems has been analysed. The attention was paid to the state of oak forests in the country and the sustainability of provided services. The focus of the discussions is

* Corresponding Author's Email: ryana_l@yahoo.com.

also on the protection of habitats and the opportunities for their sustainable use as a natural resource.

Keywords: oak ecosystem, biological diversity, functionality, ecosystem services, ecosystem health, sustainability, Bulgaria

1. Introduction

In Bulgaria, forests cover over 37,4%, or 4 230 825 ha of the territory and broadleaved forests are over 70% of the forest area. The total volume is over 680.10^6 m^3, the annual increment is 14.10^6 m^3 and annual harvesting - 8 415 300 m^3. The average timber supply per capita is about 1.5 m^3 (Agricultural report, 2014).

Control and management of forests in Bulgaria are entrusted to the Ministry of Agriculture and Food and 6 State Forest Enterprises according to Art. 163 (LF) and only 4.13% is managed by the Ministry of Environment and Waters. Important legislation acts for the forestry sector in Bulgaria and related to the services offered by the forest ecosystems are: National Strategy 2013 – 2020 (NSDFSRB) and Strategic Plan 2014 – 2023 (SPDFSRB) for Forest Sector Development, as well as the Low on Forests (Lyubenova ed., 2019). According to the forest management data of 2010, the area of oak forests amounts to 1 325 959 ha, which is almost 35% of the total afforested area in the country. The dominant species of genus *Quercus* that determine the appearance of oak forests are *Q. frainetto* Ten. (Hungarian oak), *Quercus cerris* L. (Turkish oak), *Q. dalechampii* Ten. (Dalechamps oak), and *Quercus petraea* Liebl. (s.l., sessile oak). Some botanists believe that *Q. dalechampii* Ten. is predominant in the sessile oak forests as its southern variety, and *Q. petraea* Liebl is involved in these forests of varying abundance (Bondev, 1991). In-depth genetic and field studies have not yet been conducted to prove this statement. These oak species grow slowly, fruit periodically. The undergrowth is light loving, capricious and only grows intensively when thinning the forest. Due to the slow growth of oak forests, they reach an operating age for longer periods. However, these local forest ecosystems are of significant ecological importance for the adjacent territories - climate, water and soil protection, recreational and for the natural components purification from the pollution produced by the nearby settlements. Under the good management, they can offer a variety of provisional benefits and services.

The still preserved oak communities of the Robur section, although isolated and fragmented, have a negligible area of 1332.5 ha (*Quercus robur*/*Quercus hartwissiana*) (Tsavkov, 2016).

Based on the detailed typological studies of the oak forests in Bulgaria, Marinov et al. (1995) discussed 3 groups according to the forests specifics and the sylvicultural activities conducted in them.

The first group includes types of oak forest growing on deep, fresh to moist fertile soils, distributed mainly in Strandja, Eastern Stara Planina and Ludogorie. These are high yielding, monodominant and mixed *Quercus daleshampii* Ten. stands, which produce large timber for veneer and furniture production.

The second group includes types of oak forest growing on medium-deep, fresh, relatively fertile soils. These are pure and mixed stands of *Quercus daleshampii* Ten., *Q. cerris* L. and *Q. frainetto* Ten. with a middle wood extraction.

The third group includes low-productive forest types growing on stony, shallow, dry and poor soils. The forest types is composed of *Quercus pubescens* Willd., *Quercus cerris* L., *Quercus frainetto* Ten., *Carpinus orientalis* Mill. and various species of dry shrubs. The occupied habitats are not generally suitable for the oak farming, but these ecosystems are most resilient to those habitats and they are noteworthy in the changing climatic conditions.

The emergence and development of civilization is linked precisely to areas, where the various species of oak were dominant, and this is why many of the extensive oak forests in the past have been largely reduced, e.g., Ludogorie, Dobrudzha, Thracian Plain (Stefanov, 1964; Ganchev, 1966). Their multifunctional use and location near settlements are the reason for the reduction of the areas and the change of their natural structure and floral composition. The current state is primarily the result of intense human impact. The deterioration of the growth conditions, as well as the biological and the different oak species environmental requirements do not allow the natural regenerative process to proceed normally. As a result, over 80% of the oak forests in Bulgaria are shoots, some of them low-stemmed. Many of the areas are intended for reconstruction or for the conversion to high-stemmed. At a later stage, the management of coppice oak forests is commented by Hinkov (2002) and Kostov et al. (2017).

In an advanced stages of degradation, the xerothermic oak forests turn into bushes with very low canopy - "kayrak" or "shiblyak." These communities are the sad fate of the forests of the Balkan Peninsula as a result of intensive grazing and felling. The strong anthropogenic pressure on communities of different oak species requires the conservation of some communities.

Oak ecosystems are a source of diverse ecosystem services and benefits for the population - provisional, regulatory, supporting and cultural. The importance of xerothermic oak forests as a source of services and benefits will increase with the climate warming in Bulgaria.

Since the middle of 1980s, the tendency of the average annual air temperature in Bulgaria have changed towards warmer climate. In the period 1988-2016, the average annual air temperature for the lower part of the country (for areas up to 800 m altitude) was increased on average with 0.8°C relative to the reference period 1961-1990 climatic normal and ranged between 10.6°C and 13.0°C. The climate in Bulgaria became not only warmer, but also drier at the end of 20th century. During the last decade however, the precipitation totals have increased, as 2014 was the rainiest year for 1988-2016. The average annual precipitation amount was 1013 mm for the areas up to 800 m altitude that is more than the previous reached maximum of 924 mm in 2005 (Marinova et al., 2017).

The chapter deals with the analysis of information on oak forests in Bulgaria - biodiversity, structure and functions, as well as their potential to provide services and benefits in terms of adaptation of their management in the context of climate change.

2. BIODIVERSITY AND DISTRIBUTION OF OAK FORESTS IN BULGARIA

2.1. Species Diversity

In Europe, the genus *Quercus* L. is represented by 20 species (Schwarz, 2010), and in Bulgaria the number of species is 17 (*Quercus brachyphylla* Kotschy, *Q. cerris* L., *Q.coccifera* L., *Q. longipes* Steven, *Q. dalechampii* Ten., *Q. mestensis* Bondev & Gančev, *Q. erucifolia* Steven, *Q. pedunculiflora* C. Koch, *Q. frainetto* Ten., *Q. petraea* (Mattuschka) Liebl., *Q. hartwissiana* Steven, *Q. polycarpa* Schur, *Q. pubescens* Willd., *Q. robur* L., *Q. thracica* Stef. & Nedjalkov, *Q. trojana* Webb, *Q. virgiliana* (Ten.) Ten. (Assyov et al., 2012).

Q. thracica and *Q. mestensis* are protected species, included in the Red Data Book of Bulgaria (Rusakova, 2015) as critically endangered (CR) Bulgarian endemic (more precisely local Rodope endemic) species. In the Bulgarian Red list of vascular flora (Petrova and Vladimirov, 2009), *Q. coccifera* and *Q. hartwissiana* are also included as respectively endangered (EN) and vulnerable (VU) species. In 2017, the species *Quercus protoroburoides* (Rila oak) was validated (Tashev and Tsavkov, 2017). Its communities are also included in the Red Data Book of Bulgaria (Rusakova, 2015).

In online sources like http://oaks.of.the.world.free.fr/ and http://www.theplantlist.org/, *Q. brachyphylla* Kotschy and *Q. virgiliana* (Ten.) Ten. are the synonyms of *Quercus pubescens* subsp. *pubescens*; *Q. longipes* Steven, *Q. erucifolia* Steven and *Q. pedunculiflora* C. Koch are the synonyms of *Q. robur* subsp. *pedunculiflora* (K.Koch) Menitsky and *Q. polycarpa* Schur is in the sessile oak group.

2.2. Diversity of Oak Communities

The conspectus of plant communities in Bulgaria (Apostolova and Slavova, 1997) lists 22 formations, 19 groups of associations and 156 associations (Gogushev, 2009). Many syntaxa with oak dominance have been described in later studies conducted in accordance to the methodological principles of dominant approach. The phytocenological studies of oak forests in accordance with the floristic method of Brown-Blanke were conducted by Dimitrov and Glogov (2003), Tsonev (2002), Roussakova and Tzonev (2003), Tzonev (2009), Lyubenova et al. (2011) and Tzonev et al. (2019). The thermophilous oak communities in Bulgaria are classified and characterized that led to the identification of 19 associations and subassociations and two communities (shown below). Six new associations and three new subassociations were proposed (Tzonev et al., 2019).

- *Cotino-Quercetum cerridis* (*Quercus cerris* Forest-Steppe forests)
- *Quercetum pubescenti-roboris* (Forest-Steppe *Quercus robur* subsp. *pedunculiflora* forests)
- *Quercetum frainetto-cerridis* (Xeromesophilous mixed oak forests on the foothills and low mountains)
- *Hedero helicis-Quercetum cerridis* (Mesothermophilous oak forests on the slopes of the ravines and valleys)
- *Genisto carinalis-Quercetum petraeae aceretosum monspessulani* (Dry thermophilous *Quercus petraea* subsp. *petraea* forests)
- *Genisto carinalis-Quercetum petrae*ae (Mesophilous and thermophilous *Quercus petraea* subsp. *petraea* forests)
- *Moehringio pendulae-Quercetum petraeae* (Acidophilous high-altitudinal *Quercus petraea* subsp. *petraea* forests)
- *Haberleo-Quercetum petraeae* (Calciphilous to neutrophilous *Quercus petraea* subsp. *petrae* forests on rocky slopes)
- *Rusco aculeati-Quercetum frainetto* (Coastal *Quercus frainetto* forests)
- *Symphyto ottomani-Quercetum frainetto* (Submediterranean *Quercus frainetto* forests on deep soils)

- Community of *Quercus frainetto* and *Hypericum cerastoides* (Xeromesophilous mixed oak forests on the foothills and low mountains of East Rhodope)
- *Genisto januensis-Quercetum pubescentis carpinetosum orientalis* (Species poor *Quercus pubescens* woods)
- *Genisto januensis-Quercetum pubescentis* typicum (Typical temperate *Quercus pubescens* woods)
- *Genisto januensis-Quercetum pubescentis dianthetosum gracilis* (Transitional temperate *Quercus pubescens* woods)
- *Cyclamini hederifolii-Quercetum frainetto* (Submediterranean mixed *Quercus pubescens* and *Quercus frainetto* forests on shallow soils)
- *Cisto incani-Quercetum pubescentis* typicum (Submediterranean *Quercus pubescens* wood)
- *Cisto incani-Quercetum pubescentis juniperetosum deltoidis* (Submediterranean *Quercus pubescens* woods with evergreen trees and shrub)
- *Carpino orientalis-Quercetum cocciferae* (*Quercus coccifera* woods)
- Community of *Jasminum fruticans* and *Quercus pubescens* (Coastal *Quecus pubescens* woodlands)
- *Calluno-Quercetum polycarpae* (Dry Euxinian *Quercus petraea* subsp. *polycarpa* forests)
- *Trachystemono orientalis-Quercetum polycarpae* (Mesophilous *Quercus petraea* subsp. *polycarpa* forests).

2.3. Diversity of Habitats Formed by Genus *Quercus*

The oak stands are found in all three forest vegetation areas in the country - the Moesian, the Thracian and the Southern border regions (Figure 1). The natural vegetation that include representatives of genus *Quercus* L. is listed below (Bondev, 1991).

1 - Moesia region; 2 - Thracian region; 3 - Southern border area

Figure 1. Distribution of oak forests in Bulgaria (by Marinov et al., 1995).

2.3.1. Mesophytic and Xeromesophytic Microthermal Vegetation in the Horn-Oak Belt

The vegetation is characteristic for the horn-oak forest belt and widespread in all mountains. In the most cases, the belt covers 600-700 m and up to 1000 m altitude slopes. The lower boundary is limited by the xerothermic oak forests belt and the upper one - by the mesophyte beech forests. The vegetation includes *Querceto-Carpineta betuli* forests; mixed forests of *Carpinus betulus* L., *Quercus cerris* L., *Q. dalechampii* Ten. and *Acer campestre* L.; *Querceta daleshampii* forests; mixed forests of *Quercus daleshampii* Ten. and *Ostrya carpinifolia* Scop. with *Fraxinus excelsior* L., *F. ornus* L., *Carpinus orientalis* Mill.; mixed forests of *Q. dalechampii* Ten., *Q. cerris* L. and *Q. frainetto* Ten; mixed forests of *Q. dalechampii* Ten. and *Q. frainetto* Ten; mixed forests of *Q. dalechampii* Ten. and *Carpinus orientalis* Mill.; *Fraxineta excelsioris* forest often with *Q. dalechampii* Ten., *Q. cerris* L. and *F. ornus* L.

2.3.2. South Euxinian Vegetation

The South Euxinian vegetation is widespread in Strandzha and the Eastern Balkan Mountains, with a predominance of indigenous formations of *Fagus orientalis* Lipsky and *Quercus polycarpa* Schur. From the oak communities here are common *Querceta policarpae forests*; mixed forests of *Quercus polycarpa* Schur. and *Carpinus betulus* L.; mixed forests of *Quercus polycarpa* Schur., *Q. frainetto* Ten. and *Q. cerris* L.; mixed forests of *Quercus polycarpa* Schur. and *Quercus frainetto* Ten.; mixed forests of *Q. polycarpa* Schur. and *Carpinus orientalis* Mill.

2.3.3. Xerophytic and Mesophytic, Microthermal and Mesothermal Vegetation in the Xerothermic Oak Belt and Hilly Plains

This vegetation is spread from 10 m to 500-700 m, rarely up to 800 m altitude. The climate is temperate continental. The soils are Chernosems and Gray Luvisols in the Northern and Vertisols and Chromic Luvisols in the Southern Bulgaria. The oak vegetation is composed by *Quercus cerris* L. forests; mixed forests of *Q. cerris* L. and *Q. frainetto* Ten.; mixed forests of *Q. cerris* L., *Q. frainetto* Ten. and Mediterranean elements; mixed forests of *Q. cerris* L., *Q. frainetto* Ten. and *Carpinus orientalis* Mill. occurring in places as secondary vegetation; mixed forests of *Q. cerris* L., *Q. frainetto* Ten., *Q. hartwissiana* Stev. and *Carpinus orientalis* Mill.; mixed forests of *Q. cerris* L., *Q. pubescens* Willd. and *Q. virgiliana* (Ten.) Ten.; mixed forests of *Q. cerris* L., *Q. pubescens* Willd. with undergrowth *of Cotinus coggygria* Scop.; mixed forests of *Q. cerris* L. and *Fraxinus ornus* L.; mixed forests of *Q. cerris* L. and *Carpinus orientalis* Mill.; *Q. frainetto* Ten. forests; *Q. frainetto* Ten. forests with Mediterranean elements; mixed forests of *Q. frainetto* Ten., *Q. pubescens* Willd. and *Q. virgiliana* (Ten.) Ten.; mixed forests of *Q. frainetto* Ten. and *Fraxinus ornus* L.; mixed forests of *Q. frainetto* Ten. and *Carpinus orientalis* Mill.; mixed forests of *Quercus frainetto* Ten. and *Carpinus orientalis* Mill. with Mediterranean elements; *Q. pubescens* Willd. and *Q. virgiliana* (Ten.) Ten. forests; *Q. pubescens* Willd. and *Q. virgiliana* (Ten.) Ten. forests with Mediterranean elements; mixed forests of *Q. virgiliana* (Ten.) Ten., *Q. pubescens* Willd. and *Acer campestre* L.; mixed forests of *Q. virgiliana* (Ten.) Ten., *Tilia tomentosa* Moench. and *Fraxinus ornus* L.; mixed forests of *Q. pubescens* Willd., *Q. virgiliana* (Ten.) Ten. and *Carpinus orientalis* Mill.; mixed forests of *Q. pubescens* Willd., *Q. virgiliana* (Ten.) Ten. and *Carpinus orientalis* Mill. with Mediterranean elements; evergreen shrubs with a predominance of

Q. coccifera L., *Q. pubescens* Willd., *Carpinus orientalis* Mill. and xerothermic grass formations (Bondev, 1991).

2.3.4. Mesophytic and Hygrophytic Microthermal Vegetation in the Lowlands, in the Valley Fields and Near Rivers

Mesophytic and hygrophytic vegetation are located throughout the country from 0 to 200 m, and in the basins - from 300 to 1000 m altitude. The soils are alluvial and alluvial-meadow, deluvial and deluvial-meadow. The vegetation is composed of *Querceta pedunculiflorae*, *Querceta roburis* and *Acer campestre* L. mixed forests; forests of *Q. pedunculiflora* C. Koch with *Quercus cerris* L. and *Q. virgiliana* (Ten.) Ten. (Bondev, 1991).

Table 1. Reserves and localities for the oak communities' protection (ExEA site)

№	Name	Area (ha)	Location	Oak vegetation
1	Protected area "Tulovska koria"	75.7	Stara planina 42.3349°N, 25.3312 E	An age-old oak forest
2	Reserve "Beli Lom"	773.0	East Danube plane 43.5337° N, 26.5411° E	Oak-lime forests
3	Reserve "Vitanovo"	1112.4	Stranja mountain 43.5337° N, 26.5411° E	*Q. cerris* L., *Q. frainetto* Ten.
4	Reserve "Lopushna"	2581.5	Stranja mountain 41.952222°N, 27.27°E	Oak-beech forests
5	"Tisovitca"	749.3	Stranja mountain 42.1415° N, 27.7020° E	Oak-beech forests
6	Reserve "Silkosia"	396.0	Stranja mountain 42.0580° N, 27.7800° E	Oak-beech forests
7	Borovets Maintained reserve	35.9	East Rhodopes 41.5273° N, 25.4094° E	*Quercus pubescens* Willd., *Quercus frainetto* Ten.
8	Reserve "Sokolna"	1250.0	Stara planina 42.7031° N, 25.1278° E	The forests are composed of oak, hornbeam, maple
9	Maintained reserve "Kirchov dol"	54.2	East Stara planina 42.9529° N, 27.5914° E	150-year-old forest of *Quercus frainetto* Ten., *Q. cerris* L., *Quercus dalechampii* Ten. and *Fagus sylvatica* L.
10	Maintained reserve "Varbov dol"	70.6	East Stara planina 42.9232° N, 27.3829° E	Mixed forests of *Quercus frainetto* Ten., *Quercus dalechampii* Ten. and *Q. cerris* L.
11	Reserve Sokolata	219.6	Maleshevska mountain 41.5307° N, 23.0980° E	*Quercus frainetto* Ten. forests

The system of Natura 2000 sites was applied in Bulgaria, as a country of the European Union. More than 55% of the forest territories are in NATURA 2000 and the Regime for Sustainable Management of these forests has been developed in 2011. The network also includes habitats with oak representatives, as: 91M0 "Pannonian-Balkanic turkey oak-sessile oak forests"; 91AA* "East pubescent oak forest"; 91F0 *Quercus robur*, *Ulmus laevis* and *Fraxinus excelsior* or *Fraxinus angustifolia* forests near big rivers (*Ulmenion minoris*); 9170 "*Galio-Carpinetum* oak-hornbeam forests"; 91G0* "Pannonian forests with *Quercus petraea* and *Carpinus betulus*"; 91H0* "Pannonian forests with *Quercus pubescens*"; 91I0* "Euro-siberian steppe forests with *Quercus* spp" (Kavrakova et all., 2009).

The territory declared as protected areas in Bulgaria is 34.3% (3 National parks, 11 Natural parks and more than 700 other protected areas). The protected areas including oak stands are presented on Table 1.

3. *Quercus cerris* - *Quercus frainetto* Forests in Bulgaria

Mixed forests with a predominance of Turkish oak and Hungarian oak are widespread and of greater economic importance among xerothermic oak forests in the country. They refer to the native, xerophytic vegetation, a major element of forest vegetation in the plains, hilly areas and Foremountains areas of the xerothermic oak belt. The limiting ecological factors for the oak vegetation in this belt are the available periods of drought, low average annual rainfall, relatively high summer and low winter temperatures, as well as the intensive anthropogenic pressure. The climate is temperate continental. The average annual air temperature is 10-11°C, in January - from 0 to -3°C, the average July temperature is 20-21°C. The humidity in summer is insufficient and the droughts are well expressed. Annual rainfall is 500-600 mm in average with May-June maximum and January-March minimum – Figure 2.

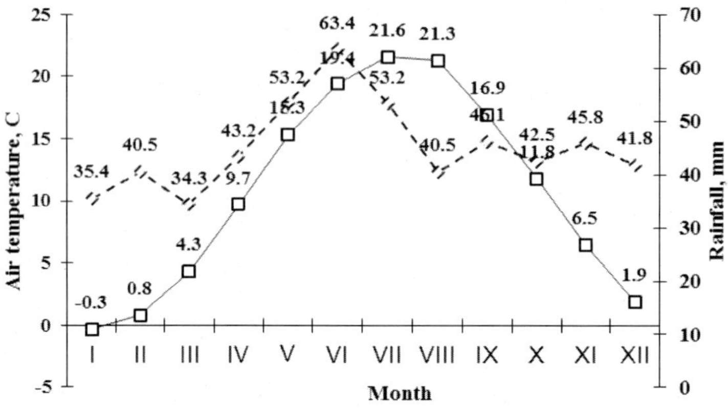

Figure 2. Walter climatic diagram (G. Toshevo, 230 m; 1953 - 2016) – Lyubenova et al. (2019).

(A) (B)

Figure 3. (A) Coppice Turkish oak (B) Hungarian oak forests.

Figure 4. Overgrown *Pteridium aquilinum* L. on the deforested oak forest areas.

Figure 5. Age-old Turkish oak from the area of Dragoman.

The xerothermic oak forests have been subjected to centuries-old anthropogenic effects. Most of the occupying land is arable land today, and the rest of communities are mostly coppice, low-productivity and do not meet good habitat conditions (Figures 3-4).

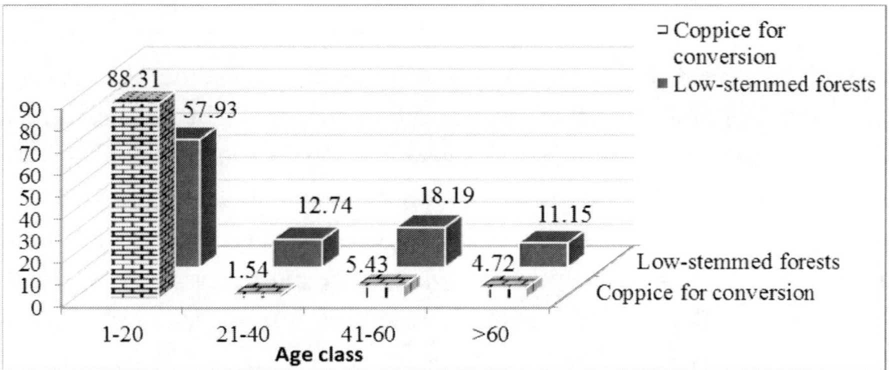

Figure 6. Distribution of coppice for convertion and low-stemmed stands by the age (%).

According to the inventory conducted in 2015, the total area of oak forest ecosystems was 8214072 ha, as coppice conversion forests was 94.77%, high-stemmed forests and low-stemmed forests – respectively only 3.32% and 1.92%. *Q. cerris-Q. frainetto* forests are respectively 26.79% and 28.86% of the area covered by the high-stem (281810 ha) and low-stem coppice forests (162930 ha) and 1.92% of the coppice forests for conversion

(8051142 ha). The distribution of coppice forests for conversion and low-stemmed oak forests by age is presented on Figure 6 - in both types, the first age class is predominated. For high-stemmed forests, the first, second and third age classes are predominated - Figure 7.

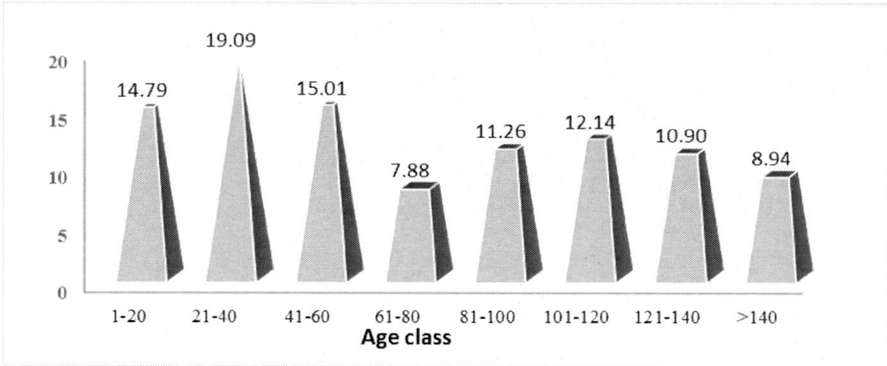

Figure 7. Distribution of high-stemmed stands by the age (%).

Hungarian oak is a light- and heat-loving species, attached to dry and deep soils, with slower growth and lower production than other oak species, but with better timber. Turkish oak is demanding for light and heat, most often the communities are spread over well developed and rich, low to moderate moist soils formed on a diverse rock base. Turkish oak is the fastest growing species of all oaks. As a result of the selected logging of Hungarian oak, the participation of Turkish oak increased and many of Turkish oak forests are now of secondary origin. The seed oak forests are now preserved in some of the reserves, e.g., Sokolata Reserve in Malashevska Mountain – Table 1, in the area of Staro Oryahovo State Forestry, where the communities are used as seed fund, etc.

In the total wood stock of 142822088 m^3, coppice forests for conversion are the biggest contributors - 71.61%, followed by high-stemmed stands - 25.02% and low-stemmed forests - only 3.37% participation – Table 2. *Q. cerris – Q. frainetto* forests take part with 39%, 35.98% and 23.75% in the wood reserves respectively of low-stemmed, coppice for conversion and high-stemmed forests.

Table 2. Wood reserves of oak forests (m^3)

Forest type	Total
High-stemmed forests	35739631
Other oaks	27249881
Q. cerris - Q. frainetto	8489750
Coppice for conversion	102269522
Other oaks	65475791
Q. cerris - Q. frainetto	36793731
Low-stemmed forests	4812935
Other oaks	2935699
Q. cerris - Q. frainetto	1877236
Total	142822088

3.1. *Quercus cerris* L. Forests (*Querceta cerris*)

The formation of *Quercus cerris* L. (900 – 1000 m altitude; about 234 000 ha) has larger areas in Northern Bulgaria, especially in the Ludogorie region, smaller ones in Targovishte, Pleven, Troyan and Vidin, also the Eastern Stara Planina and limited in Southern Bulgaria (the region of Tran and Kraishte, Sredna Gora, Western Border Mountains and Strandja Mountain) – Figure 7. The soils are Chernosems, Grey Luvisols, Chromic Luvisols and Rendzic Leptosols. Generally, soils occupied by *Quercus cerris* L. forests are heavier and more clayer than soils over which *Quercus frainetto* Ten. - *Quercus cerris* L. forests are distributed.

The *Quercus cerris* L. forests are monodominant. They are extremely rich in tree, shrub and grass species. In tree layer, *Ulmus minor* Mill., *Acer campestre* L., *A. tataricum* L., *Fraxinus excelsior* L., *F. ornus* L., *Carpinus betulus* L., *C. orientalis* Mill., *Sorbus torminalis* (L.) Crantz., *Quercus pedunculiflora* C. Koch., *Q. daleshampii* Ten., *Q. frainetto* Ten., *Q. pubescens* Willd., *Q. virgiliana* (Ten.) Ten, *Tilia tomentosa* Moench., *Malus sylvestris* Mill., *Pyrus pyraster* Burgsd. and others participate with single individuals.

Crataegus monogyna Jacq., *Ligustrum vulgaris* Scop., *Viburnum lantana* L., *V. opulus* L. and other shrubs are the most common.

The herb cover is often well developed. *Brachypodium sylvaticum* (Huds.) P. B., *Festuca heterophylla* L., *Dactylis glomerata* L., *Poa nemoralis* L., *P. angustifolia* l., *Buglossoides purpurocaerulea* (L.) L. M. Johnst., *Asparagus tenuifolius* Lam., *Lathyrus niger* (L.) Bernch., *Carex muricata* L., *Geum urbanum* L., *Clynopodium vulgare* L. and others prevail in it (Bondev, 1991).

3.2. *Quercus frainetto* Ten. Forests (*Querceta frainetti*)

Quercus frainetto Ten. forests (up to 800 – 900 m altitude; 50% of total oak area) are relatively widespread in Northen Bulgaria – between ForeBalkan and the meadle part of Danube Plain (Lovech, Pleven, Vidin), North-Eastern Bulgaria (Sowth of Ludogorie) in the Thracian Plain (mainly between Haskovo and Asenovgrad), the Eastern Rhodopes, Mesta and Struma River valley, etc. In Strandzha Mountain, the mixed forests of *Q. frainetto* Ten. and *Q. polycarpa* Schur are widespread. The forests grow mainly on silicate terrains, on Gray Luvisols and Chromic Luvisols soils. In the composition of tree floor, *Quercus frainetto* Ten., *Q. cerris* L., *Q. virgiliana* (Ten.) Ten.), *Q, pedunculiflora* C. Koch, *Fraxinus ornus* L, *Pyrus pyraster* Burgsd., *Sorbus domestica* L., *Acer tataricum* L, *A. campestre* L. and others take part.

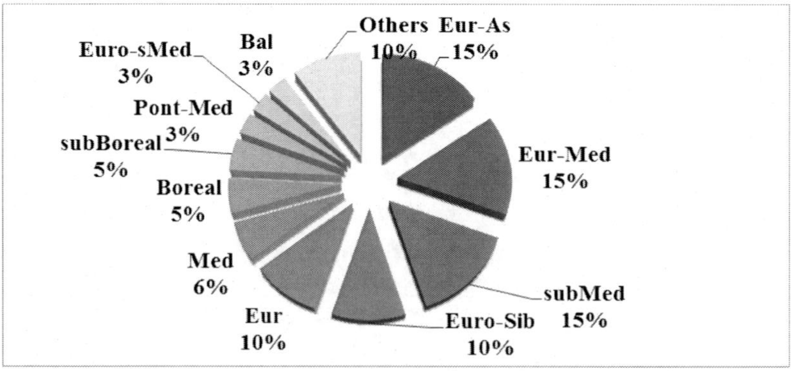

Figure 8. Participation of geoelements (%) in Turkish oak – Hungarian oak forests (Lyubenova et al., 2009).

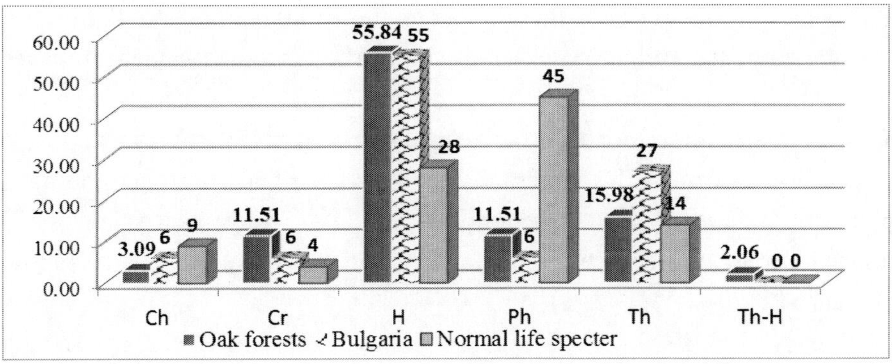

Figure 9. Participation of life forms (%) in Turkish oak – Hungarian oak forests compared to Bulgarian vegetation and the norm for the earth (Lyubenova et al., 2009).

Crataegus monogina Jacq., rarely *Cotinus coggygria* Scop. and *Prunus spinosa* L. are the most common shrubs. *Rhamnus catharticus* L., *Rosa gallica* L., *R. canina* L., *Ligustrum vulgare* L., *Cornus mas* L., *Euonymus europaea* L., *Clematis vitalba* L. and others also participate.

The herb cover is relatively well developed, composed by *Dactylis glomerata* L., *Buglossoides purpurocaerulea* (L.) I. M. Johnst, *Teucrium chamaedrys* L., *Lychnis coronaria* (L.) Desr., *Asparagus tenuifolius* Lam., *Helleborus odorus* W. et K. ex Willd., *Clynopodium vulgare* L., *Carex flacca* Schreb., less often *Asparagus officinalis* L., *Lathyrus niger* (L.) Bernch., *Agrimonia eupatoria* L., *Veronica chamaedrys* L., *Doronicum herbaceum* Vill., *Brachypodium pinnatum* (L.) P. B., *Festuca heterophylla* Lam. and others.

Investigations for the structure of mixed forest communities of *Quercus cerris* and *Quercus frainetto* in Bulgaria were conducted by Lyubenova et al. (2009). About 598 species of vascular plants and about 24 species of bryophytes have been established during the investigation. The participation of geoelements and life forms in the species composition of xerothermic oak forests is presented on Figures 8 and 9.

Georgieva et al. (2013) identified the floristic diversity of xerothermic oak forests with predomination of *Quercus cerris* L. and *Quercus frainetto* Ten. in Site of Community Importance "Zapadna Stara planina i Predbalkan." The inventory of the floristic complex resulted in more than

100 species of higher plants that belong to almost 80 genera and 40 families represented in the Bulgarian vascular flora. The biological spectrum based on Raunkiaer's life forms shows the predominance of hemicryptophytes and takes an intermediate position between the spectrums of the forests in the temperately warm and temperately cold zone. The phytogeographical spectrum shows the predominance of European-Asian and subMediterranean floral elements, which are represented equally (about 20% each).

The relev′es of oak communities (*Quercus frainetto* Ten., *Q. cerris* L., *Q. dalechampii* Ten.) from different parts of Bulgaria collected since 1980s and with different origin were analysed (Lyubenova et al., 2011). Five groups resulted from TWINSPAN classification were obtained. They belong to the alliances *Quercion confertae* (Horvat, 1949) and *Quercion petraeae* (Z′olyomi et Jakucs, 1963). *Quercion confertae* forests are typical thermophilous oak forests distributed in the xerothermal oak belt in Bulgaria. The results obtained are one of the first generalizations of the available data for the widespread oak vegetation in our country and gives some information of the most distributed alliances in lowlands and low mountain belt.

3.3. Functional Parameters Studied

Each ecosystem service is based on a specific functional characteristics of the ecosystem, with one function being able to provide several services and one service to be the result of several functions performed. Ecosystem is the basic functional units of the biosphere, a self-regulating and developing unity between biocenosis and biotope, ie. the living and non-living part, through which energy is fixed and transmitted, the small biological cycle takes place, based on productive and destructive processes and processes of energy transformation (Lyubenova, 2011).

Table 3. General characteristics of four xerothermic oak ecosystems

Objects	W. Balkans, v. Cheparlinci	W. Balkans, v. Govejda	W. Balkans, v. Chuprene	Sofiiska Planina, t. Kremikovci	Lyulin planina, q. Suhodol
G. coordinates	22.9070°	23.0439° - 23.0412°	22.4601° - 22.4549°	23.4998°	23.2240°
	43.0186°	43.2245° - 43.2338°	43.3342° - 43.3330°	42.7843°	42.6961°
Altitude, m	740	284 - 223	307 - 360	1070 - 1100	910
Exposition	W	SE - N	NE - NW	SW	NE
Slope, °	12 - 14	10 - 25	10 - 15	9 - 15	5 -10
Soil	Luvisols /Rendzic	Luvisols /Rendzic	Luvisols	Luvisols	Leptosols
Dominated trees	Qc-Qf (+Q.p., Q.d)	Qf /Qf-Qc	Qf /Qf-Qc	Qd - Qc	Qc-Qf (+Qd)
Canopy	0.4-0.6	0.6 0.9	0.5-0.7	0.7-0.8	0.4-0.5
Number of trees, $n.ha^{-1}$	1102	836 - 600	596 - 668	1133	1101
Age, y	70	50	60	95	80
DBH, cm	19.9-28.1	15.8-19.2	18.9-21.5	17.8/33.5-15.8/40.2	26.1-31.7
Av H, m	17.9	19.2	21.7	18.2	17.9
Bush layer, %	50-60	10-20	30-40	30 - 40	40-60
Grass layer, %	50-80	30-50	30-50	60 - 80	60-70
Litter-fall, $t.ha^{-1}$	3.597	4.059	3.323	2.384	2.432
Mulch, $t.ha^{-1}$	13.713	11.284	12.264	14.823	18.408

Lubenova (2009) reviewed published data on oak ecosystem functions in Bulgaria for the period 1985-2009. The general habitats characteristics of four studed oak communities are presented on Table 3.

The average stocks of mulch is 14.098 (STDEV 2.763) $t.ha^{-1}$, as they replenish annually on average with 3.159 (STDEV 0.734) $t.ha^{-1}$ litter-fall. The biological turnover intensity is inhibited (a = 4.5) of 6 degree (Lyubenova, 2007).

The oak leaves functional characteristics as total content of chlorophyll "a", chlorophyll "b" and carotenoids, determined by "Spectol 10" spectrophotometer applying standard procedure (Arnon, 1949) and the amount of proline determined by Bates et al. (1973) are presented on Table. 4. The values are average over three vegetation seasons (June - October) (Любенова, 2009).

The underground and aboveground biomass of tree floor and undergrowth in representative *Quercus frainetto* - *Quercus cerris* ecosystems in the “Western Balkan and Predbalkan” protected area have been identified. According to the biomass data collected, the studied xerothermic forest vegetation would be given a rating of 8 class (300.1–400 t·ha^{-1}) by the Rodin et al. (1968) scale for broad-leaved forest ecosystems. The aboveground plant mass distribution were in the range of other cited data: respectively 60–81% perennial organs and 1–3% leaves. The estimated percentage of the belowground biomass was higher than the cited data (17–29%) (Lyubenova et al., 2014).

Table 4. Some average functional parameters of oaks leaves

Species	Leaf area (cm^2.g^{-1})	Prolin, μM. g^{-1} DW	Chlorophyll “a” (mg.g^{-1} DW)	Chlorophyll “b” (mg.g^{-1} DW)	Carotenoids (mg.g^{-1} DW)
Q. frainetto Ten.	279.918	0.442	3.171	2.620	3.223
Q. cerris L.	275.584	0.438	3.295	2.933	2.341
Q. pubescens Willd.	258.286	0.440	2.328	1.817	2.383
Q. dalechampii Ten.	311.410	0.883	3.103	3.077	2.737
Av	281.300	0.551	2.974	2.612	2.671
STDEV	22.142	0.222	0.438	0.563	0.409

Table 5. Underground biomass reserves for two oak ecosystems (t.ha^{-1}, %)

Fractions	*1	2	Av	%
Thin roots	11.833	14.542	13.188	20,862
Thick roots	31.181	42.233	36.707	58,07
Sum	43.014	56.775	49.895	78,932
Dry mass	25.325	1.31	13.318	21,068
Total	68.339	58.085	63.212	100

*1 Q.frainetto+Q.cerris-Festuca heterophylla+Poa nemoralis ecosystem on Luvisols.

2. Q.pubescens+Q.frainetto-Cotinus coggygria-Brachypodium pinnatum ecosystem on Chromic cambisols.

The published data for underground biomass of xerothermic oak ecosystems is presented on Table 5. The underground distribution of phytomass is shown on Figures 10 and 11 (Ninov et al., 1983; Lyubenova, 1987, 2009).

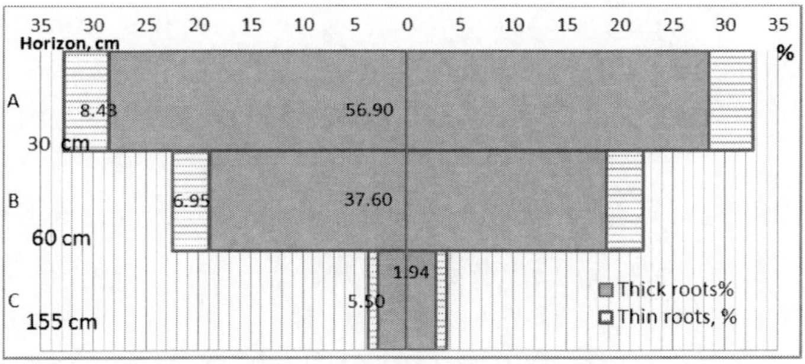

Figure 10. Distribution of *Q. frainetto+Q. cerris-Festuca heterophylla+Poa nemoralis* ecosystem underground phytomass by Luvisols soil horizons (155 cm thickness of soil layer).

Table 6. Average energy content (kJ.g^{-1}) in biomass fractions of xerotermic oak ecosystems

Biomass fractions	*Q. cerris* L.	*Q. frainetto* Ten.	*Q. dalechampii* Ten.	Av.
Aboveground biomass				18.95
Leaves				19.68
seedlings	19.67	17.51	19.22	18.80
tree layer	21.84	19.66	20.34	20.61
bushes				19.63
Branches				19.80
seedlings	17.58	17.32	17.53	17.48
tree layer				
perrenual	20.70	20.31	20.64	20.55
annual	21.34	21.25	21.31	21.30
bushes				19.89
Wood				
tree layer	18.24	18.22	18.25	18.24
Bark				
tree layer	17.99	18.01	18.20	18.07
Herb layer				18.55
legumes species				19.23
mixo-herbosa				18.60
grass sp.				18.36
sedges				18.01
Underground biomass				15.40
Tree layer				15.35
thick roots				15.56
thin roots				15.13
Bush layer				15.40
Herb layer				15.48

Because of underground biomass later studies, the following quantitative data were obtained for *Quercus frainetto* Ten., aged 65 y -11.6 kg.m^{-2} and accumulated carbon - 2.6 kg.m^{-2} (Dimitrova et al., 2014).

The published results for the biomass of *Quercus frainetto* Ten. model tree (BDH 2.6 mm, height 14.9 m, forest density of 22 n.m^{-2} and mean age of 6 years, the following results were obtained: aboveground biomass – 20.451 g.m^{-2} (12.902 g.m^{-2} stem and 7.549 g.m^{-2} leaves) and underground biomass – 30.509 g.m^{-2} (Dimitrova et al., 2014a).

The average energy equivalent of different biomass fractions from xerothermic oak forests, obtained by burning in a Gallencamp CBB01L calorimeter, is presented on Table 6 (Ninov et al., 1983; Lyubenova, 1994; 2009).

The average chemical elements composition of different phytomass fractions and soil from xerothermic oak forests is presented in Tables 7 and 8. Total nitrogen was determined by the Kjeldahl method (Kjeldahl, 1983) and the total elements forms by Perkin-Elmer 3030 atomic absorption spectrophotometer, only Pb was determined by PYE UNYCAM SP 1950 air-acetylene flame spectrophotometer (AAS). All samples were processed by standardized method ISO 11466.

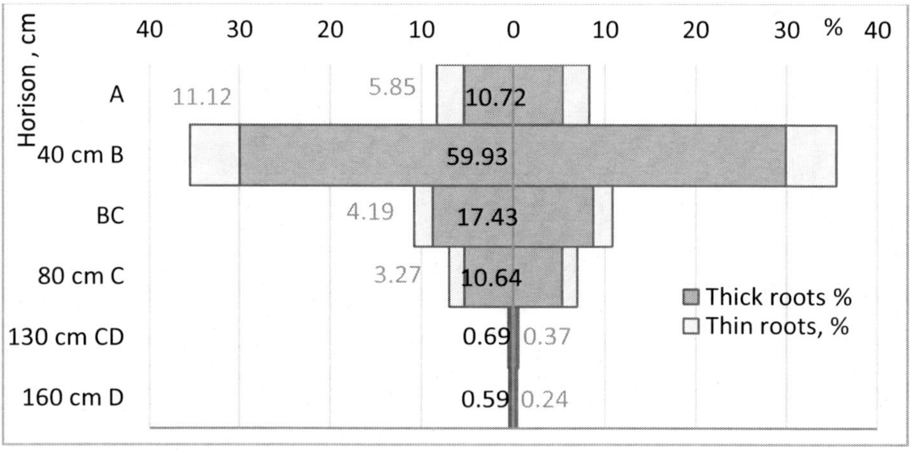

Figure 11. Distribution of *Q. pubescens+Q. frainetto-Cotinus coggygria-Brachypodium pinnatum* ecosystem underground phytomass by Chromic cambisols soil horizons (160 cm thickness of soil layer).

Table 7. Average chemical content in oaks biomass fractions (%, $mg.kg^{-1}$)

Species/ Fractions	C %	N %	P, $mg.kg^{-1}$	Mg, $mg.kg^{-1}$	Ca, $mg.kg^{-1}$	K, $mg.kg^{-1}$	Na, $mg.kg^{-1}$
Q. cerris L.							
Leaves	42.08	1.3122	1966	1873	8730	8892	613
Ann. branches	53.99	0.7873	1322	2042	14135	5762	520
Acorns	56.95	0.6415	1070	946	2841	8922	482
Branches <5 cm	50.12	0.3791	818	922	12900	5271	449
Branches >5 cm	56.28	0.2624	426	1041	1911	5831	409
Bark	42.88	0.3158	426	624	20120	2998	414
Wood	55.34	0.5807	821	565	1685	2290	297
Q. frainetto Ten.							
Leaves	42.08	1.3122	2358	1792	8416	10767	702
Ann. branches	51.40	0.6124	1042	1875	11076	5910	515
Acorns	49.58	0.5248	874	1061	4303	6708	492
Branches <5 cm	39.28	0.2916	846	1072	9955	5005	451
Branches >5 cm	53.60	0.1166	426	374	1251	3972	444
Bark	45.56	0.4374	678	1189	30086	3116	412
Wood	43.42	0.3062	566	365	4310	3264	430

Table 8. Average chemical content in litter-fall and mulch (%, $mg.kg^{-1}$)

Elements	Litter-fall	Mulch	Soil
C %	55.78	37.75	3.19
N %	0.75	0.58	0.12
P, $mg.kg^{-1}$	1059.00	736.50	374.00
Mg, $mg.kg^{-1}$	1892.25	1964.25	3572.50
Ca, $mg.kg^{-1}$	10794.25	11518.00	3438.50
K, $mg.kg^{-1}$	6508.00	2231.75	6733.25
Mn, $mg.kg^{-1}$	2309.75	266.00	1354.00
Cu, $mg.kg^{-1}$	13.85	20.48	29.33
Fe, $mg.kg^{-1}$	334.25	8303.75	25759.25
Zn, mg.kg-1	34.35	43.20	59.40

Ecological status of xerothermic oak ecosystems in SCI "Zapadna Stara Planina i Predbalkan" has been investigated (Georgieva et Lyubenova, 2014). The investigated xerothermic oak ecosystems are in a good ecological status according to the obtained data and calculated indices for tree layer (regeneration capacity, mulch and annual litter-fall reserves, intensity and chemistry of biological turn-over). Nevertheless a couple of negative trends have been established: low average age of seedlings, e.g., low survival rate; turfing in herbaceous layer for about the 50% of communities; trend of xerophytisation in the part of studied area – increasing the number of shrubs seedlings, degradation of tree layer and increasing of dried trees; the high content of Mn (over 0,05 $g.kg^{-1}$) and Fe (over 0,1 $g.kg^{-1}$), also the little higher content of Cu (over 0,006 $g.kg^{-1}$) in plant mass due to the human influence; the very low intensity of biological turnover of Fe and high intensity – of K have been also found.

3.4. Provided Ecosystems Services and Benefits

The representatives of genus *Quercus* L. occupy vast territories in Bulgaria and this determines their ecological and economic importance. In the temperate zone, they are the most important forest producers (Krahl-Urban, 1959).

Oak forests offer a wide range of ecosystem services and benefits - provisional, regulatory, supportive and cultural - that can provide a meaningful life for the human communities that have created their settlements in the vicinity or in their occupied territories. The variety of provisional services provided is related to the quality of the various phytomass fractions. Oakwood is rot-resistant (contains up to 5% tannins), durable and beautiful, and is a high-quality fuel. The wood is used in construction, furniture production, production of small wooden products, and for heating. As a building material, wood of the Hungarian oak is especially appreciated, it is also used for the production of barrels for the maturation of alcoholic beverages. The wood of Turkey oak is very heavy, solid and has high water resistance, whence can be used for the construction of water facilities – bridges, enclosures and others. Unlike the Hungarian oak wood, it decomposes faster and can be used for construction purposes after impregnation. The Dalechamps and sessile oak wood is lighter than the wood of *Q. pedunculiflora* C. Koch. and is used for making chairs, armchairs, stools and other woodworks. The wood of the kermes oak (*Q. coccifera* L.) is good-quality, heavy and solid, but is used mainly for fuel. The wood is not of particular resource importance in the country due to the limited distribution of the species.

The oak bark contains up to 12% tannins (up to 20% at *Q. dalechampii*) and can be used for animalskins treatment. The bark (at *Q. cerris*) and the oak-galls are extremely rich in natural dyes and can be used for painting the wool and silk in gray, beige (*Q. dalechampii bark*), brown (*Q. cerris* bark) and black color.

Leaf phytomass contains up to 30% protein, up to 31% starch and up to 3% fat, so it is a valuable fodder for animals. Only the summer oak dry leaves are of fodder importance, since fresh ones cause the intestinal disorders of animals. The acorns also serve as food for some domestic and wild animals. The flour obtained by grinding the *Q. dalechampii* acorns is used as a fodder and medicine in poultry farming.

The oak forests are rich in species diversity, expressed floor structure (tree, bush, grass and sometimes moss layers) and are the source of many other biological resources - mushrooms, herbs, berries, bee grazing and honey production (more honey dew is produced by the summer oak), hunting resources (deers, roedeers, wild boars, rabbits, pheasants, etc.) and more. Due to its developed grass layer, oak communities have traditional using for grazing.

In the natural medicine, oak is popular as an ingredient in many recipes. Baked and ground cupolas of Hungarian oak are a natural remedy for intestinal disorders, and leaf extract has antiprotista action; bark of summer oak (containing 10-20% quercetin and gallic acids) and oak - galls (up to 40% tannins, sugars, starch, etc.) are used as antiseptic agents - disinfection of the mouth and throat, washing of wounds, etc. The bark and roots of kermes oak are used as cure for dysentery. In the Mediterranean sides, the aphid species is cultivated on the leaves of kermes oak for the carotene production.

Oak forests provide the full range of regulatory and supporting services, which is very important for the adjacent territories, where the settlements are located. The deep and intensive root systems provide pronounced water protection - regulation of water runoff and maintenance of water flow, soil protection, mitigation of the climate and purification of air, soils and water from pollutants. This effect is of great importance for the public health, as well as for increasing the yields of agro-ecosystems, it also affects the quality of production.

The range of cultural services provided is extremely diverse. The oak ecosystems with their species richness, structure and functions create wonderful conditions for aesthetic enjoyment, for tourism and recreation, ie. have a pronounced recreational importance. These communities are deeply connected to the lifestyle, traditions and culture of the Bulgarian population. They have been the shelter of national heroes - fighters for freedom and independence. They are the source of inspiration for the creation of works of art – literature works, folk songs, stitches, fairy tales, stories and more (Lubenova, 2014; 2019).

CONCLUSION

The diversity of forest-forming species in the xerothermic oak belt, the large species diversity of communities, the complex structure of these forests, the potential for ecosystems to function at high temperatures, relatively low rainfall, and prolonged periods of drought make them promising for growing under warming climate.

The traditional use of the large range of services and benefits provided creates the conditions for the development of different industries, jobs creation for the population and improvement of its well-being, as well as for the development of a sustainable economic environment in the regions.

In the face of climate change, the importance of xerothermic oak forests and the services they offer will increase as communities adapted to the periods of long summer drought, while maintaining their ecological functions.

Proper management of oak ecosystems will be essential for the ecology and economy of the regions. Adequate and effective payment schemes for the services provided by oak ecosystems are needed, which will provide sufficient resources for sustainability of the management and flow of services and benefits that ecosystems can provide.

REFERENCES

Apostolova, I., Slavova L. 1997. *A conspectus of the plant communities in Bulgaria. S.,* 340 p. (in Bulgarian).

Arnon, D. J. 1949. "Copper enzymes in isolated chloroplasts. Polyphenoloidase in *Beta vulgaris.*" *Plant Physiol.*, 24: 1-15.

Assyov, B., Petrova, A., Dimitrov, D., Vassilev, R. 2012. Conspectus of the Bulgarian vascular flora. *Horology and floral elements.* Fourth revised and enlarged edition. Bulgarian Biodiversity Foundation, Sofia, 490 p. (in Bulgarian).

Bates, L. S., Walden, R. P., Teare, L. D. 1973. "Rapid determination of free proline for water stress studies," *Plant Soil*, 39, 205-207.

Bondev, I. 1991. Vegeta*tion in Bulgaria. Map 1: 600 000 with explanatory text,* St. "Kliment Ohridski" University Press, Sofia, 183 p. (in Bulgarian).

Dimitrov, M., Glogov P.. 2003. Characteristic syntaxes of forest phytocoenoses in the Lozen mountain. In: *Proccedings of International Scientific Conference "50 Years of University of Forestry."* Department of Forestry and Landscape Architecture, S., pp. 15-20.

Dimitrova, V., Zhyanski, M., Lyubenova, M. 2014. *The role of underground phytomass in the annual fixation of CO_2 from forest ecosystems of major forest species in Bulgaria.* UF Publishing House, Sofia, ISBN 978-954-332-111-7, 112 p. (in Bulgarian).

Dimitrova, V., Lyubenova, M., Dimitrov, D. 2014a. "Biomass of young tree saplings of main forest species in Bulgaria and its role as a carbon depot." *Silva Balcanica*, ISSN 1311-8706, 15(2), p. 35-42.

ExEA, *Executive Environment Agency* – http://eea.government.bg.

Ganchev, I., Bondev, I. 1966. Genus 220 (3) Oak - *Quercus* L. In: Yordanov D (Ed.) *Flora of the People's Republic of Bulgaria*, vol. 3: 105-145. (in Bulgarian).

Gogushev, G. 2009. Floristic classification of oak forests in the Western border Mountains. *Jubilee Scientific Conference 35 Years of EEA Training* at the University of Forestry, Sofia, 54-78 pp. (in Bulgarian).

Georgieva N., M. Lyubenova. 2014. "Ecological status of xerothermic oak ecosystems – SCI" Zapadna Stara Planina I Predbalkan." *Bulgarian Journal of Agricultural Science*, 20 (Supplement 1), 105–109 pp.

Georgieva, N., Pachedjieva K., Lyubenova M.. 2013. "SCI "Zapadna Stara Planina and Predbalkan" – floristic studies on xerothermic oak forests." *Bulg. J. Agric. Sci.,* Supplement 19(2), 218–221pp.

Kavrakova, V., Dimova D., Dimitrov M., Tzonev R., Belev T., Rakovska K. (Eds). 2009. Guidance for determining of the habitats of Europe significance in Bulgaria. Second edition. Sofia, WWF, Danube – Karpatian program, federation "*Green Balkans*," 131 p. (in Bulgarian).

Kjeldahl, J. 1883. "Neue Methode zur Bestimmung des Stickstoffs in organischen Körpern" (New method for the determination of nitrogen in organic substances), *Zeitschrift für analytische Chemie*, 22 (1) : 366-383.

Kostov, G., Aleksandrov, N., Borisov, M., Tonchev, T. 2017. "*Condition and guidelines for management of the coppice oak forests in Bulgaria*." *Gora* 2: 8–9 (in Bulgarian).

Ljubenova, M., I. Bondev. 1987. Underground phytomass of the ass. *Quercus frainetto+Quercus cerris - Festuca heterophylla+Poa nemoralis* tree floor near Gabra village, Sofia. - In: *Proceeding of the IV th National Conference on Botany*, vol. 3, 72-76 (in Bulgarian).

Ljubenova, M. 1994. "Above - ground energy reserves and annual accumulation of energy in shrub and herb layers in ass. *Quercus frainetto-Quercus cerris- Festuca heterophylla +Poa nemoralis* in the

Vakarel Mountains." *Ekol. Zast. Zivot. Sred.*, vol. 2, №1, 29-35. ISSN 0354-2491.

Lyubenova, M. 2004. *Phytoecology*. Acad. M. Drinov Publishing House, Sofia, 574 p. ISBN 954-322-026-3 (in Bulgarian).

Lyubenova, M. 2009. *Functional Biocenology*. Sofia, An-Di Publishing house: 368 p. (in Bulgarian).

Lyubenova, M., Tzonev R., Pachedjieva K.. 2009. "Floristic investigation of *Quercus cerris* and *Q. frainetto* communities in Bulgaria." *Biotechnol.& Biotechnol. Eq.* 23/2009/SE ON-LINE 314-317 pp.

Lyubenova, M., Tzonev R., Pachedjieva K.. 2011. "Syntaxonomy of *Quercetea pubescentis* (Oberd, 1948) Doing Kraft, 1955, in Bulgaria." *Comptes rendus de l'Acad'emie bulgare des Sciences*, tome 64 (4), 565-580 pp.

Lyubenova, M. 2011. "About Ecological Fundaments of New Generation Global Models." Comptes rendus de l'Academie Bulgare des Sciences. *Biologie, ecologie*, t. 64, № 11, 1589-1601.

Lyubenova, M., Dimitrova, V., Georgieva, N., Dimitrov, D. 2014. "Biomass of Tree Xerotermic Oak Ecosystem on a site of community importance, BG0001040 "Zapadna Stara Planina i Predbalkan," Bulgaria. *Drewno*, Vol. (57): 89-106, No. 192 DOI: 10.12841/wood.1644-3985.S10.06. ISSN: 1644-3985.

Lyubenova, M. (Ed.). 2019. *Forest ecosystem services and Payment schemes* (Case study). St. Kliment Ohridski University Press, Sofia, p. 131. ISBN 978-954-07-4778-1

Marinov, M., Kostadinov, K., Popov, G., Stiptsov, V., Bozhinov, H., Dinev, D., Denev, D., Horozov, S. 1995. *The oak forests in Bulgaria*. Zemizdat, 254 p., Sofia, (in Bulgarian).

Marinova, T., Malcheva, K., Bocheva, L., Trifonova, L.. 2017. "Climate profile of Bulgaria in the period 1988-2016 and brief climatic assessment of 2017." *Bul. J. Meteo & Hydro* 22/3-4, 2-15.

Ninov, N., Bondev, I., Georgiev, N. 1983. "Undergroung phytomass, annual production and energy reserves of oak phytocoenose in the zone of Chromic cambisols." *Soil science and agrochemistry*, vol. XVIII, № 6, 87-95. (in Bulgarian).

Rodin, L., Remezov, N., Bazilevich, N. 1968. Методы изучения динамики биомассы и биологического круговорота в растительных сообществах [*Methodological guidelines for the study of the dynamics and the biological cycle in plant communities*]. 145 p. (in Russian).

Rusakova, V. 2015. 20G1 Forests of the Rila oak (*Quercus protoroburoides*). In: Biserkov, V. et al., (ed.) 2015. *Red Data Book of the Republic of Bulgaria. Volume 3.* Natural habitats. BAS & MOEW, Sofia. (in Bulgarian).

Roussakova, V., Tzonev R.. 2003. "Syntaxonomy of the oak forests in the region of Pleven (Danube Plain in Bulgaria)." *Fitosociologia*, 40, 1, 23–31.

Stefanov, B. 1964. The oak and oak forests in Bulgaria /dendrological - ecological characteristic/. In: *On the problems of forestry in the oak forests of the NRB. S.* (in Bulgarian).

Tashev, A., Tsavkov, E. 2017. Validation of the name *Quercus protoroburoides* (Fagaceae). *Phytotaxa* 308(2): 232–238.

Tzavkov, E. 2016. "*Morphological and genetic variability of oaks from Robur Rchb. section in Bulgaria*." Ph D thesis. 129 p. (in Bulgarian).

Tzonev, R. 2009. "Syntaxonomy of the natural and semi-natural vegetation of the Middle Danube Plain in Bulgaria." *Biotechnol. & Biotechnol. EQ.* 23/2009/SE, Special edition – on line, 354–359.

Tsonev, R. 2002. "*Flora and vegetation in the Middle Danube plain between the valleys of the Vit and Studena rivers*." Ph D thesis. Sofia University, (in Bulgarian).

Tzonev, R., Dimitrov, M., Gussev, Ch., Pachedjieva, K., Gogushev, G., Apostolova-Stoyanova, N., Nikolov, I., Alexandrova, A., Glogov, P. 2019. "Phytosociological classification of the thermophilous oak forests in Bulgaria: new interpretations and gaps in knowledge." *Phytocoenologia,* vol. 49, Issue 4, 369–391.

BIOGRAPHICAL SKETCHES

Mariyana I. Lyubenova

Affiliation: Sofia University, Faculty of Biology, Department of Ecology and Environmental Protection, 8 Dragan Tzankov Bld., 1164 Sofia, Bulgaria. E-mail: ryana_l@yahoo.com.

Education:

1984–1988, PhD, Institute of Botany and Mycology, Bulgarian Academy of Science. Title of dissertation: "Primary biological production and energetic reserves of *Quercus frainetto* - *Quercus cerris* forest ecosystems in Vakarelska Mountain."

1973–1978, MSc, Faculty of Biology, Sofia University. Title of thesis: "Influence of water-soluble bioactive substances from *Nicotiana tabacum* L. on the quantity and quality of *Matricaria hamomilla* L. hamazulen oil"

Business Address: Faculty of Biology, Department of Ecology and Environmental Protection, 8 Dragan Tzankov Blvd., 1164 Sofia, Bulgaria.

Research and Professional Experience: 40 years.

2016 to Present - Professor; 2000–2015 - Assoc. Professor, Department of Ecology and Environmental Protection, Sofia University "St. Kliment Ohridski"; 1993–2000 - Chief Assistant; 1989–1993- Senior Assistan; 1986-1989 - Assistant; 1981 – 1986 - Biologist.

1979-1981 – Biologist, Institute of Botany and Microbiology, Bulgarian Academy of Science, Sofia; 1978–1979 Biologist, Institute of Plant Physiology, Bulgarian Academy of Sciences, Sofia.

Professional Appointments: Lectures and Practice.

For bachelor degree:

- Ecology and environmental protection

- Biological resources and their protection
- Methods of ecological researches
- Ecological monitoring

For master degree:

- Functional biocoenology
- Phytoecology
- Ecotoxicology
- Chronoecology
- Natural ecosystems

Head of 53 master – graduates, Leader and advisor of five PhD students.
Forest ecology, Dendrochronology, Functional Biocoenology; Plant monitoring, Ecosystem Services / Ecotoxicology, Modeling of Forest Ecosystems, Plant Resources Protection
H index 7; 185 publications incl. 12 books and 3 handbooks; 250 citations.
Certificate for Excellence in Competition, 2013.
22 International and 19 National projects.
Member of 4 scientific organizations, 3 editorial boards.

Publications from the Last 3 Years:

Báliková K., Červená T., De Meo I., De Vreese R., Deniz T., El Mokaddem A., Kayacan B., Larabi F., Lībiete, Z., Lyubenova M. I., Pezdevšek Malovrh Š., Potocki K., Pelyukh, O., Rugani B., Sarvasova Z., Šálka, J., Stevanov, M, Stojnic, S., Jarský, V., Vuletić D., Zahvoyska, L., Paletto A. 2019. How Do Stakeholders Working on the Forest–Water Nexus Perceive Payments for Ecosystem Services? *Forests* 2020, 11(1), 12; https://doi.org/10.3390/f11010012 ISSN 19994907.

Boteva, S., M. Lyubenova, H. Kalcheva. 2019. Growth Test with *Lepidium sativum* (L.) for Soil Monitoring of Sites, Part of the Bulgarian National Soil Monitoring System. *Bulgarian Journal of Soil Science*, v. 4, iss. 1, 46-54. ISSN 2367-9212 ISSN 2534-8787 DOI: https://doi.org/10.5281/zenodo.3250851.

Bruelheid, H., J. Dengler, O. Purschke, J. Lenoir, B. Jiménez-Alfaro, M. Lyubenova 2018. Global trait–environment relationships of plant communities. Natute. *Ecology & Evolution* https://doi.org/10.1038/s41559-018-0699-8 www.nature.com/natecolevo l.

Chikalanov, A., M. Lyubenova, S. Peteva, Y. Petkov, I. Todorova. 2019. Cost Effectiveness Assessment of Forest Ecosystem Services for Water Bodies Protection.- In: *Reflections on the Mediterranean.* K. JurËeviÊ, L. Kaliterna LipovËan, O. Ramljak (Eds.), Publishers: Institute of Social Sciences Ivo Pilar, VERN' Group Croatian, Academy of Sciences and Arts, Scientific Council of Anthropological Research (HAZU), Euro-Mediterranean Academic Network (EMAN), 637 p. Mediterranean Issues, b.2, 481-492. ISBN 978-953-7964-77-1 (Institute of Social Sciences Ivo Pilar) ISBN 978-953-8101-04-5 (VERN' Group); National and university Library in Zagreb under the number 001034095.

Dineva, S., M. Lyubenova, J. Olsen. 2018. *Trees under polluted environment.* Lap Lambert Academic Publishing, Mauritius, 169 p. ISBN: 978-620-2-07811-5.

Lyubenova, M., A. Chikalanov, Y. Petkov. 2019. Structural Model of Forest Ecosystem Services Payment for Water Quality Impruvment. – In: Proceedings Book "*International Symposium "The Environment and The Industry," SIMI 2019.* Pascu, L., C. Lehr, G. Vasile, A. Constantin (Eds.), Publisher: National Research and Development Institute for Industrial Ecology ECOIND, 238-245. ISSN-L: 1843-5831 DOI: http://doi.org/10.21698/simi.2019;http://doi.org/10.21698/simi.2019.fp31.

Lyubenova, M. (Ed.). 2019. *Forest ecosystem services and Payment schemes* (Case study). St. Kliment Ohridski University Press, Sofia, p. 131. ISBN 978-954-07-4778-1.

Lyubenova, M., A. Chikalanov, S. Peteva, Y. Petkov, I. Todorova. 2019. Development of Schemes for Provided Services of Forest Shelterbelts in Dry Climate. – In: *Agrofood Confference E-BOOK*, 14-27.

Lyubenova, M., S. Dineva, K. Cala, B. Dinich, S. Boteva. 2019. Ecotoxicity of Purified Industrial Waste Water. *Environment and Ecology Research* 7(4), 208-219. doi: 10.13189/eer.2019.070402 http://www.hrpub.org.

Lyubenova, M., S. Nedkov, M. Zhiyansky, P. Nikolov, G. Popchev, P. Petrov. 2019. Assessment of Pollination Ecosystem Service Provided of Urban Ecosystems in Bulgaria. Nedkov St. et al., (Eds.) Smart Geography. Key Challenges in Geography, 405-416. *EUROGEO Book Series.* Koutsopoulos K., González R., Schmeinck D. (Eds.). Springer International Publishing. ID: 476867_1_En ISBN: 978-3-030-28190-8 doi 10.1007/978-3-030-28191-5.

Lyubenova, M. 2019. *Ecosystem services and benefits. Cost, Efficiency, Payments.* University Publishing House "St. Kliment Ohridski," Sofia, 250 c. ISBN 978-954-07-4729-3.

Peteva, S., M. Lyubenova. 2018. Investigation of Dendroflora in the City of Sevlievo. *Journal of Balkan Ecology*, vol. 21, №4, 425-442.

Peteva, S., M., Lyubenova. 2019. Green System and Air Quality in town Sevlievo, Bulgaria. *Ecologia Balkanica*, Vol.11, Issue 2. Union of Scientists in Bulgaria - Plovdiv, ISSN 13140213, 13139940.

Tzvetkova, P., M. Lyubenova, S. Boteva, E. Todorovska, S. Tsonev, H. Kalcheva. 2019. Ecotoxicology of herbicides Paraquat and Glyphosate on the early development of two plant biotests. *IOP Conf. Ser.:* Earth & Environmental Sciences, vol. 221012137 https://www.mess-earth.org/gallery.html.

Violeta G. Dimitrova

Affiliation: University of Forestry, Faculty of Forestry, Department of Dendrology, 10 Kliment Ohridski Blvd., 1756 Sofia, Bulgaria. E-mail: vilydi@abv.bg.

Education: PhD

Business Address: University of Forestry, Faculty of Forestry, Department of Dendrology, 10 Kliment Ohridski Blvd., 1756 Sofia, Bulgaria.

Research and Professional Experience:

08.1999-01.2000 – ecologist, Sofia Municipality, District of Kremicovtci.

02.2000 -11.2007 - Senior expert, Ministry of Environment and Water.

1.11.2007 – 06.2018 - Assistant Professor, University of Forestry.

06.2018 – till now – associated Professor, University of Forestry.

From 2003 to 2007- PhD in Ecology and Ecosystem Protection.

Professional Appointments:

Lectures and practice:

- Botany;
- Medical botany;
- Phytocoenology;
- Ornamental plants in the Bulgarian flora

Publications from the Last 3 Years:

Dimitrova, V. (2016). Forest habitats in Natura 2000 protected zone BG0000211 “Tvardishka planina” – floristic composition and nature conservational status. *Ecologia Balcanica*, v. 8, issue 1, pp. 47-55.

Dimitrova, V. (2017). Floristic inventory and nature conservational status of the forest nature habitats in “Karvav kamak,” Bulgaria protected zone from Natura 2000. *Comptes rendus de l’Academie Bulgare des Sciences* t. 70 (1), 91-102.

Dimitrova, V., Dimitrov, D., Velizarova, E. (2017). Biomass and carbon content in undergrowth from areas affected by fires. *J. BioSci. Biotechnol.* SE/ONLINE:81-84.

Dimitrova, V. (2018). Stocks of dead biomass of beech (*Fagus sylvatica* L.) forest ecosystems in *West Balkan Range*, Bulgaria, Šumarski list, 7–8: 363–370.

Dimitrova, V. (2018). Flora in Region of Studena Quarry and Measures for Conservation of Protected Species. *Journal of Balkan Ecology*, vol. 21, No. 4, pp. 393-399.

Dimitrova, V. Tashev, A. (2019). Medical plants on the territory of Petrohan Training and Experimental Forest Range, Bulgaria. *Agriculturae*

Conspectus Scientificus, vol. 84, No. 1 pp. 37-46. ISSN: 1331-7768, e-ISSN: 1331-7776.

Tashev, A., Dimitrova, V. (2019). Medicinal Plants of Bulgaria. *Current Perspectives of Medicinal and aromatic plants*, ISSN 2619- 9645, e-ISSN: 2667-5722, v. 2, 29-39.

In: Quercus: Classification, Ecology and Uses ISBN: 978-1-53618-026-8
Editor: Bent J. Steffensen

Chapter 8

STUDY ON SEEDLING POPULATION RESPONSE TO DISTURBANCE OF A GARHWAL HIMALAYAN OAK (*QUERCUS LEUCOTRICHOPHORA* A. CAMUS) FOREST

V. Singh[1,*], D. S. Chauhan[2] and S. Dasgupta[3]
[1]Department of Forestry, Dolphin PG Institute of Biomedical and Natural Sciences, Dehradun, Uttarakhand, India;
[2]Department of Forestry and Natural Resources, HNBGU, Srinagar Garhwal, Uttarakhand, India
[3]Department of Forestry and Biodiversity, Tripura University, Suryamaninagar, Tripura (West), India

ABSTRACT

An extensive study was conducted to access the population status of *Quercus leucotrichophora* seedlings in a part of Garhwal Himalaya Oak forest. Permanent plots were established in different altitudes and aspects.

---

* Corresponding Author's Email: vikaspals@gmail.com.

Among all disturbance stands the seedling density of *Quercus leucotrichophora* was found higher in the north aspect at moderately disturbed stand followed by highly and undisturbed stands. The *Quercus leucotrichophora* seedlings were best represented in all root collar diameter classes at north aspect as compared to south aspect. In the south aspect under root collar diameter class of 7.5 – 10.0 mm in undisturbed stand, < 10.0 mm root collar diameter class in moderately disturbed stand and 0 - 2.5 mm root collar diameter class at highly disturbed stand there were no seedling of *Quercus leucotrichophora*. In the rest of seedling species, among all disturbed stands, the seedling density was found higher again at north aspect in most cases.

Keywords: soil, north facing aspect, south facing aspect, altitude

INTRODUCTION

Banj Oak (*Quercus leucotrichophora*) is the most common broadleaf tree in the mid-elevational ranges of central Himalaya in India. This broad-leaved genus of family Fagaceae has many superior characteristics for its wide acceptance as fuel, timber and nutritious fodder throughout the Himalaya. Past studies on this tree have reported that regeneration is sparse to absent promoting speculation that this species will soon be replaced by others (Thadani and Asthon, 1995). *Quercus* is a large genus with many of its five hundred species being canopy dominants. The broad leaf oak forests are considered to provide the most effective water and soil conservation services (Saxena and Singh, 1982).

One important objective in forestry is to establish and maintain natural regeneration of young trees with good height growth and a future timber quality (Drexhage and Colin, 2003). Regeneration is a key process for the existence of species in a community. It is also a critical part of forest management, because regeneration maintains desired species composition and stocking after biotic disturbances (Khumbongamyum et al., 2005). Recruitment of tree species on the forest floor is governed by various factors and microhabitat conditions (Kitajima, 2007). Likewise, germination of seeds in nature is strongly influenced by internal as well as external environmental factors (Baskin and Baskin, 1998). Regeneration of a species is dependent on internal community processes and exogenic disturbance (Barker and Patrik, 1994). It has been recorded that regeneration of a species is effected by fire (Sukumar et al., 1997), grazing, light, canopy, density, soil moisture and nutrients and anthropogenic pressure (Murthy et al., 2002).

The process of seedling growth and development of forest trees largely depends on gaps/canopy openings in the forest created due to natural disturbance, thus influencing the regeneration and species composition of the forest (Khumbongmayum et al., 2005). The forest regeneration is that vital process, which in the due course, leads to silvigenesis and build-up of the forest stand. Though the regeneration of forest is possible through both natural as well as artificial means, in many situations the natural regeneration is the predominant process (Tripathi and Khan, 2007). Regeneration of a particular species is poor if seedlings and saplings are much less than the mature trees. The study of regeneration of forest trees has important implications for the management of natural forests, and is one of the thrust areas of forestry (Pokhariyal et al., 2010).

Study Area

This study was conducted in Chandrabadani Oak forest. The Chandrabadani Oak forest forms a moist-temperate area falling in the jurisdiction of Tehri district of Uttarakhand State. The study area was assessed along three altitudes viz; upper, middle and lower in both north facing and south facing aspects. The studied altitude in north facing aspect were in the range of 1500 m to 2100 m and in south facing aspect from 1500 m to 2200 m altitudes (Table 1). The study area dominated by broad leaved species including *Quercus leucotrichophora, Rhododendron arboreum, Myrica esculenta, Lyonia ovalifolia, Pyrus pashia, and Benthamidia capitata* etc., with dominant shrub species of *Woodfordia fruticosa, Rubus ellipticus, Pyracantha crenulata, Berberis aristata,* and *Colebrookea oppositifolia.*

Table 1. Location of studied altitudinal ranges in Chandrabadani oak forest

Studied Altitudes	Altitude (m.)	Latitude	Longitude
North Upper	2092	30 ⁰ 19’ 03.7”	78 ⁰ 36’ 55.1”
North Middle	1781	30 ⁰ 19’ 15.3”	78 ⁰ 37’ 10.1”
North Lower	1597	30 ⁰ 19’ 26.5 ”	78 ⁰ 37’ 20.1”
South Upper	2103	30 ⁰ 18’ 48.0”	78 ⁰ 37’ 02.7”
South Middle	1829	30 ⁰ 19’ 42.3”	78 ⁰ 36’ 36.5”
South Lower	1563	30 ⁰ 20’ 13.5”	78 ⁰ 37’ 14.2”

METHODOLOGY

The study area was assessed along three altitudes viz; upper, middle and lower in both north facing and south facing aspects. The studied altitude in north facing aspect were in the range of 1500 m to 2100 m and in south facing aspect from 1500 m to 2200 m altitudes. Vegetation study was conducted in both north-facing and south-facing aspects. Sampling was done in stratified random manner along the altitudinal transects. A total of six transects were laid down along altitudinal gradient (03 each in north-facing and south-facing aspects). Transects were spatially distributed so as to minimize the autocorrelation among the vegetation. Quadrats were laid down along each transect in stratified random manner and placed at an interval of 100m. Species area curve was used to determine minimal sample area which is based on quantitative variation of the vegetation in terms of species number. Quadrats of 10 m x 10 m were used for tree layer, 2m x 5m for shrubs and 1mX1m for herb species. About 5% of total forest area was sampled.

The whole study area on each aspect was divided into three altitudinal ranges viz. upper altitudinal range (> 1900 m asl), middle altitudinal range (1700 to 1900 m asl) and lower altitudinal range (< 1700 m asl). In each altitudinal range permanent plots were established for enumeration of seedlings. All seedling individuals were tagged for assessing the growth and development of seedlings. The classification of disturbance classes were determined on the basis of tree density, canopy cover, cut stumps and lopping intensity as per methodology given by Mishra et al., (2004) (Table 2). The seedling population status was determined by classifying different root collar diameter ranges at each permanent plots among all studied altitudes. The root collar diameter classes were setup as 0 – 2.5 mm., 2.5 – 5.0 mm, 5.0 – 7.5 mm, 7.5 – 10.0 mm and < 10 mm. In each permanent plot the seedlings were tagged and monitored periodically. During each observation the recruitment, survival, growth and development of the tagged seedling were recorded.

Table 2. Disturbance index categorization of Chandrabadni Oak forest

Disturbance parameters	Undisturbed	Moderately disturbed	Highly disturbed
Tree Density (h^{-1})	> 1200	800-1200	< 800
Canopy cover (%)	> 50	30-49	< 30
Lopping (%)	< 20	21-29	> 30
No. of cut stumps (h^{-1})	< 80	81-149	> 150

RESULT

Undisturbed Stand

In the north aspect of undisturbed stand, maximum seedling density (1.2 plants/100m2) of *Quercus leucotrichophora* was found under 0 – 2.5 mm and 7.5 – 10.0 mm root collar diameter class and minimum seedling density (0.8 plants/100m^2) was recorded under 2.5 – 5.0 mm and 5.0 – 7.5 mm root collar diameter classes. The seedling density of < 10mm collar diameter class was found as 1.0 plants/100m^2. At south aspect of the undisturbed stand, *Quercus leucotrichophora* showed highest seedling density (5.4 plants/100m^2) was recorded under 2.5 – 5.0 mm root collar diameter class and lowest seedling density (1.4 plants/100m^2) was under 5.0 – 7.5 mm root collar diameter class. The root collar diameter classes of 0 – 2.5 mm and 5.0 – 7.5 mm showed the seedling density 1.8 plants/100m^2 and 1.4 plants/100m^2 respectively. While, under 7.5 – 10.0 mm root collar diameter class no seedling of *Quercus leucotrichophora* was found (Figure 1).

Moderately Disturbed Stand

In the moderately disturbed stand the north aspect showed higher seedling density as compared to the south aspect for *Quercus leucotrichophora* seedlings. In north aspect the highest Quercus seedling density (3.2 plants/100m^2) was recorded under 5.0 – 7.5 mm root collar diameter class and lowest density (0.2 plants/100m^2) in 7.5 – 10.0 mm root collar diameter class. The root collar diameter class of 0 – 2.5 mm showed 1.4 plants/100m^2 seedling density, 2.5 – 5.0 mm as 0.8 plants/100m^2 and < 10.0 mm root collar diameter showed 0.4 plants/100m^2 seedling density. In the south aspect maximum seedling density (2.6 plants/100m^2) was observed under 5.0 – 7.5 mm root collar diameter class followed by 2.5 – 5.0 mm root collar diameter class (2.0 plants/100m^2), 0 – 2.5 mm root collar diameter (0.4 plants/100m^2). While, in the root collar diameter class of < 10 mm no seedlings were recorded (Figure 1).

Highly Disturbed Stand

The seedling density of *Quercus leucotrichophora* was found maximum (2.8 plants/100m^2) in 2.5 – 5.0 mm root collar diameter class and minimum seedling density (0.2 plants/100m^2) was in 0 – 2.5 mm root collar diameter class. The root collar diameter class of 5.0 – 7.5 mm showed 1.8 plants/100m^2 seedling density, 7.5 – 10.0 mm collar diameter class as 1.0 plants/100m^2 and < 10.0 mm root collar diameter class showed 0.6 plants/100m^2 seedling density. In the south aspect of highly disturbed stand, highest seedling density (3.8 plants/100m^2) was recorded in 2.5 – 5.0 mm root collar diameter class followed by 5.0 – 7.5 mm (0.8 plants/100m^2), < 10 mm root collar diameter class (0.6 plants/100m^2) and 7.5 – 10.0 mm root collar diameter class (0.4 plants/100m^2). While, under 0 – 2.5 mm root collar diameter class not a single seedling of *Quercus leucotrichophora* was found (Figure 1).

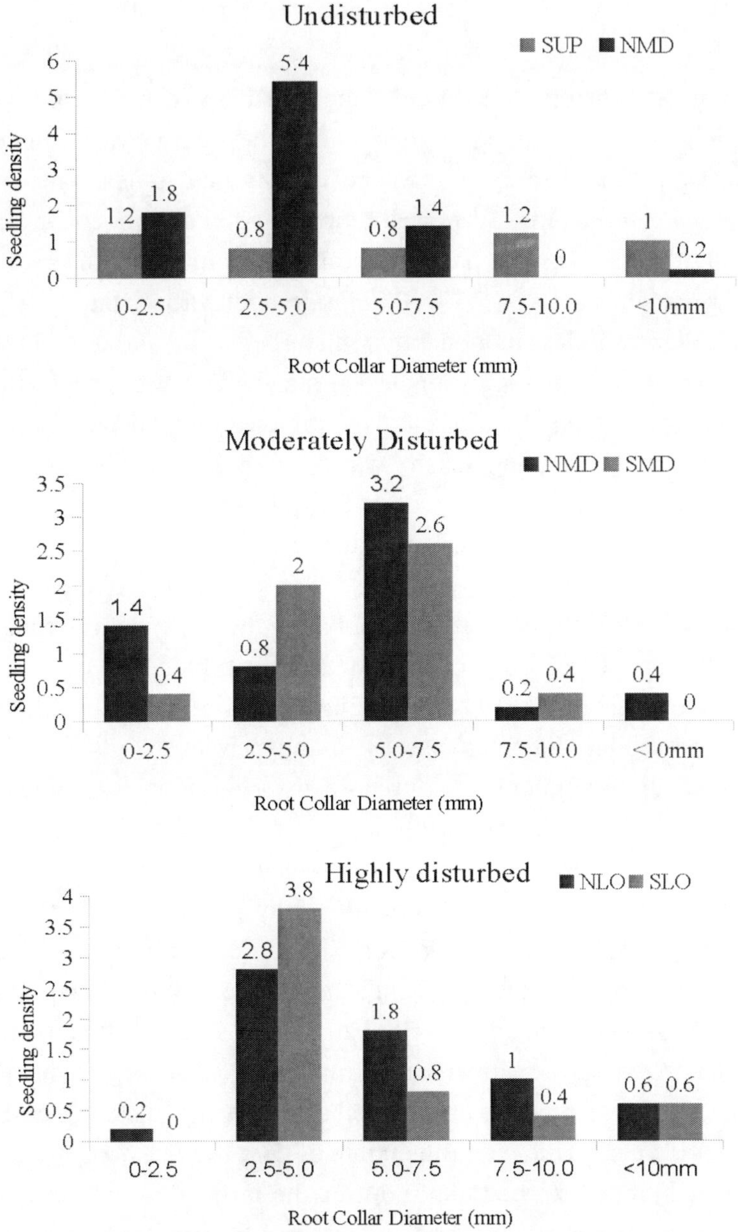

**NUP-North Upper, SUP- South Upper, NMD-North Middle, SMD- South Middle, NLO-North Lower, SLO-South Lower*

Figure 1. Population density of *Quercus leucotrichophora* seedlings observed in different disturbance stands at studied permanent plots.

DISCUSSION

Among all disturbance stands the seedling density of *Quercus leucotrichophora* was found higher in the north aspect at moderately disturbed stand followed by highly and undisturbed stands. The *Quercus leucotrichophora* seedlings were best represented in all root collar diameter classes at north aspect as compared to south aspect. In the south aspect under root collar diameter class of 7.5 – 10.0 mm in undisturbed stand, < 10.0 mm root collar diameter class in moderately disturbed stand and 0 - 2.5 mm root collar diameter class at highly disturbed stand there were no seedling of *Quercus leucotrichophora*. In the rest of seedling species among all disturbed stands the seedling density was found higher again at north aspect in most of cases. Regeneration of species is greatly influenced by the interaction of different biotic and abiotic factors of the environment (Akasmit and Iring, 1984). These factors may affect the recruitment, survival and growth of tree regeneration. Unfortunately, the regenerative capacity of Oak forest is poor not only in the Himalayan region but also in North America (Lorimer et al., 1994) and Europe (Andersson, 1991). Among different root collar diameter classes in the permanent plots, the major portion of seedling population was recorded for the root collar diameter of 2.5 – 5.0 mm and 5.0 – 7.5 mm.

In general regeneration of species is effected by anthropogenic factors (Barik et al., 1996, Khan and Tripathi, 1989) and natural phenomena (Welden et al., 1991). The greater contribution of seedlings, saplings and young trees to the total population indicates successful regeneration (Khan et al., 1987, Saxena et al., 1984, Bhuyan et al., 2002) of most of the tree species. The seedling recruitment in permanent plots among all disturbance stands was varied with the seasons as well as the level of disturbance. During the observation period the recruitment rate of *Quercus leucotrichophora* was higher in the highly disturbed stands during the spring season (April). It may be due to low canopy cover, and low stem density which make canopy openings in the highly disturbed stands facilitating the suitable environment for seedling recruitment. Shrestha et al., (2007) observed similar findings for regeneration status of *Betula utilis*; in which they revealed that the open forest left after clear cutting became suitable for seed germination and seedling establishment. *Quercus leucotrichophora* with seedfall at the end of autumn (September - October) and the emergence of seedlings start in February and peak by the April.

CONCLUSION

The recruitment of seedlings was greater at highly disturbed stands as compare to undisturbed and moderately disturbed stands. It may be due to scattered canopy present at highly disturbed stands which causes suitable microenvironmental conditions for the recruitment of new seedlings. Extraction of mature trees, collection of fuel-wood and fodder by the villagers and grazing by domestic animals were the major factors of anthropogenic disturbance in the study area. The low and medium level of disturbances will be helpful in maintaining the high plant biodiversity. Furthermore regeneration of *Quercus leucotrichophora* was found severely influenced by insect attack on the fallen seeds, seed predation by birds, overcrowding of herbaceous layer and excess litterfall.

REFERENCES

Aksamit, S.E. and Irring, F.D. (1984). Prescribed burning for lowland black spruce regeneration in northern Minnesota. *Cac. J. For. Res.* 14: 107-113.

Andersson, C. (1991). Distribution of seedlings of Quercus robur in a grazed deciduous forest. *Journal of vegetation Science*. 2: 279-237.

Barik, S.K., Tripathi, R. S., Pandey, H.N. and Rao, P. (1996). Tree regeneration in a subtropical humid forest: effect of cultural disturbance on seed production, dispersal and germination. *Journal of Applied Ecology* 33: 1551-1560.

Barker, P.C.J. and Krik Patrik. J.B. (1994). Phyllocladus asplenifolius: variability in the population structure of the regeneration niche and dispersion patterns in Tasmanian forest. *Australian Journal of Botany* 42: 163-190.

Baskin, C.C. and Baskin, J.M. (1998). *Seeds: Ecology, Biogeography and Evolution of Dormancy and Germination*. Academic Press, San Diego, CA, 666 pp.

Bhuyan, P., Khan, M. L. and Tripathi, R. S. (2002). Regeneration status and population structure of Rudraksh (*Elaeocarpusganitrus* Roxb.) in relation to cultural disturbances in tropical wet evergreen forest of Arunachal Pradesh. *Current Science*, 83: 1391-1394.

Drexhage, M. and Colin, F. (2003). Effect of browsing on shoots and roots of naturally regenerated sessile Oak seedlings. *Ann. For. Sci.* 60: 173-178.

Khan, M.L. and Tripathi, R.S. (1989). Survival and growth of transplanted nursery seedlings of three sub-tropical trees at burnt and unburnt sites in dense and sparse forest stands. *Tropical Ecology* 30: 20-30.

Khan, M.L., Rai, J.P.N. and Tripathi, R.S. (1987). Population structure of some tree species in disturbed and protected sub-tropical forests of northeast India. *Acta Ecology*. 8: 247-255.

Khumbongamyum, A.D., Khan, M.L. and Tripathi, R.S (2005). Survival and growth of seedlings of a few tree species in the four sacred groves of Manipur, Northeast India. *Current Science* 88(11): 1781-1788.

Kitajima, K. (2007). Seed and seedling ecology. In: F.I. Pugnaria and F. Valladares (eds.) *Functional Plant Ecology*. IInd Edition. CRC Press Tylor and Francis Group. Pp. 549-580.

Lorimer, C.G., Chapman, W. and Labmcrt, W.D. (1994). Tall understory vegetation as a factor in the poor development of oak seedlings beneath mature stand. *Journal of Ecology*. 82: 227-237.

Mishra, B.P., Tripathi, O.P.,Tripathi, R.S. and Pandey, H.N. (2004). Effects of anthropogenic disturbance on plant diversity and community structure of a sacred grove in Meghalaya, northeast India. *Biodiversity and Conservation* 13: 421-436.

Murthy, I.K., Murali, K.S., Hegde, G.T., Bhat, P.R. and Ravindranath, N.H. (2002). A comparative analysis of regeneration in natural forest and joint forest management plantations in Uttara Kannada Districts. Western Ghats. *Current Science* 83: 1358-1364.

Pokhariyal, P., Uniyal, P., Chauhan, D.S. and Todaria, N.P. (2010). Regeneration status of tree species in forest of Phakot and Pathri rao watersheds in Garhwal Himalaya. *Current Science* 98(2): 171-175

Saxena, A.K. and Singh, J.S. (1982). A phytosociological analysis of woody species in forest in forest communities of a part of Kumaon Himalaya. *Vegetation,* 50: 3-22.

Saxena, A.K., Singh. S.P. and Singh. J.S. (1984). Population structure of forests of Kumaon Himalaya: implications for management. *Journal of Environmental Management* 19: 307-324.

Shrestha B.B., Ghimire, B., Lekhak, H.D. and Jha, P.K. (2007). Regeneration of Treeline Birch (*Betula utilis* D.Don) Forest in trans-Himalayan Dry Valley in Central Nepal. *Mountain Research and Development* 27(3):250-258.

Sukumar, R., Suresh, H.S., Dattaraja, H.S. and Joshi, N.V. (1997). In: *Forest Diversity Research, Monitoring and Modeling: Conceptual Background and old World Case Studies* (eds. Dallmeier, F. and Comiskey, J.A.), Parthenon Publishing. Vol. 1 pp. 529-540.

Thadani, R. and Asthon. P.M.S. (1995). Regeneration of banj Oak (*Quercus leucotrichophora* A. Camus) in Central Himalaya. *Forest Ecology and Management* 78: 217-224.

Tripathi, R.S. and Khan, M.L. (2007). Regeneration dynamics of natural forests- A review. *Proc. Indian Natn. Sci. Acad.* 73(3): 167-195.

Welden, C. W., Hewett, S. W., Hubbell,S. P. and Foster, R. B. (1991). Sapling survival, growth, and recruitment: relationship to canopy height in a neotropical forest. *Ecology* 72:35–50.

In: Quercus: Classification, Ecology and Uses ISBN: 978-1-53618-026-8
Editor: Bent J. Steffensen

Expert Commentary

SOCIETY, ENVIRONMENT AND FORESTS FUTURE

Mariyana I. Lyubenova
Faculty of Biology, University of Sofia "St. Kliment Ohridski",
Sofia, Bulgaria

An observable global trend since the mid-20th century is that anthropogenic pressure and anthropogenic transformations in the biosphere and its transformation into the noosphere are tangential to its degradation. Technology development, including innovative technologies related to the consumption of resources cannot be stopped and limited, as it is embedded in the development of human knowledge. Human society's dependence on ecosystem services, including natural resources, is steadily increasing, while practices for managing and using these services and resources are leading to the environmental degradation, the ecosystems and the biodiversity loss (de Groot et al., 2002).

The only viable strategy for the conservation of natural resources is the Sustainable development strategy. This means assessing the ecosystem services provided by the natural communities, valuing them and incorporating this cash flow into national budgets as part of it, to spend on restoring the resources expended, using ecosystem services and benefits, ie. "The resources protection in the processes of their utilization."

In 2009, the World Network - Ecosystem Services Partnership (ESP) was created to enhance the science, policy and practice activities related to the ecosystem services for protection and sustainable development (https://www.es-partnership.org/). The network includes 48 environmental organizations, universities, companies and more as members. The International Ecosystem Economics and Biodiversity Initiative (IEB) also published a series of reports, calling for an ever change in national and international environmental policies in separate "areas" - air, water, soil, biodiversity - towards the more integrated approach based on the comprehensive ecosystem assessments. The Conference of the Parties to the Convention on Biological Diversity (CBD) adopted a Strategic Plan aimed at "maintaining ecosystem services, maintaining a healthy planet and providing benefits for all people "(CBD 2010). The EU also accepts its goal of halting biodiversity loss and degradation of ecosystem services by 2020 and restoring them where possible. In 2011, the European Commission adopted "Biodiversity Strategy," in which the concept of ecosystem services is extremely integrated. In 2012, the vision for a "Green Economy" outlined in Rio + 20 assumes that economic indicators depend on the effective management of ecosystems and biodiversity. In the same year, the International Finance Corporation published its revised Implementation Standards (IS) on environmental and social sustainability.

The Environmental Program, using the UN-initiated ecosystem services concept - The Economics of Ecosystems and Biodiversity (TEEB2, http://www.teebweb.org/about/the-initiative/) continues its environmental activities. The development of the EU CICES (Towards a Common Classification of Ecosystem Services, https://cices.eu/) has been published on the website of the European Environment Agency. The projects of Seventh Framework Program (FP7) - OpenNESS (2015), OPERA (2015) and ESMERALDA (http://www.esmeralda-project.eu/) have contributed to the development of the concept and methodologies for evaluating the ecosystem services and the implementation of concept through real-world case studies. In 2012, the European Commission established a Special

Working Group on Mapping and Assessment of Ecosystems and Their Services (MAES, https://biodiversity.europa.eu/maes). In 2012, the Intergovernmental Platform on Biodiversity and Ecosystem Services (IPBES, https://www.ipbes.net/) was established, providing mechanisms for evaluating, integrating and critically analyzing different knowledge and information generated by scientific, governmental and non-governmental organizations, etc. institutions that are consistent with both the scientific community and the governance structures (Maes et al., 2012; Lyubenova et al., 2015; 2017). There are an extremely large number of organizations at national and international level that are relevant to environmental, economic or social aspects of biodiversity and ecosystem services.

Gómez-Baggethun et al., (2010) has made a detailed literature review of the evolution of the concept of ecosystem services in the scientific literature since its emergence in the second half of the 20th century, when the benefits of natural functions to society began to be emphasized.

The current definition of ecosystem services as "the contribution of ecosystem structures and functions, combined with other elements, to human well-being" (Burkhard et al., 2017; Assenov et al., 2016, 2017) includes both the contribution of the natural environment and anthropogenic impacts and human contribution to the well-being of society. The explanation "in combination with other elements" covers changes in nature as a result of human activities - engineering facilities, agricultural crops, agroforestry systems (including forest belts), changes in land use, intensification the human impacts etc., which nowadays have reached considerable proportions and lead to a significant change in the natural functions of nature, especially in industry and agriculture. Biophysical assessment of ecosystem goods and services in most EU countries has been completed, followed by the monetary evaluation projects.

The implementation of Sustainable Development Strategy requires an aggregate knowledge of the structure and functioning of ecosystems, which will enable them to assess the potential benefits and services they provide and to monitor the status and dynamics of services to ensure their sustainable use. Forest ecosystems are of paramount importance to human society among other ecosystems. They are the source not only of a variety provisional services (Lyubenova, 2004; 2019), but also of regulatory and supporting services set that are important for maintaining equilibrium in nature, maintaining gas balance, soil fertility and soil integrity, water balance and the flow of surface water bodies and drinking water supplies, as well as for the removal of contaminants from all natural components. They are a source of diverse cultural services. The development of civilization is directly related to forest ecosystems. They are embedded in the minds, traditions and folk psychology of the local communities. Forests are a place for relaxation and aesthetic enjoyment, inspiration for the creation of works of art, a tool for the personality educating, etc.

In the warming climate conditions, particular attention must be paid to the forest ecosystems that are adapted to a more continental climate and are of economic importance for the regions. In Bulgaria, these are xerothermic oak forests located on the foremountains, hills and hilly plains. Aggregation of existing information on the structure and functioning of these forest ecosystems is essential in valuing the ecosystem services and benefits they offer, creating adequate management solutions and appropriate payment schemes that will enhance their capacity to influence the environment and the economy of local communities as well as the sustainable development of regions.

REFERENCES

Assenov, A., A. Chikalanov, M. Lyubenova, S. Kostadinova. 2017. Ecosystem services: a function of natural capital, Stephenson Library, Newcastle upon Tyne, NE6 2PA, UK. In: Recep Efe Münir, Öztürk (Eds). *Contemporary Studies in Environment and Tourism,* Ch. 4. ISBN (10): 1-4438-7283-0 ISBN (13): 978-1-4438-7283-6.

Assenov, A., K. Vassilev, N. Pedashenko, B. Koulov, E. Ivanova, B. Borisova. 2016. Research of the Biotope Diversity for the Purposes of Economic Valuation of Ecosystem Services in Chepelare Municipality (The Rhodopes Region of Bulgaria). *European Journal of Sustainable*

Development, 5 (4), 409-420 ISSN: 2239-5938 doi: https://10.14207/ejsd.2016.v5n4p409.

Burkhard B., Maes J. (Eds.). 2017. *Mapping Ecosystem Services*. Pensoft Publishers, Sofia, 374 p.

de Groot, R. S., Wilson, M., & Boumans, R. 2002. A typology for the classification, description and valuation of ecosystem functions, goods and services. *Ecological Economics,* 41(3), 398-408.

Gómez-Baggethun E., R. Groot, P. Lomas, C. Montes. 2010. The history of ecosystem services in economic theory and practice: From early notions to markets and payment schemes. *Ecological Economics,* 69, 1209-1218.

Lyubenova, M. 2004. *Phytoecology*. Acad. M. Drinov Publishing House, Sofia, 574 p. ISBN 954-322-026-3 (in Bulgarian).

Lyubenova, M., Y. Pavlov, A. Chikalanov, K. Spassov. 2015. A Decision Making Support Model for a Smart Forest Ecological Management System. – In: *Proceedings International Conference "Big Data, Knowledge and Control Systems Engineering - BdKCSE'2015,"* 5-6 November 2015, Sofia, 119-131. http://conference.ott-iict.bas.bg/wp-content/uploads/2015/12/BdKCSE2015_Proseedings1.pdf.

Lyubenova, M., K. Stoyanova, Cv. Ivanova. 2017. Biological Diversity, Ecosystem Services and Their Payment in Bulgaria. *Journal of Balkan Ecology,* vol. 20, №3, 229-235.

Lyubenova, M. (Ed.). 2019. *Forest ecosystem services and Payment schemes (Case study).* St. Kliment Ohridski University Press, Sofia, p. 131. ISBN 978-954-07-4778-1.

Maes, J., B. Egoha, L. Willemena, C. Liquetea, P. Vihervaarab, J. Philipp, S. Agnera, B. Grizzettic, E. G. Drakoua, A. La Nottea, G. Zuliana, F. Bouraouia, M. Luisa Paracchinia, L. Braatd, G. Bidoglioa. 2012. Mapping ecosystem services for policy support and decision making in the European Union. *Ecosystem Services* 1, 31-39. Open access under CC BY-NC-ND.

SEEA EEA 2015. *System of Environmental Ecosystem Accounting. Experimental ecosystem accounting* (Technical Recommendations). United Nations, New York.

TEEB Synthesis. 2010. *Mainstreaming the Economics of Nature: A Synthesis of the Approach Conclusions and Recommendations of TEEB*. Earthscan, London, Washington.

TEEB. 2011. *The Economics of Ecosystems and Biodiversity in National and International Policy Making. Edited by Patrick ten Brink.* Earthscan, London and Washington.

TEEB. 2012. *The economics of ecosystems and biodiversity in local and regional policy and management*. Edited by Heidi Wittmer and Haripriya Gundimeda. Earthscan, London, ISBN 978-3-9812410-2-7.

INDEX

C

D

E

F

G

H

I

K

L

M

N

O

P

T

V

W